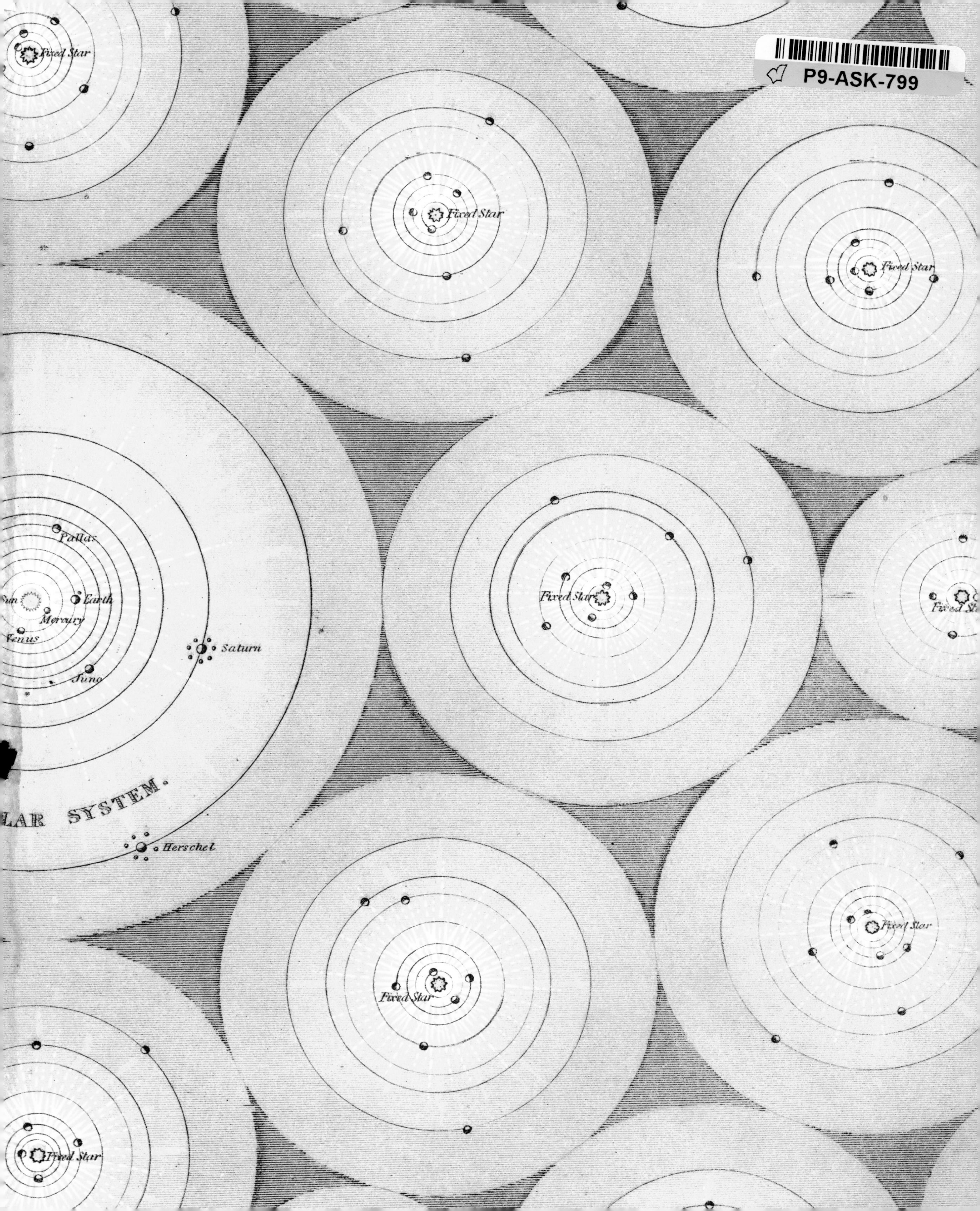
Fixed Star
Fixed Star
Fixed Star
Pallas
Sun
Earth
Mercury
Venus
Saturn
Juno
Fixed Star
LAR SYSTEM.
Herschel
Fixed Star
Fixed Star
Fixed Star

For Melita

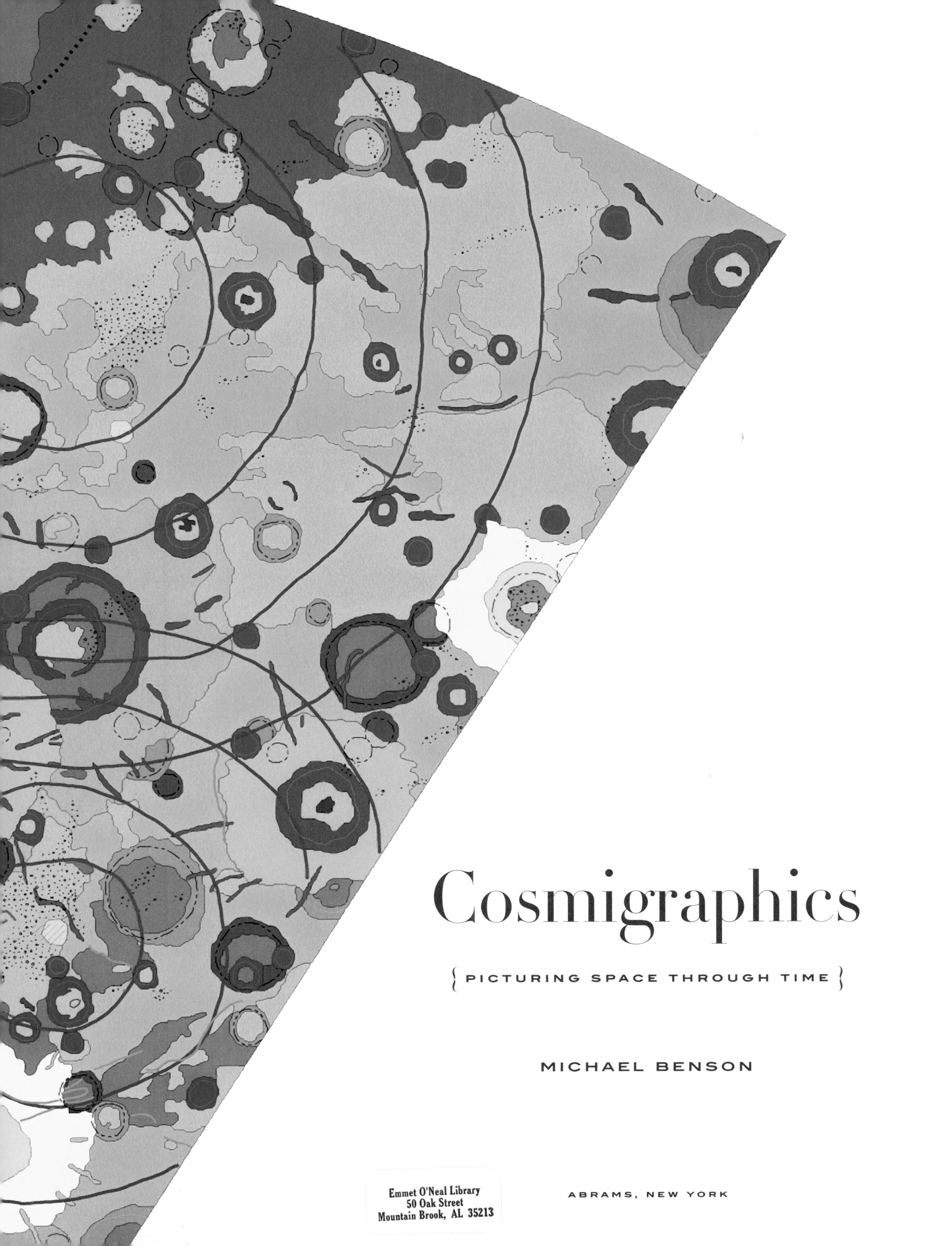

Cosmigraphics

{ PICTURING SPACE THROUGH TIME }

MICHAEL BENSON

ABRAMS, NEW YORK

TIMETABLE OF ILLUSTRATIONS

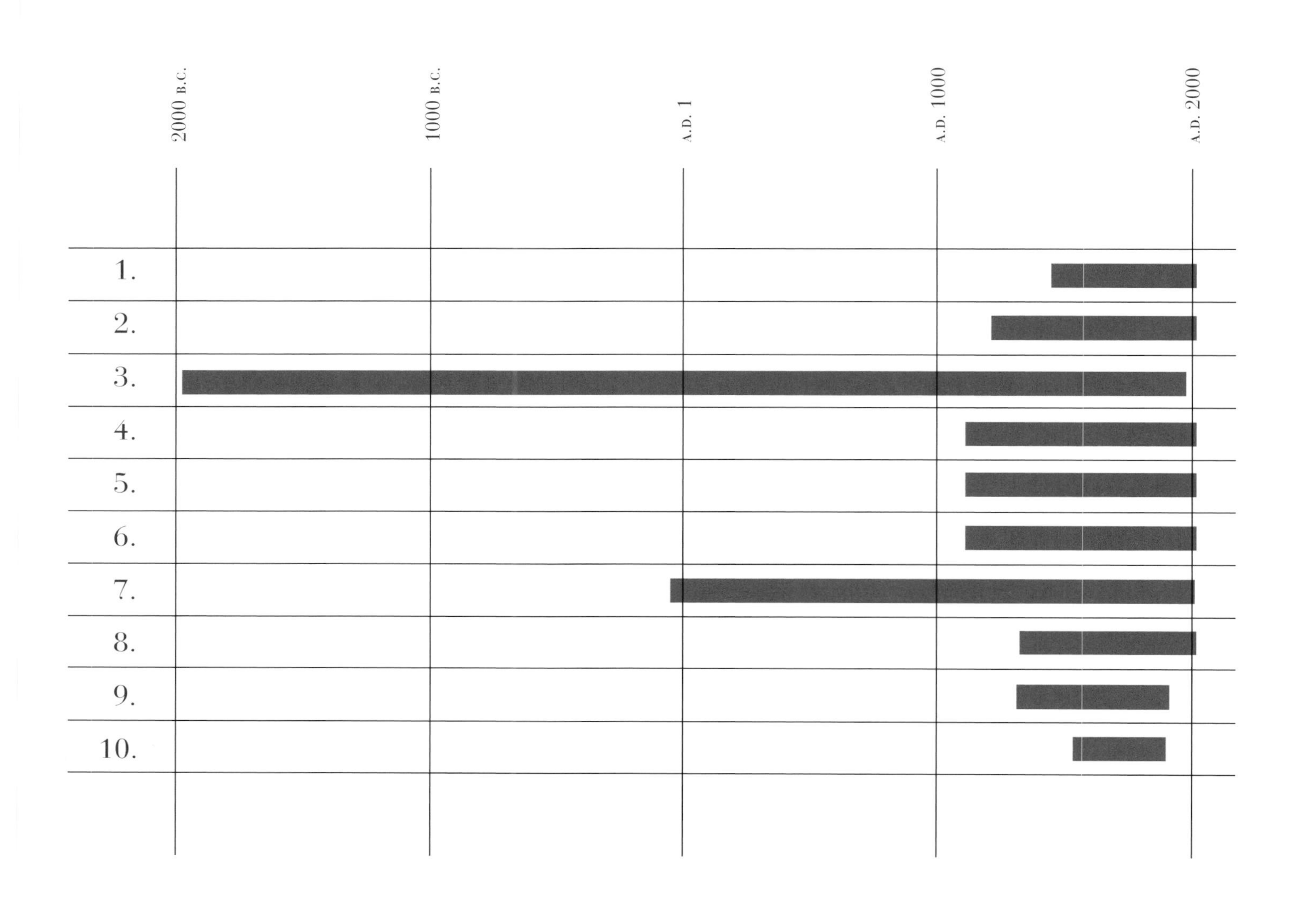

TABLE OF CONTENTS

I

Foreword by Owen Gingerich

BEHOLD! YOU HAVE IN YOUR HANDS AN EXTRAORDINARY visual sampling of the human response to the beauty and mystery of the heavens!

For millennia, night was a fearsome time when wild beasts prowled, but the endless march of the moon and stars across the sky also imparted a calm serenity. These nocturnal rhythms inspired awe and wonder. "When I behold the stars which thou hast ordained, what is man that thou are mindful of him?" exclaimed the psalmist.

These rhythms also inspired curiosity. The sun's annual passage back and forth along the horizon appeared to be linked to the changing seasons, and hence to the biological cycles of plants and animals. The moon with its phases gave a shorter and more convenient interval, but why did it have to be so complicated? The number of "moonths" in a solar year never came out quite even, though 19 years and 235 lunations differ by only a few hours. Even more complex was the way the moon conspired to produce those ominous eclipses. And thus, out of fear and curiosity, astronomy was born.

Eventually, astronomy began to leave a paper trail—papyrus, parchment, rag paper, not just in black and white, but with a rainbow of colors. And that is what *Cosmigraphics* is all about, a colorful record of wonderment, discovery, and understanding. *Cosmigraphics* is not intended as a history of astronomy. It nevertheless sometimes works that way, because the images within each of its thematic sections are largely arranged chronologically, and Michael Benson's essays and reflections place these images in the story of human knowledge. It also contains a history of graphic techniques, from the fastidious painting on parchment in medieval monasteries to modern computer-aided design. And it is more simply a paean to the aesthetic seeing eye.

Today, there are more astronomers alive than all that had ever lived before 1950. With some exceptions, this volume largely honors their predecessors, who, through science and art, revealed the universe. Nevertheless, there are precious few pages prior to the tenth century, an indicator of just how rare these early manuscripts are. If we take just the last millennium, and slice it roughly in two around 1540, then for more than half of those centuries, educated people firmly believed that Earth was solidly fixed in the middle of the cosmos and was composed of just four elements: earth, water, air, and fire. At that temporal divide, America had just been discovered. Printing with movable type was in its infancy. No one knew that blood circulated through the heart, and smallpox was a scourge with no vaccine. Keeping healthy involved periodic bloodlettings set on astrologically determined dates.

In 1540, the most spectacular book of its century was published, Peter Apian's *Astronomicum Caesareum* (Caesar's Astronomy). Standing nearly 18 inches high with 114 brilliantly hand-colored pages, many with complex layers of moving parts, it was truly a worthy gift for Holy Roman Emperor Charles V of Spain. For his efforts, Apian won a new coat of arms, the right to appoint poets laureate, and the privilege to pronounce children born out of wedlock as legitimate. As for the moving volvelles, they contained all the bells and whistles of the ancient geocentric system of Claudius Ptolemy from A.D. 150. With these paper wheels and the attached threads it was possible to determine the positions of the planets to within a degree (that is, within the accuracy of Ptolemy's system, which could at bad times go out by five degrees or more). Apian's *Astronomicum Caesareum* is among the most popular sources in *Cosmigraphics*, appearing in six full-page illustrations scattered throughout this volume.

Apian's luxurious production represented a high-water mark for the Earth-centered Ptolemaic system. Beautiful, even dazzling, it stood at the cusp of a radical change in human outlook, for just three years later one of the greatest highlights of astronomical publishing appeared, Nicolaus Copernicus's presentation of a heliocentric cosmology. (On the rare book market, Apian and Copernicus are still in competition—*Astronomicum Caesareum* fetches nearly a million dollars today, but *De revolutionibus orbium coelestium*—On the Revolutions of the Heavenly Spheres—has passed the two-million mark.) Unusually, the original manuscript of *De revolutionibus* survives because Copernicus retained his handwritten leaves and a copy was made for the typesetters in Nuremberg (who presumably wore out that copy in the process). It is the famous autograph sheet with the diagram of a heliocentric system that appears in this volume on page 146, a trophy of one of the greatest unifications the scientific community ever witnessed.

Copernicus's revolutionary system did not immediately win the day. The idea of being on a globe spinning at a thousand miles an hour at its equator seemed preposterous—wouldn't people be spun off into space? And didn't Psalm 104 say that the Lord God laid the foundations of the Earth that it not be moved forever? So for a century and a half, Copernicus's sun-centered system was viewed as a recipe book for computing the positions of planets, but not as a description of the physical universe. This began to change with the work of Kepler and Galileo, which you will find embedded in this volume.

You will wish to stroll through this gallery at your own pace, but let me mention a few of the images that particularly capture my attention and my imagination.

The lunar gallery, with an intriguing array of telescopic images across five centuries, is especially challenging with respect to details that may not jump out at first. Galileo was not so much interested in cartography as in topography, the heights and depths that made the moon Earthlike. Nevertheless, his lunar image (see page 79) is sufficiently accurate for us to be sure it was drawn on December 18, 1609. In constructing his telescope Galileo required that his instrument show images right-side up, so north is at the top of his drawing. Likewise, the facing map prepared by Thomas Harriot is also a north-side-up image, but, drawn at full moon, it shows no shadows and hence no mountains. Although Harriot made his first lunar sketch well before Galileo, his first instrument was inadequate to reveal the craters, and he did not map any craters until after he had improved his telescope and had seen Galileo's published drawings. Harriot's best map of the full moon (which he never published) is cartographically superior to anything Galileo made, but unlike Galileo, he made no description of lunar topography or the height of mountains on the moon.

In the years immediately following Galileo's pioneering telescopic work, most astronomers adopted the alternative Keplerian arrangement of lenses, which offered a much wider field of view but an inverted image. At first, astronomers continued to depict the moon right-side up, as shown in the 1693–98 "plenilunum" by Maria Clara Eimmart on page 87, where the brilliant-rayed southern crater Tycho appears at the bottom, not far from the South Pole. However, by the time of the monster 1878 lunar map by Wilhelm Gotthelf Lohrmann on pages 96–97, serious charts had south at the top, and the large northern dry seas (Serenitatis and Imbrium) at the bottom.

When Étienne Trouvelot drew the lovely detail of Mare Humorum (the Sea of Humors, page 99) in 1875, the large crater Gassendi was at the lower edge, that is, on the north side, and on the next page Humorum and Gassendi are visible near the edge of the full moon in a two o'clock position. But when Mare Humorum appears again (see page 108) in a modern geological map, Gassendi, on the north side of the crater, is back on top. Astronauts need real maps of the moon, with north safely on top!

Galileo grabs my attention again in the section on the sun, with a finely textured engraving of the sun with sunspots (see page 121). Galileo was not the first to turn the new telescope to the sun, but with his 1613 book that showed solar images for twenty-five consecutive days, he convinced his readers that the "perfect" sun was actually blotched and rotating. This image, with the delicate penumbral areas ringing the dark umbral nuclei, has proved a real challenge for others to reproduce, and it is seldom seen so perfectly in facsimiles as it appears here. In Galileo's book, on the image for the next day, July 8, the sunspot labeled *B* is greatly foreshortened as it revolves over the edge of the sun, showing decisively that the spot is on the surface of the sun and not an intervening cloud, thereby settling a fierce contemporary debate.

An often-reprinted graphic appears on page 59. Until recently, most reproductions carried a caption calling it a late-medieval woodblock, with a traveler who has found where the Earth and sky meet. Chaucer would have known better than to link a flat Earth with a spherical sky, and he would never have imagined this imaginary celestial machinery. Such curiosity would not have been welcomed in the Middle Ages. Altogether it is anachronistic, but charmingly memorable nevertheless. Repeated searches to find an early source for this graphic have come to naught, and it now appears to have been fashioned by a famous French popularizer, Camille Flammarion, who published it in his *L'Atmosphère: météorologie populaire* (The Atmosphere: Popular Meteorology; Paris, 1888).

Among the "Structure of the Universe" pages is a pairing I particularly admire (see pages 156–57). On the left is a classic view of the Whirlpool nebula, drawn by Lord Rosse, who discovered its spiral structure with his giant telescope in Ireland. The earl showed his sketch at the 1845 British Association for the Advancement of Science meeting, and J. P. Nichol, the astronomy professor in Glasgow who was preparing a popular book on astronomy, asked for permission to use it. The image shown here is reproduced from Nichol's book, the first published picture of a spiral galaxy, though in 1846 no one had a clue what a spiral galaxy was. Four decades later (in 1889) Vincent van Gogh painted *The Starry Night*, a picture that appears to include an allusion to the Whirlpool nebula, and it has been suggested that he may have got the vision from one of Flammarion's popular books. Van Gogh's own reed pen study, done after his oil painting, flanks Rosse's picture on the right side of the spread.

A cluster of images beginning on page 237 raises my pulse. The first one is an inside view of a spiral galaxy, although Thomas Wright in 1750 wouldn't have known what a spiral galaxy was. His vertical slab of stars was a cross section of the Milky Way. Though Galileo's telescope had revealed that the milkiness of the Milky Way was created by myriad stars too faint to be distinguished by the naked eye, Galileo had never suggested that it was a flat disc-like structure seen on edge.

A swath of blues splashes across the sky of the Skalnate Pleso atlas plate on page 245, a trailblazing celestial atlas of the twentieth century. They delineate the ambiguous boundaries of the Milky Way, perforated by the purplish dark nebulae where interstellar dust restricts the view. Cassiopeia's W sits in the Milky Way at top right. The Pleiades crowd together near the lower-left boundary, a touch of green indicating a faint veil of nebulosity illuminated by the stars themselves. The two most conspicuous red ellipses are spirals in our local family of galaxies: to the right, our twin, the Andromeda spiral galaxy, and to the left, the rather smaller Messier 33, in the inconspicuous constellation Triangulum. And notice how the sprinkling of tinier and more distant galaxies avoid the dust-laden Milky Way.

As Michael Benson and I were discussing the vast sweep of this book and the wonderful variety of images, many of which are not commonly seen, it occurred to me that I had someplace in my files an image he might like as a final book end. It was a fresco of an angel rolling up the scroll of the heavens at the end of time. I couldn't remember the name of the church, but I knew it was in Istanbul. Now, Michael Benson is nothing if not indefatigable, and by the next morning he had found it from that brief description. It's in the early-fourteenth-century Church of the Chora. And there it is at the end of this volume, rolling up time itself.

Cambridge, Massachusetts

II Introduction by Michael Benson

As Italo Calvino slyly observes in the epigraph to this book, we understand through signs, be they images or texts. Our way to interact with the greater universe is to a great extent mediated. We create worlds of words and universes of pictures. Without them, their subjects may as well not exist.

Cosmigraphics presents a visual record of our attempts to visualize the universe and our place within it. The images here span a period from almost two thousand years B.C. to contemporary high-definition visualizations of interactions between extraordinarily far-flung galaxy groups. They're made of materials running the gamut from the hammered-copper and gold Nebra Sky Disc—the oldest-known realistic depiction of the cosmos in history—to point clouds of supercomputer-generated pixels, with each pixel cluster representing an individual galaxy containing hundreds of billions of stars. In between are hand-painted manuscript illustrations on calfskin parchment, multiple methods of printmaking from woodcuts to copper and steel-plate etchings, and detailed geographical maps of planets and moons originally released as offset prints or in contemporary digital file formats. Photographic images aren't represented here, unless they document handmade objects or serve as the basis for maps.

This material is organized in chapters covering ten themes, with each ordered chronologically (apart from the first). While the result is that the same image sources sometimes feed into the book in multiple places, these thematic divisions were the best way to structure a pictorial archive amassed over the course of several years of research.

This is an entirely subjective survey. I have felt free to include material that would not necessarily figure in a presentation of exclusively astronomical images. I'm interested in innovative approaches to the conundrum of how to present such a vast subject within the frame of a graphic image, even if they aren't directly associated with scientific research and occasionally represent conservative reactions against astronomical findings. I'm biased toward the striking and unusual, even if it restates a case that has previously been made with less visual flair. While this isn't an objective visual history of astronomy, I do believe that sometimes a subjective approach reveals cultural or historical truths better than a dutifully comprehensive method.

Most of the definitions we take as self-evident arrived at their current meanings after meandering journeys through time. Even if we exclude some of the more idiosyncratic material presented here, a majority of the individuals standing behind the images in this book weren't scientists or artists the way we understand those words today; rather they were scholars, or "natural philosophers"—even theologians. Many doubled as astrologers, alchemists, or priests—sometimes all three in one. At least one is now a saint.[1] Their motivations were in many cases very different from that of contemporary researchers or image makers. After more than a century of exponential scientific and technological innovation, it's hard to remember that *science* is itself a new term, and wasn't understood as an autonomous field of endeavor until the nineteenth century. Astronomy and physics were profoundly linked to theology and astrology for much of history.

The greatest astronomer of antiquity, Claudius Ptolemy, author of the single most influential work of astronomy in human history, the *Almagest*, also wrote a keystone text of astrology. The most important physicist of all time, Isaac Newton, formulator of the laws of universal gravitation and inventor of infinitesimal calculus, spent half his precious time attempting to achieve alchemical transformations of matter. Newton reportedly absorbed so much lead and other toxic substances attempting the transmutation of base metals to gold that he had a nervous breakdown. John Maynard Keynes observed that he was "not the first of the age of reason, he was the last of the magicians."

Motivated by theological convictions, German astronomer and astrologer Johannes Kepler devised a complex polyhedral cosmology in which platonic solids defined the distance between the planets. He spent a lifetime searching for God's geometric key to the harmony of the celestial spheres, discovering his revolutionary laws of planetary motion on the way. The inexplicably obscure eighteenth-century astronomer Thomas Wright envisioned a multi-galaxy cosmos, arriving at an epiphanic understanding of the Milky Way's shape, and yet he saw the eye of providence at the center of his "celestial mansions," or galaxies. We'll return to Wright a bit later.

The visual side of their attempts to understand the cosmos were made by the astronomers themselves, or by professional artists and illustrators, or frequently a combination of the two, in which the astronomer made the initial detailed drawings and the engraver converted these to prints suitable for mass reproduction. The invention of the telescope predated the arrival of photography by two hundred years, and photographic emulsions sensitive enough for use in astronomy came a century after that, or a good three hundred years after Galileo. In the absence of photography, skill in depicting what they were seeing was a substantial asset to working astronomers.

Is *Cosmigraphics* an art or a science book? Yes, and yes. As with science, the role of art was far from its contemporary meaning for most of the history sampled here. Until the seventeenth century or later, the arts and sciences were essentially fused. The great Renaissance painters advanced the science of optics, and were prized for their ability to convey realistic depictions of nature. Many were as much scientists and engineers as they were artists, even if their art is what we most celebrate now. The natural philosophers of the Enlightenment also developed their mimetic abilities, the better to depict natural phenomena. When astronomer John Herschel traveled to South Africa in 1833, it was to catalog the southern stars, and document the 1835 return of Halley's Comet from an observatory he set up for his giant twenty-one-foot telescope. But he and his wife, Margaret, also became engrossed in the flora of the Cape, producing 132 exquisite full-color illustrations still used by botanists today. (His rendition of the comet is on page 292.)

Before the Renaissance and as late as Romanticism, artists were essentially considered craftspeople, artisans involved in a relatively low-ranking guild devoted to embellishing church architecture, manuscripts, and civic buildings. Their names weren't necessarily important, even if the greatest rose to prominence. Sienese master Giovanni di Paolo belongs to the latter category. While today he's considered one of the greatest painters of the High Renaissance, it took some serious detective work by twentieth-century art historians to confirm that he was the artist behind many of the illuminations in the most spectacular fifteenth-century manuscript of *The Divine Comedy*. He's represented here by nine works.

Just as scientific endeavor wasn't autonomous from theology and was viewed as a way to comprehend God's

1 Hildegard von Bingen; see page 39.

design, so art operated in service as a form of illustration. Some of the most striking images presented in the first chapter of the book illustrate the biblical creation narrative, combining the Aristotelian-Ptolemaic multisphere geocentric universe with the iconography of Old Testament monotheism. These include an awe-inspiring series of paintings by Portuguese artist and philosopher Francisco de Holanda, a student of Michelangelo, as well as di Paolo's *The Creation of the World and the Expulsion from Paradise*.

The visual legacy encompassed by *Cosmigraphics* documents the stages of our evolving understanding as a species—a gradually dawning, forever incomplete situational awareness about the cosmos and our position within it, rising across millennia. If there's one overarching subject, it's the enigma of our emergence as conscious beings within an unspeakably vast and cryptic universe, one that doesn't necessarily guard its secrets willfully—actually it strews them promiscuously around in the form of hints, indications, clues, and manifestations—but doesn't exactly hand out codebooks either.

There's something very human about attempting to create meaningful depictions of such extraordinarily vast and complex subjects—be they planets, nebulae, galaxies, galaxy groups, or, for that matter, the grand totality of space-time—in two-dimensional pictures produced in scales small enough to open between two hands. It would reek of hubris if it wasn't so daring, and necessary, and without an alternative plan. In fact, it's as integral to the species as building rattletrap flying machines, parsing nature down to the atomic level and then splitting what was found, and converting ballistic missiles into rockets capable of taking human beings to the moon. To quote architectural theorist Dalibor Vesely, in his book *Architecture in the Age of Divided Representation*, "The rather limited mode of representation is, owing to our finite abilities, the only way to come to terms with the inexhaustible richness of reality." And: "The primary purpose of representation, we may conclude, is its mediating role, which can also be described as participatory because it enhances our ability to participate in phenomenal reality."

If it enhances, it also provides designs. Some of the images here played the role of tools, no less than flint arrowheads, stone wheels, or space telescopes. (Some still perform that part.) However incomplete the understanding behind them, they're one way by which we've implicated ourselves in the universe, insinuated ourselves into its design. Just as the creation of a tool via the agency of human intelligence forces evolutionary development by in effect necessitating new neuronal pathways—as the "wetware" learns to use the tools it's made—so presentations of the universe's putative design in hand-illustrated manuscripts, wood-block prints, or supercomputed galaxy clouds spur follow-on refinements to these conceptual structures, be they evolutionary or revolutionary. Nothing is ever complete, no field theory ever quite unified.

The reliance on mediating images—on Calvino's "general thickness of signs superimposed"—brings with it the danger that we'll mistake our construct for reality, like Pygmalion falling in love with his milk-white statue of Galatea. (A name that shares etymological roots, incidentally, with the comparatively new term *galaxy*, also associated with "milk"; "Milky Way" being a translation, by way of the Latin *via lactea*, of the Greek *galaxías kyklos*, or "milky circle"—itself a perfect example of the genius of the Ancient Greeks, who correctly saw a circle where others saw only a meandering *via*.) According to this viewpoint, the endless reification of a geocentric multisphere Aristotelian-Ptolemaic cosmology across fifteen centuries may have blocked progress, because the universe was considered already solved. Vesely quotes physicist Werner Heisenberg: "Contemporary thought is endangered by the picture of nature drawn by science. This danger lies in the fact that the picture is now regarded as an exhaustive account of nature itself so that science forgets that in its study of nature it is studying its own picture."

And yet images have two edges: They're dialectical. Without those endlessly iterative depictions of an Earth-centered universe, each as similar to its predecessor and successor as that Virgin-and-Child chain of Orthodox icons descending down through time (and each as subject to individual artistic skill), Copernicus would have had nothing to react to, and nothing on which to base his alternative cosmology. The propagation of the established narrative brought the story to the plot's turning, however long it took. Copernicus and his followers clearly recognized the danger Heisenberg later warned of; the need for a different picture, or at least a modification.

In its way, it was all a form of natural selection. As he worked in Cape Town to add southern observations to the star catalogs of the northern heavens that his father, the great discoverer of Uranus, and his aunt Caroline, the great comet hunter, had started in the previous century, John Herschel took to meditating about the "replacement of extinct species by others," and compared it to the ongoing evolution of language through time. Charles Darwin, who visited when HMS *Beagle* docked in Cape Town in the summer of 1836, was clearly influenced by

Herschel's ideas, later referring to him obliquely as "one of our greatest philosophers" in the opening of *On the Origin of Species*.

The images in *Cosmigraphics* are sequenced in ways calculated to allow for these evolutionary threads to be seen as each chapter subject unfolds. Sometimes they can seem very incremental, until one looks more closely. Take the case of English astronomer Thomas Digges, who in 1576 printed the first defense of Copernicanism in England. At first glance, his "perfit description of the Coelestiall Orbes," a woodcut of a heliocentric solar system, seems more or less indistinguishable from similar depictions, including Copernicus's own from 1543—until one notices that the stars at the periphery are no longer organized in a neat outer circle (see page 228).

Instead they extend in all directions: Digges had discarded the concept of fixed stars mounted like glittering ornaments on a rotating outer celestial sphere in favor of an infinite number scattered liberally across an infinite universe. This was no Copernican concept, though it was certainly inspired by him. It could have been adapted from the radical cosmology of fifteenth-century German philosopher-astronomer-cardinal Nicholas of Cusa, who considered Earth a star among innumerable others distributed throughout the universe. While Digges was a contemporary of Giordano Bruno, he probably didn't know of the Italian visionary's similar ideas, as Bruno didn't come to England until 1583.

Other images already seem revolutionary even though they date back to the Middle Ages and start the timeline of an individual chapter. One example is the extraordinary diagram representing celestial motion from the 1121 encyclopedia *Liber floridus* (Book of Flowers). (See page 177.) Although hand-painted on animal-skin parchment, its grid filled with the zigzagging lines of planetary motion through time—a classic early "visual display of quantitative information"—has a startlingly contemporary effect. A whiz-bang infographic in a manuscript illuminated three hundred years before Gutenberg, it's like the sudden materialization of a Mies van der Rohe skyscraper at the center of a half-timbered medieval village—an early artifact of what Vesely has called the "mathematization" of reality.

Another excellent example is the astonishing curvilinear waveforms visible in two prints on pages 285–86 from Kepler's 1619 treatise *De cometis libelli tres* (The Comets Trilogy). Other published works by Kepler, some seventeen books across five decades, largely contain graphics very much of their time, or the Baroque period (see, for example, the plate on page 150). But Kepler's Gehryesque comet depictions, which track the changing angles of the comet's tail as it progresses past the orbits of the inner planets, seem to predict an entire universe of twentieth- and twenty-first-century technology, architecture, and design. They're computer graphics centuries before computer processing. (They're also a great example of what Heisenberg was warning about: Although these images appear absolutely analytical and self-assured, in fact Kepler depicted comets traveling in straight trajectories—something they never do.)

This foreshadowing of styles and methods that would only become commonplace centuries later is no accident, because Kepler's way of seeing his cometary flight paths was specifically mathematical, and his stripped-down plates are thus completely utilitarian and not subject to the frippery of extraneous ornamentation. In attempting to render only the data that had been acquired, they belong to what Vesely calls an "instrumental" point of view. It follows that they're also predictive of a technological way of thinking.

A symbiotic relationship between representation and understanding can be seen throughout *Cosmigraphics*, with the latter not necessarily preceding the former. Many of these images were produced under the conviction that comprehension comes through the making—that a depiction of a working model of nature is itself a form of knowledge. That some of these models contradicted others is of course no surprise: During the Copernican Revolution that extended from the mid-sixteenth century to Newton's death in the first quarter of the eighteenth, the cosmologies of Ptolemy, Aratus, Copernicus, Brahe, Kepler, Riccioli, Newton, and others sometimes seemed to duel across the pages of rival tracts. Their opposing or complementary concepts were also anthologized in sumptuous detail in such celestial atlases as Andreas Cellarius's 1660 *Harmonia macrocosmica* (Cosmic Harmony), which attempted to show all sides of the argument. Seven plates from that atlas are reproduced here.

The seriousness with which these graphics were taken is an artifact of their seamless incorporation within the scientific theories being presented. In most cases, the visual accompaniment wasn't a separate and lesser element of a hypothesis but was integral to the argument. These images also frequently rose to pinnacles of aesthetic achievement. Athanasius Kircher's ideas concerning the hydrology and magma flows of the subterranean Earth were as reliant on his riveting double-page plates as they were on the accompanying text (see pages 48–51). French artist-astronomer Étienne Trouvelot's superb chromolithographs of sunspots, comets, and lunar features were the direct result of his work as a staff member of the Harvard College Observatory. (Eleven of his pieces are scattered throughout the book.) Danish painter Harald Moltke's participation in two arctic expeditions was funded by the Danish Meteorological Institute and dedicated to studying the northern lights; his seething electric skyscapes were intrinsic to the research goals of the expeditions (see pages 313–15).

One could even say that although these images were working in service of theories or observations—or better, because of it—they were a precursor form of contemporary conceptual art, in which concept rules form and has priority over purely aesthetic concerns.

Vesely discusses why this may be so in a way that deserves to be quoted at length:

> *We must keep in mind that the traditional understanding of art includes every kind of making—from the making of shoes and tools to arithmetics and geometry. They were distinguished by their degree of involvement with matter and manual labor and were placed in broad categories, which were most often expressed only by adjectives—the mechanical arts (*artes mechanicae*), usually situated at the bottom of the hierarchy because of the labor involved; the liberal arts (*artes liberales*), which include the trivium (grammar, rhetoric, and logic) and the quadrivium (arithmetic, music, geometry, and astronomy); and, finally, the theoretical arts, sometimes known as* scientiae, *consisting of theology, mathematics, and physics. That the arts represented not only experience and skills but also an important mode of knowledge is reflected in the ambiguity of their relation to science.*

The conviction that the construction of mimetic models amounts to understanding nature's behavior is reflected in the variations on circular shapes visible throughout *Cosmigraphics*. Ancient Greek astronomers recognized not just that celestial movements unfold in complex cycles but also that they're manifestations of a circularity in nature's design. As I argue later in the book, all of our contemporary technologies can be traced back to what the ancients conceived as the urgent need to understand these circular mechanisms. This resulted in both two-dimensional representations such as can be seen here, and the construction of working machines like that astonishing first-century B.C. analog-geared computer, the Antikythera Mechanism (see chapter 3 essay). Both were rehearsals of comprehension.

By forcing graspable matter to behave in a manner directly analogous to the movements of unreachable celestial objects, a kind of transubstantiation was brought about. Rather than bread and wine becoming body and blood, in this case tangible terrestrial matter *here* behaved in the same way as unattainable celestial matter *there*—or at least enough so to facilitate accurate predictions of celestial movement. You could say that metaphysics became physics.[2]

Although they're rooted in a Ptolemaic cosmological model, the volvelles, or paper dials, in German mathematician and cosmographer Peter Apian's 1540 book *Astronomicum Caesareum* (Caesar's Astronomy) provide a great example of this. The first depicts a polar planispheric projection of the northern constellations centering on Polaris, the North Star (see page 227). It's designed to be turned one full revolution every *thirty-six thousand years*—or Ptolemy's rate of precession. (Precession, the movement of the axis of a spinning body around another axis due to torque, gradually causes the terrestrial poles to trace the shape of a cone over a long period of time: picture a top oscillating as it spins.) Each of the tabs visible at the edge of the volvelle represent a planet, and can be set to compensate for the Earth's incremental precession. After being correctly set, the positions established are meant to be exported to successive volvelles throughout the book. The first paper dial, in other words, rules the rest.

As of this writing, Apian's 474-year-old book has sailed through one-seventy-sixth of a full turn of its key volvelle.

Cosmic matter's inexorable circularity and also sphericity was clearly very much on the mind of English astronomer and mathematician Thomas Wright, whose 1750 book *An Original Theory or New Hypothesis of the Universe* provides one of the best-case studies of scientific reasoning through image. It contains thirty-two plates, many of them full-page, in a book of eighty-four pages. Although Wright made a series of audacious conceptual breakthroughs concerning the form of galaxies and the structure of the universe that have proven to be largely correct, he remains an obscure figure. In part, this is because he wasn't a physicist, and didn't come up with elegant laws to back his theories, like Kepler or Newton. Nor did he discover anything directly through observational astronomy.

Although lavishly produced, *An Original Theory* attracted very little attention, and Wright's ideas might have vanished without a trace if philosopher Immanuel Kant hadn't read a detailed review of the book in a German journal. Although he apparently never tried to get a copy of the book itself, Kant was prompted to write his own work on the subject. It was published in 1755 and

2 Something Vesely also implies in *Architecture in the Age of Divided Representation*.

contained many of Wright's core ideas—which, however, Kant very correctly attributed to Wright. While Kant's book, which appeared fairly early in his career, also wasn't much noticed—in part because his publisher was going bankrupt at the time—as his reputation grew it became more widely read. As a result, Kant is frequently and mistakenly given credit for conceiving of a universe made of multiple galaxies.

In retrospect, Wright's cosmological meditations were a logical extrapolation on Copernican heliocentrism and Newton's laws of universal gravitation. But it bears remembering that they came only 150 years after Galileo had first transmitted to the world that the Milky Way was comprised of innumerable stars, too faint to differentiate with the naked eye. Also, *An Original Theory* was published decades before William Herschel started his systematic campaign to catalog non-stellar deep-sky objects, or nebulae, many of which were eventually determined to be galaxies in their own right. During Wright's time, and all the way to the end of the first quarter of the twentieth century, our galaxy *was* the universe, and its shape was a complete mystery.

While his prose style is sometimes opaque, in combination with his illustrations Wright's meanings are clear. *An Original Theory* is a superb example of the image as a mode of knowledge. He wrote that, in his view, the stars are all in motion, and the sun is a star similar to the rest. He referred to the organizational principle of the solar system, in which the planets orbit a far more massive sun, and theorized that it and other stars are also orbiting another center of mass. He then proposed two alternative ways in which they might move around that hypothetical center, "but which of the two will meet your Approbation, I shall not venture to determine."

In his first galaxy model, the sun and the rest of the stars are "all moving the same Way, and not much deviating from the same Plane, as the planets in their heliocentric Motion do round the solar Body." He illustrated this concept with a series of plates, accompanied by text explaining what was being depicted. In one, the Milky Way is seen from inside, as "a perfect Zone of Light"—exactly as it does from Earth (see page 237). In two others, the galaxy is viewed from outside, with the stars surrounding the central nucleus arranged in concentric rings (see page 153). Wright had taken the flattened disc shape of the solar system and extrapolated his way to the Milky Way's actual form—the first human being ever to do so.

In the completed book, Wright did not conceive of spiral structure (though in his preparatory drawings he did, albeit somewhat ambiguously[3]). He does, however, compare his hypothetical galactic form to Saturn and its rings—and in a brilliant insight, writes, "I cannot help being of the Opinion, that could we view *Saturn* thro' a Telescope capable of it, we should find his Ring no other than an infinite Number of lesser Planets, inferior to those we call his Satellites." His assessment was entirely sound; Saturn's rings consist of innumerable chunks of ice and rock.

Wright's second possible galactic shape was globular, with stars "all moving with different Direction around one common Center, as the Planets and Comets together do round the Sun, but in a kind of Shell, or concave Orb." (See page 236. Unlike the planets, which all revolve on the same plane, long-period comets dive into the solar system from all directions; they originate in a spherical cloud named after Dutch astronomer Jan Oort.) In general form, if not in the particulars, Wright's second shape described what we now call elliptical galaxies—many of which are close to spherical. Just as in his first concept, where he imagined the stars organized in concentric rings around a central nucleus within a flattened disc, so with this second galaxy shape he imagined (and illustrated) several discrete shells of stars within the outermost layer.

Although in both cases his internal organizational scheme was wrong—galaxies shaped like flattened discs don't have concentric rings around their nucleus, but rather spiral arms; and spherical galaxies are not made of concentric shells, though the stars do move in different directions around a common center—Wright had through pure deduction visualized both of the galactic types that appeared, almost two centuries later, in astronomer Edwin Hubble's widely distributed 1936 "tuning fork" galaxy morphologies chart. It was a breathtaking performance.

Having explicated the two main galactic forms, Wright turned his attention to the mysteriously indistinct nebulae that eighteenth-century telescopes were just beginning to discern. Many actually *were* nebulae—clouds of interstellar gas and dust within our galaxy. A few, however, were well outside the Milky Way's plane. These were what we now know to be distant galaxies, bright

3 In his preparatory drawings for plate XXI, which depict Earth orbiting the sun, which in turn orbits the galactic center, Wright inscribes a spiral shape at that center. Within it are the words "The universal center of gravitation." The sentence corkscrews down to a black star shape at the actual galactic center. (We now know that black holes anchor the center of spiral galaxies.) Wright's prescient spiral was lost in translation to the plate. We don't know who made Wright's mezzotint prints, only that they are "By the Best Masters," as per his title page.

enough to show up in the telescopes of the day as fuzzy gray patches against the blackness of space. (The term *galaxy* was not yet in wide use; Wright was still formulating a first draft of what goes after the word, as its definition, in the dictionary.[4])

The astronomer concludes his book by positing that the "many cloudy Spots, just perceivable by us, as far without our starry Regions, in which tho' visibly luminous Spaces, no one Star or particular constituent Body can possibly be distinguished; those in all likelihood may be external Creation, bordering on the known one, too remote for even our Telescopes to reach." Wright was saying that we live in a universe comprised of many galaxies. He illustrated his idea with the remarkable plate reproduced on page 154. In a single book written prior to the American Revolution, an unrecognized astronomer from Durham in northern England had presented a coherent picture of a universe virtually indistinguishable from what we recognize today.

Wright's vertiginous vision was a seed from which billions of galaxies sprouted. In the spring of 2014, R. Brent Tully, an astronomer specialized in the astrophysics of galaxy groups, submitted a paper to the journal *Nature*, accompanied by a mesmerizing supercomputer visualization encompassing thirty thousand galaxies. These were depicted flowing dynamically across a volume of space-time more than 500 million light-years in diameter (see pages 172–73). Tully, whose 1987 announcement of the discovery of the Pisces-Cetus Supercluster Complex coincided with his publication of his pioneering *Nearby Galaxies Atlas*, has been at the forefront of attempts to understand the large-scale structures of the universe for decades. (See the *Atlas* plate on pages 160–61.)

The new paper announces the discovery of an extensive, compounded swarm of galaxy superclusters, among which is the Virgo Supercluster containing the Milky Way. Its visual accompaniment, which depicts flow lines stitching together vast galaxy groups, provides an utterly contemporary example of the image as idea and vice versa. As Tully confirmed prior to submitting the paper for peer review, with large-scale galaxy mapping of this kind, discovery and visualization happen simultaneously. Rather than large-scale structures being perceived in observational data that in turn prompts the creation of a map, here the supercomputer representation itself permits the discernment of structure. Cartography is critical to the discovery.

In the paper, Tully and his collaborators for the first time define the entire extent of the galaxy flows surrounding an enormous gravitational well comparable to a terrestrial watershed. Their large-scale structure is in fact defined by its flow, with our own galaxy positioned near the border between two such curvilinear gravitational depressions. "The region deserves a name," Tully writes. "In the Hawaiian language 'lani' means 'heaven' and 'akea' means 'spacious, immeasurable.' We propose that we live in the Lanikea Supercluster of galaxies." Although the region he'd named is "not so small," the paper concludes, "there is room for 5 million such structures within the current Universe horizon, each with 100,000 large galaxies."

In November of 2013, astronomers announced that data from NASA's Kepler space telescope indicates that up to 40 billion potentially habitable Earth-size planets probably exist in the Milky Way alone. Multiply that number by four and you approach our current estimate of the number of galaxies in the visible universe. Each contains from about ten million to roughly a hundred trillion stars. All stars are now thought to have one or more planets orbiting them. The implications are staggering.

At the conclusion of the chapter in which he presents galaxies for the first time, Thomas Wright observes ruefully that the success of his theory "is a Thing will hardly be known in the present Century, as in all Probability it may require some Ages of Observation to discover the Truth of it." The truths he illustrated in thirty-two mezzotint plates are typical of the many image quests contained within *Cosmigraphics*, with Brent Tully's revelatory panoply of flowing, filamentary galaxies only the latest. So it seems right to give the last word to Wright:

> *What inconceivable vastness and magnificence of power does such a frame unfold! Suns crowding upon Suns, to our weak sense, indefinitely distant from each other; and myriads of myriads of mansions, like our own, peopling infinity, all subject to the same Creator's will; a universe of worlds, all decked with mountains, lakes, and seas, herbs, animals, and rivers, rocks, caves, and trees.... Now, thanks to the sciences, the scene begins to open to us on all sides, and truths scarce to have been dreamt of before persons of observation had proved them possible, invade our senses with a subject too deep for the human understanding, and where our very reason is lost in infinite wonders.*

4 It was first seen in English in a poem by Chaucer around 1380, where it referred to the Milky Way.

CHAPTER 1

Creation

There is something formed of chaos,
Born before heaven and earth.
Silent and void, it is not renewed,
It goes on forever without failing
It can be seen as the World-Mother.

—The *Tao Te Ching*

Apart from producing, as it says in Genesis, something "good," the act of conjuring creation from eternal darkness with a handful of words, and maybe some imperious laconic gestures, resulted in a *design*. And because after what we might call certain misunderstandings, and maybe some serpentine promptings, we'd partaken of the low-hanging fruit, we found ourselves kicked unceremoniously onto a very long road. We'd been booted out of the Garden, but we'd landed in the universe. And our thoughts turned eventually to its structure. Was it round, or square, or neither? Was it a flat disc floating in a vast ocean surrounded by a spherical sky? Was it a seven-story ziggurat? Was it resting upon a turtle, on the back of a larger turtle, supported by a stack of ever larger turtles descending down toward some kind of universal turtle bedrock?

The fruit didn't fall far from the tree, at least initially. Just northwest of Eden, around the sixth century B.C., the pre-Socratic philosophers started cooking up a number of sophisticated ideas. Things had calmed down after some eventful millennia, the sun was shining calmly down on the sea, and a time to think was at hand. They began by discarding mythological explanations concerning the universe, trying instead to understand its behavior using reason, and even experimentation. Up in Thrace, in the fifth and fourth centuries B.C., Leucippus and Democritus of Abdera, possibly influenced by the ideas of an obscure pre–Trojan War era Phoenician named Mochus, proposed that the universe was made of extremely tiny indivisible particles. They called them "atoms." In Ephesus, Hereclitus observed that everything in nature was in a constant state of change, but he also saw that logical patterns could be discerned within the perpetual flux. It was an early form of chaos theory. And on Samos, Pythagoras worked out a number of natural principles. Although little is known about him, and nothing from primary sources, the mathematical proof known as the Pythagorean theorem is credited to him. In many respects it was the E=MC[2] of antiquity.

Ancient Greece at its height extended as far north as the Crimea and as far south as Egypt, and its more observant citizens noticed differences in the sky depending on their latitude. Some constellations, for example, were clearly visible from the Nile and only partly seen from the Black Sea. This led to the idea that Earth was probably a sphere. By the third century B.C. the chief librarian at the Library of Alexandria, Eratosthenes, figured out how to measure Earth's circumference. He knew that at high noon on the summer solstice, the sun beamed directly down a deep well in the southern Egyptian city of Swenet (today's Aswan), hitting the water directly. And he had access to survey measurements dating back to the Pharaohs, demarcating distances in units known as stadia. Armed only with a sundial, some basic geometry, and the stadion distance between Swenet and Alexandria, Eratosthenes measured the angle of the sun's solstice shadow at Alexandria, muttered a bit under his breath, and deduced Earth's diameter to be 252,000 stadia. Depending on which stadion measurement he used—the Egyptian or the Attic—that's within either 1.6 percent or 16.3 percent of its true diameter. Either is remarkable for antiquity, of course, but Eratosthenes doesn't seem a man to underestimate.

Prior to Eratosthenes, Aristotle had refined an existing cosmological model in which Earth was at the center of the universe, with the sun, moon, stars, and "wandering stars" (or planets) all orbiting terra firma. Operating out of Plato's Academy in Athens, he wrote that the classical elements of fire, water, earth, and air are restricted to a spherical shell contained within the orbit of the moon. Everything above, according to Aristotle, was made of an unchangeable fifth element he called ether. The motion of the planets, the sun, the moon, and the stars was the result of each of them being mounted in an individual, rotating, ethereal, nested sphere.

Not long after Aristotle's death, a contemporary of Eratosthenes named Aristarchus of Samos produced a somewhat different hypothesis. Working probably in Alexandria, he proposed a cosmological scheme placing the sun, not the Earth, at the center of the universe. He also suggested that the universe was far larger than previously supposed. His ideas were largely rejected, sometimes quite vehemently, in favor of the philosopher from the capital. (One critic, the head of the Stoics, reportedly wanted to put him on trial for heresy, based on this attempt to disrupt a celestial order predicated on a central, immovable Earth—but cooler heads prevailed. This scenario would repeat itself more than a thousand years later with a different outcome.) Only Seleucus, an astronomer working about a century after Aristarchus, not far from Eden's shuttered gates in the city of Seleucia, agreed with the genius from Samos.

Making his own calculation of the size of the universe, Seleucus went a good deal farther than Aristarchus. Taking a break from figuring out that the sea's tides are the result of the moon, and that particularly high ones are the result of the moon's being positioned near the sun—both eminently correct ideas—Seleucus squinted at the Tigris for a long while and came up with his estimation. The universe, he said, is—infinite.

To this day it's not known by what percentage this estimate may be wrong. And it may be correct.

Unfortunately, the ideas coming out of Seleucia and Alexandria didn't make much headway against the prolific philosopher from Athens, and when the time came for the most influential astronomer of all time, Claudius Ptolemy, to take up the celestial baton in Alexandria in A.D. 90–168, he inherited a geocentric Aristotelian cosmology. A Greek with Roman citizenship living in Egypt, then a Roman province, Ptolemy was about as cosmopolitan as you could get in the ancient world; even his name was half Greek, half Roman. He had easy access to the knowledge stored in Eratosthenes's old place of employment, the Library of Alexandria. But despite the tantalizing likelihood that Ptolemy could easily have read Aristarchus and Seleucus, whose writings we now know of only indirectly, he stayed with the geocentric model.

Most of Ptolemy's major books have survived. The most important by far is a treatise, written in thirteen parts, that he called *The Mathematical Compilation*, a title that was seemingly replaced by another, *The Greatest Compilation*. In later centuries this became known as the *Almagest*, a derivation from the Arabic *Al-majisti*, which is a translation of that second title. Although Ptolemy had rejected heliocentrism, he benefited from the methods and stored data of centuries of Babylonian and Hellenistic astronomy. This he synthesized, adding the results of his own observations, and presented along with a detailed mathematical theory on the motions of the sun, moon, planets, and stars. The *Almagest* was practically synonymous with astronomy for the next fifteen hundred years—or until about a century after Copernicus, taking a discarded page from antiquity, proposed that the sun, not the Earth, was at the center of the universe.

In the *Almagest*, Ptolemy devised a geometrical model combining circular motions known as epicycles, among other techniques, to compensate for irregularities in the motions of the planets, moon, sun, and stars—irregularities that we now know are artifacts of the Earth's motion around a central sun. Apart from the use of epicycles, which he inherited from the astronomer Hipparchus, Ptolemy compensated for remaining irregularities by offsetting the centers of planetary orbits just slightly away from Earth—a small remove called an eccentricity. His combination of epicycles, eccentricities, and a third class of virtual point called an equant, did an excellent job of predicting celestial motions, and the ingenious Ptolemaic system ruled astronomy for about a millennium and a half.

A lot of creativity, in other words, was required to model creation. With the exception of the last image, most of the works in this chapter reflect the structure of an Aristotelian-Ptolemaic cosmos grafted onto the Judeo-Christian origin narrative—though there are some intriguing deviations along the way. Essentially, when the smoke clears, we're left with Aristotle's spheres—though now with a primum mobile, an outer stratum at the border between the moving and the static, the material and immaterial, a final sphere beyond which angels flutter, or gather like arrows in feathered ranks surrounding a central throne.

One of the most extraordinary images on view, by English physician and cosmologist Robert Fludd, predates Russian avant-gardist Kazimir Malevich's landmark painting *Black Square on a White Ground* by almost three hundred years (see facing page). In 1617 Fludd began work on an ambitious book, *Utriusque cosmi, maioris scilicet et minoris, metaphysica, physica, atque technica historia* (The Metaphysical, Physical, and Technical History of Two Worlds, the Macrocosm and the Microcosm). As its portentous title suggests, it was a grand attempt to synthesize and explain the entire visible and immaterial universe. Fludd's revolutionary black square opens his creation sequence.

We should remember that for most of history the concept of a vacuum was alien to cosmological thought. As a force of nature himself, Aristotle's horror of vacuums led to his ether hypothesis: something had to fill all that seemingly empty space. The existence of his fifth element wasn't finally disproved until the late nineteenth century. Fludd's image also seemingly deletes ether, by depicting the non-space that would ostensibly contain it. He insists on a vacuum.

In depicting a time before time, a space with no ground, Robert Fludd predicted the total departure from realism and representation of the abstract art of the twentieth century. His square's absence of color (because black is to colors as zero is to numbers) and its shape that's not really a shape (because as with Malevich, the square evidently resulted from a default need to put a boundary on the shapeless) signify something "born before heaven and earth," to lift a line from the *Tao Te Ching*. It's the presence of an absence.

By starting his account of the existence and behavior of everything with a depiction of its opposite—nothing, in a picture that's not a picture—Fludd initiates a trajectory into the unknown and unknowable. His successor Malevich, who was an accomplished theorist as well as an artist, positioned his *Black Square* in the corner of the room when he first exhibited it in 1915—the place where in Russian Orthodox tradition the most important icon of the house always goes. In a subsequent essay, Malevich wrote, "God exists as 'nothingness,' as non-objectivity." For his part, Fludd quotes Psalm 18: "He made darkness His secret place."

At his wake in 1935, Malevich's followers placed *Black Square* above his open coffin, like an exit portal to the empyrean. Fludd's image is bordered on all four sides by four words in Latin, meaning "And so on to infinity."

If Fludd's square is Zero to the One that would follow, the chapter's last picture, a composite image of the cosmic microwave background radiation that seethes faintly behind the galaxies in all directions, represents a distant, unmistakable echo of the Big Bang that started the whole story. Depending on your beliefs, it could be understood as affirming the universe's staggering scale, as envisioned by two ancient geniuses, Aristarchus of Samos and Seleucus of Seleucia. Or it could be perceived as a kind of affirmation-in-aftershock of the thunderous initiatory Old Testament command, *Fiat lux*: Let there be light.

And it could be seen as both. *Et sic in infinitum.*

• 1617:

This revolutionary depiction of a black void prior to the light of creation, which predates and anticipates Russian Suprematist artist Kazimir Malevich's famous painting *Black Square on a White Ground* by three hundred years, appears in English physician and cosmologist Robert Fludd's book *Utriusque cosmi, maioris scilicet et minoris, metaphysica, physica, atque technica historia* (The Metaphysical, Physical, and Technical History of Two Worlds, the Macrocosm and the Microcosm). On each side of Fludd's irregular square can be seen the words *Et sic in infinitum*, meaning "And so on to infinity." It is the first in a series of images from that book depicting the creation of the universe.

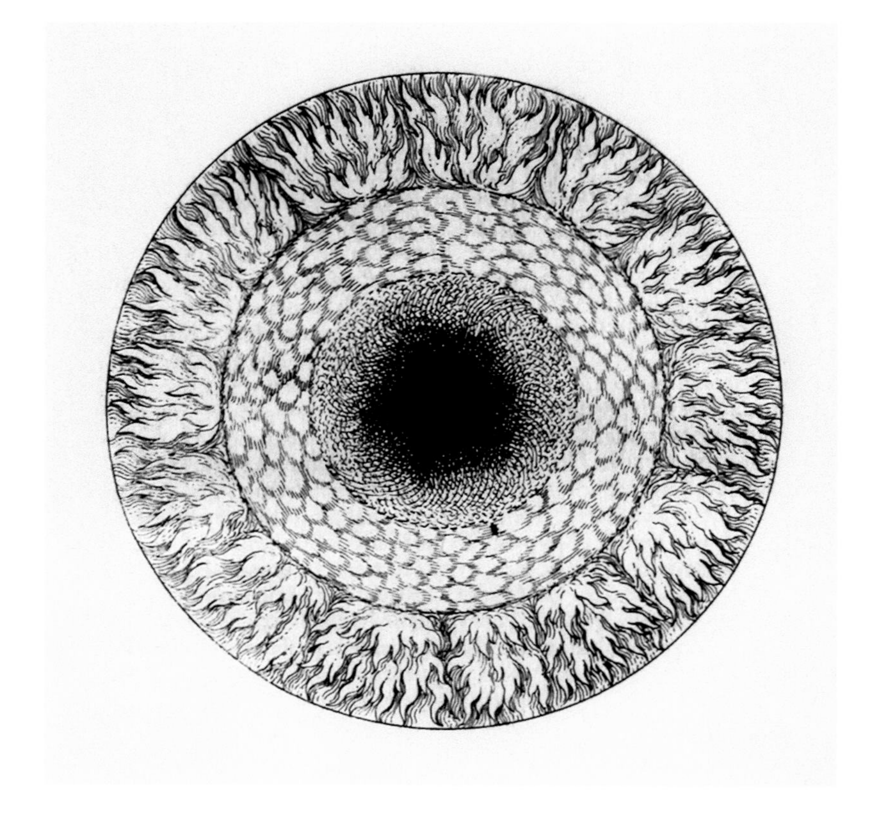

Successive images from Fludd's creation series, in which multiple chaotic fires subside until a central starlike form becomes visible, surrounded by concentric rings of smoke and debris. Intriguingly, although Fludd believed in a geocentric cosmology (see his fully realized creation on page 32), this is close to our current conceptions concerning solar system formation. In the image on the right, the first half of *Fiat lux*, or "Let there be light," is linked wordlessly to a depiction of the phenomenon of light itself, in which the Holy Spirit is represented by a dove embedded in a ring of brilliance. Apart from being an eminent Paracelsian physician, Fludd was an astrologer and mathematician. A pronounced streak of theosophy runs through his work, meaning he was engaged in an effort to provide a coherent picture of the origins and purpose of the universe, in which the human and the divine are united.

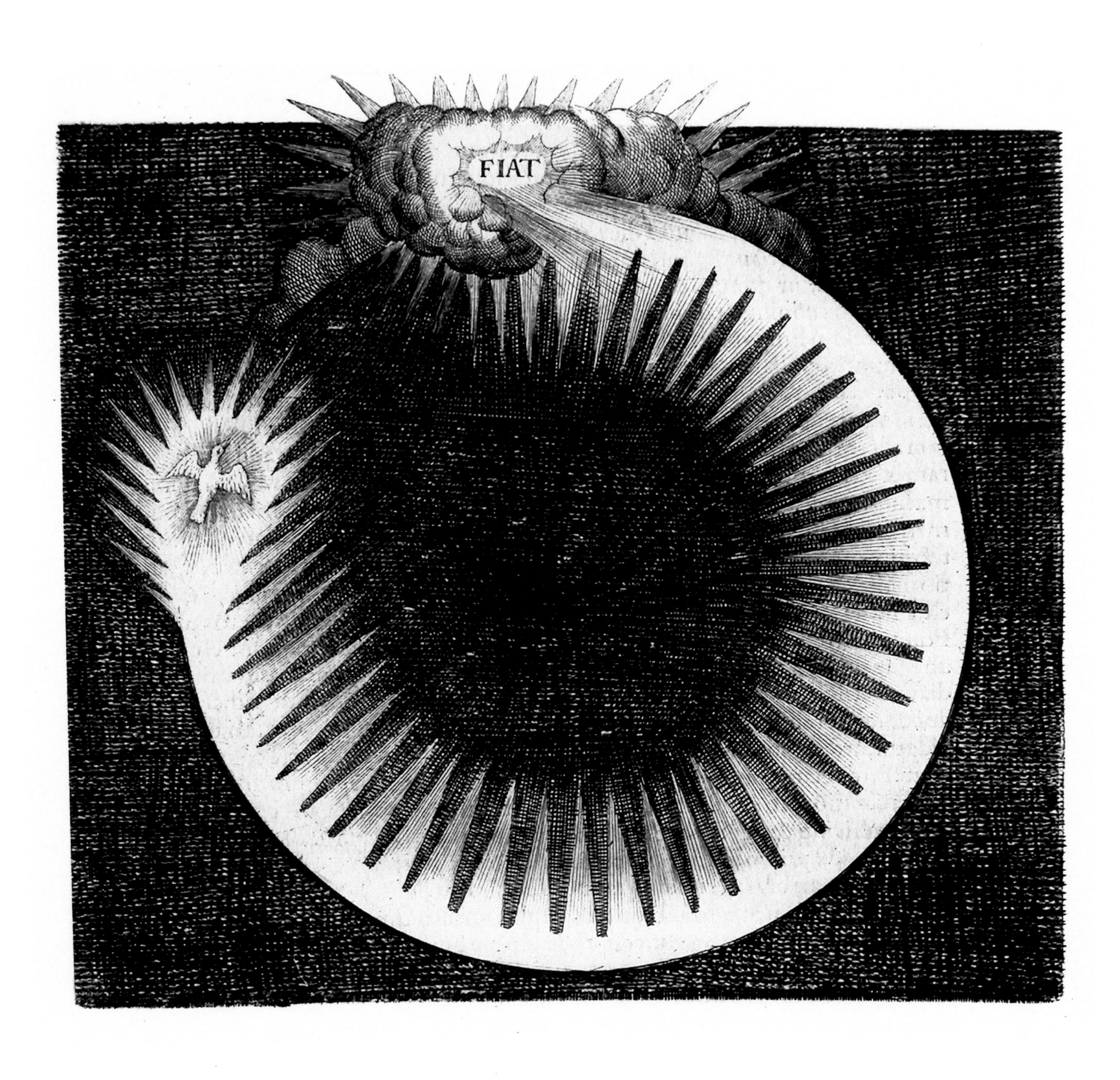
FIAT

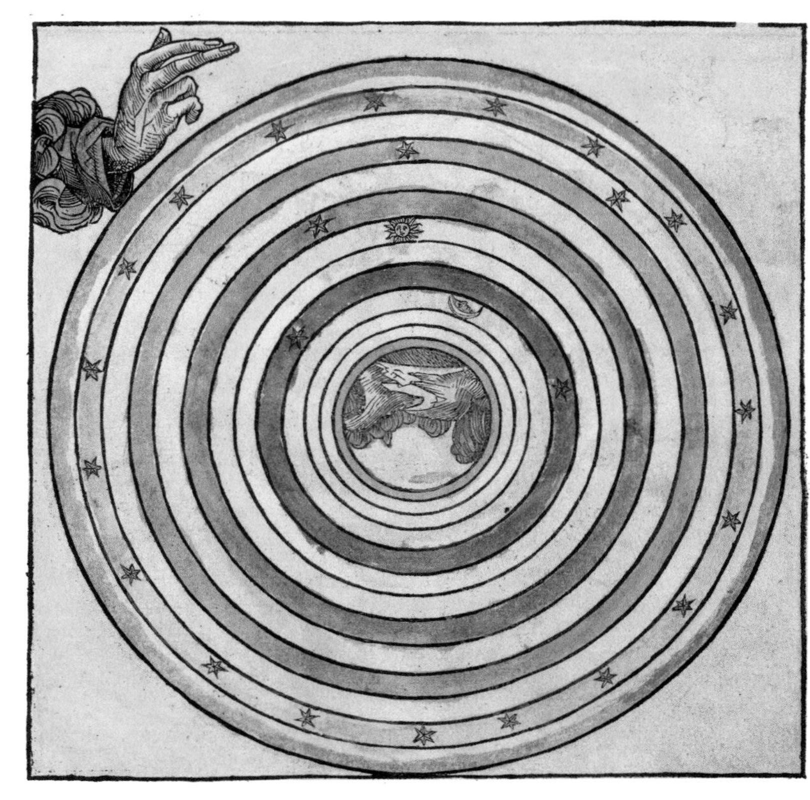

• 1493:

More than a hundred years before Fludd, German physician and cartographer Hartmann Schedel presented the six days of creation in a series of woodcuts in his *Liber chronicarum* (known in English as the Nuremberg Chronicle). Four of the days are shown above. **Top left:** The second day. **Top right:** The fourth day, in which the universe has been divided into a recognizably orderly terracentric, Ptolemaic scheme in which the Earth is upside down and surrounded by heavenly spheres, including one for each of the seven known planets at that time; a sphere containing stars; and an outermost sphere representing the primum mobile, or the "first moving sphere." **Above left:** Creation of the Garden of Eden. **Above right:** God's disembodied hand has now been replaced by a depiction of the Deity himself, who creates Adam from a lump of clay with one hand and simultaneously blesses him with the other. And the trouble begins.

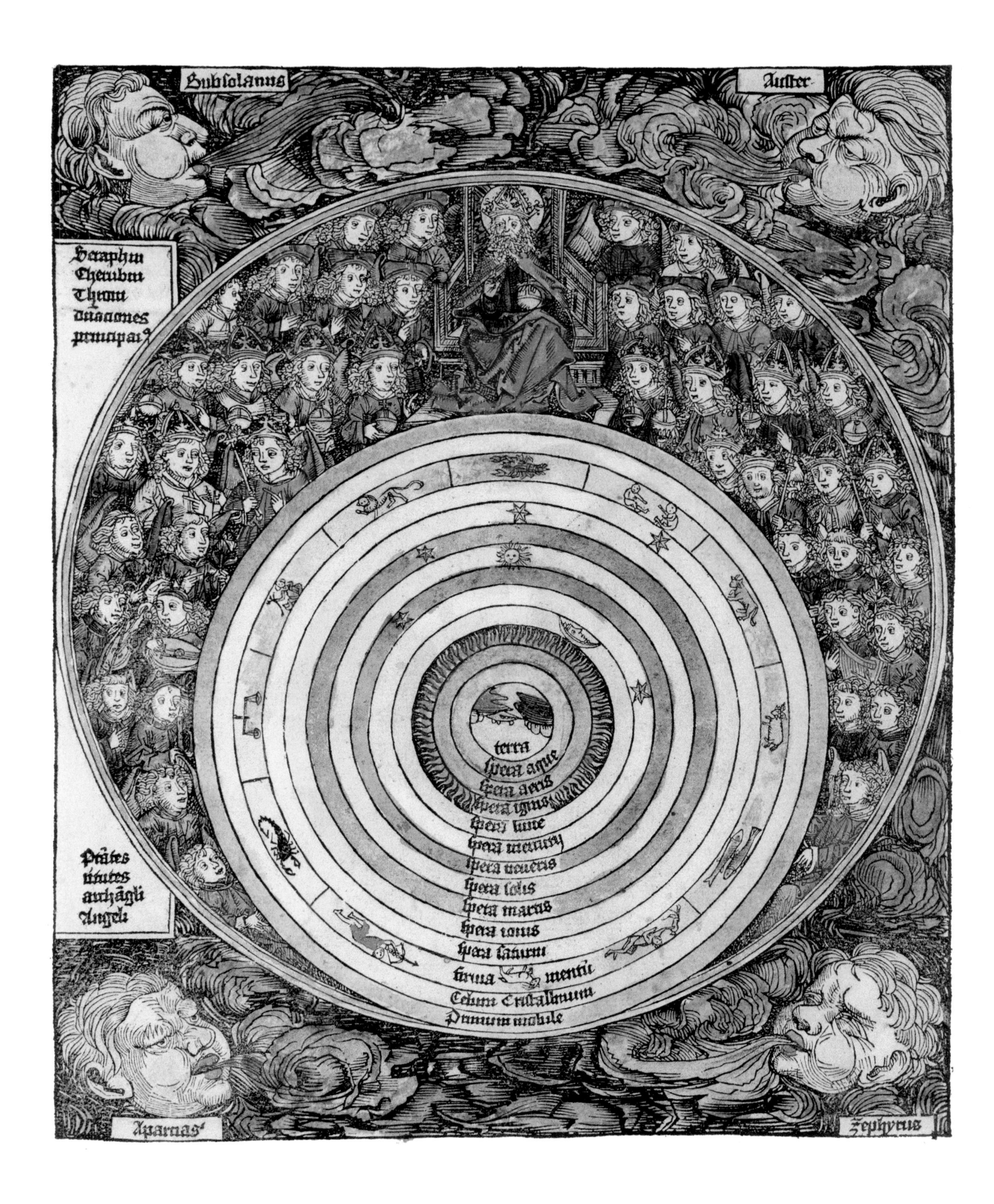

Schedel's remarkable depiction of the seventh day, or Sabbath, when God rested. (It is possible that these and other woodcuts from *Liber chronicarum* were made by a teenage Albrecht Dürer, who was apprenticed to Nuremberg printmaker Michael Wolgemut. Wolgemut's workshop produced these illustrations.) A cosmic order in which a newly minted creation spins like a complex clockwork at God's feet as he rests in his heavenly throne. At the center of the temporal universe, the Earth is surrounded by spheres representing the other elements as understood by the ancients: water, air, and fire, followed by ten celestial spheres. The latter contain the seven planets including the moon and the sun; the stars, represented by signs of the zodiac; two outer spheres, specifically an intermediate sphere sometimes called the "crystalline heavens"; and, finally, the primum mobile, at the border between material and immaterial realms. Nine orders of angels embrace the universe on either side of God's throne. At each corner, winds blow from the cardinal points of the compass. Everything is clearly labeled, and all is right with the world, which may be upside down to the mortal viewer but is correctly oriented for God's contemplation.

• 1507:

Section of an illumination from an Austrian Old Testament depicting Eve being created from Adam's rib within a familiar cosmographic scheme. Here, the ordered primary elements are readily apparent, with the rings depicting water, air, and fire encircled by a single stylized ring containing stars, as well as the moon and the sun (and doubtless planets as well, though in practice they are undifferentiated from the stars). The outermost ring of heaven is defined by angels, and, finally, in this rendering the winds seem to have redirected their attentions from terra firma and are exhaling into a great echoing void.

• 1573:

A series of paintings by Portuguese artist, historian, and philosopher Francisco de Holanda, a student of Michelangelo, depicts the creation of the universe in an unprecedentedly visionary, mystical style that seems to anticipate the work of William Blake by two hundred years. Discovered in the National Library of Spain in the mid-twentieth century, they are contained within a large volume of drawings and paintings by de Holanda illustrating the Bible titled *De aetatibus mundi imagines* (Images of the Ages of the World).

Left: A complex schematic of pyramidal forms links the material and nonmaterial spheres as a faceless God, represented by a triangle inscribed with the Greek letters alpha and omega, commands "Fiat Lux" – let there be light. Below, form materializes in a void suffused with fire and water.

Facing page: Now materialized in patriarchal glory, the omnipotent creator retains a triangular form. Belted by stars, he conjures the firmament into existence. Below, the claylike vessel seen emerging in the first painting already forms the crystalline spheres of a Ptolemaic universe, with the Earth at their center.

A
Ω
PIETAS
RATIO

FIANT LVMINARIA IN FIRMAMENTO CELI

Facing page: Francisco de Holanda's depiction of a geometry of wheeling circular shapes that has been set in motion by God's luminescent command. The largest such form outside of heaven appears to be the sun, depicted as being far larger than the Earth, which it illuminates with a powerful, shadow-casting beam. **Above:** An Earth divided into water and dry land has appeared, and the unitary God of creation has receded back into the triangle representing divine force, within which recognizable forms suggestive of the Holy Trinity have emerged.

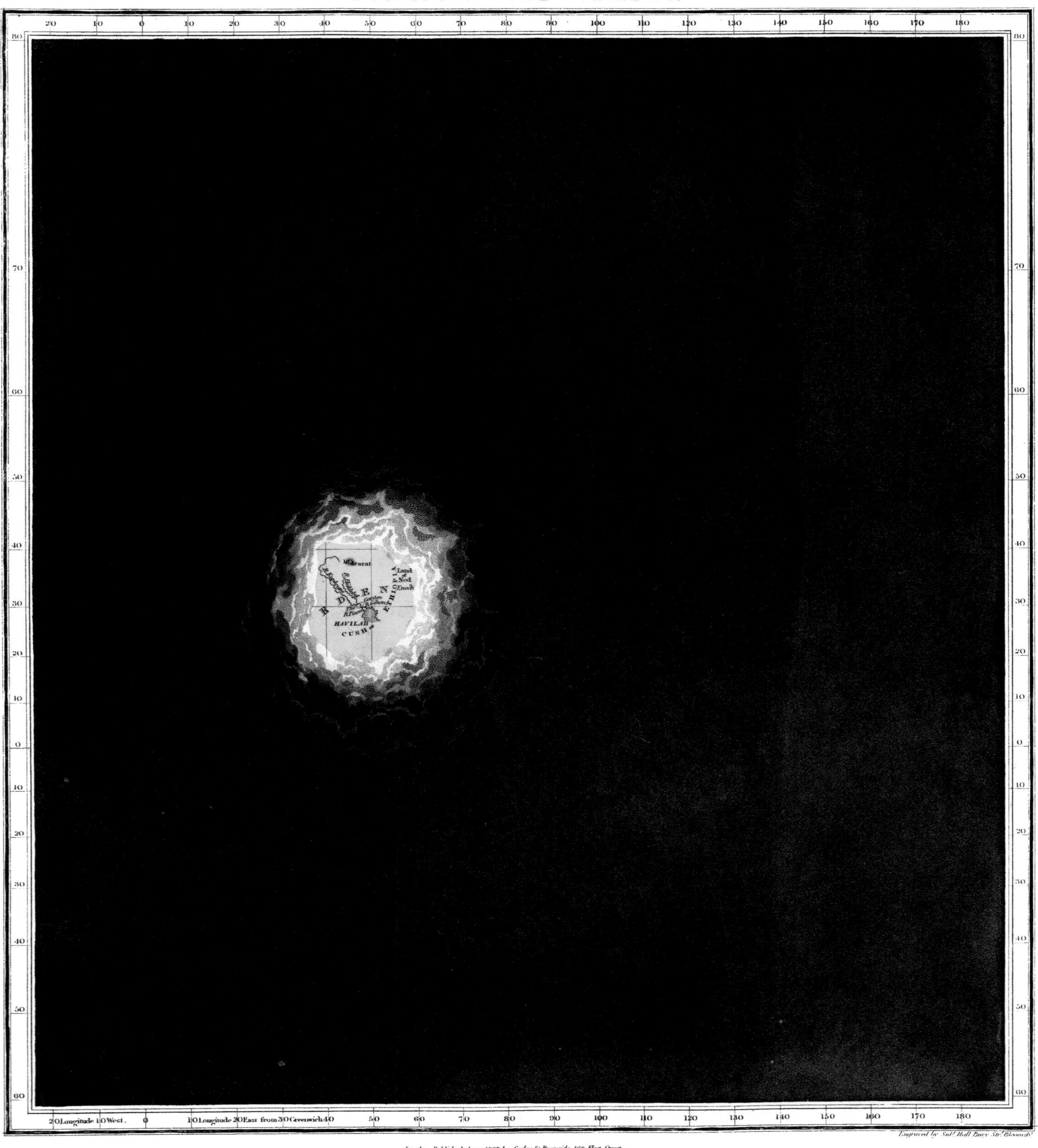

• 1830:

A parting in some very dark clouds reveals a God's-eye view of the known Earth just prior to the deluge. Below Mount Ararat and east of Eden, the Land of Nod is visible. Somewhere in this landscape the last of the antediluvian patriarchs, Noah, is presumably assembling his ark. From *An Historical Atlas* by Edward Quin.

• 1735:

Taken from Swiss doctor and natural scientist Johann Jakob Scheuchzer's astonishing four-volume *Physica sacra* (Sacred Physics), which contains more than seven hundred copperplate engravings in a work that aspired to stitch together all worldly knowledge, this engraving depicts the moment when God created the first man. On the lower left, a Latin inscription reads *"Homo ex Humo"* (Man from Dust). Intriguingly, the Wonder Cabinet–style arrangement of humanoid skeletons and depictions of fetal development seemingly functions as a harbinger of Darwin's theory of evolution, which it predates by more than a hundred years.

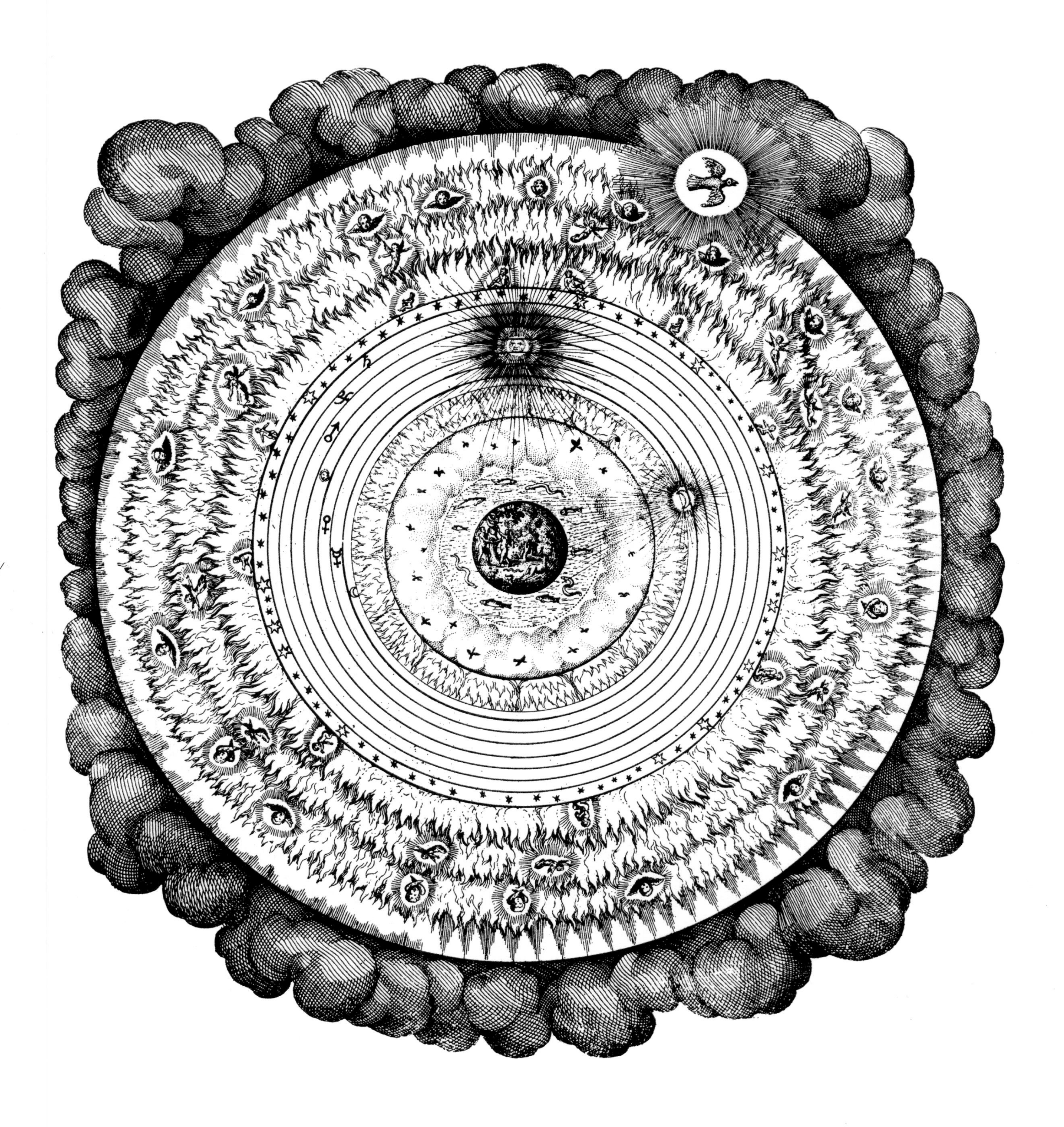

• 1617:

The completed cosmos from Fludd's "The metaphysical, physical, and technical history of two worlds." As with the illustrations by Hartmann Schedel on pages 22–23, eleven spheres rotate around the central Earth, although here, Adam and Eve can be seen, with Eve reaching toward the tree. A serpent is visible at her feet. The same dove depicted on page 21 can be seen on the upper right; here, the representation of the Holy Spirit faces away from where free will is being exercised for the first time.

• 1445:

Sienese master Giovanni di Paolo's *The Creation of the World and the Expulsion from Paradise* contains a cosmological scheme virtually identical to similar representations in this chapter, only here the central Eden, represented by a map of the world containing four rivers, has been forcibly vacated. To the right, an angel under the command of God the Father escorts Adam and Eve from the Garden. For other works by di Paolo, see pages 75, 116, 144, 178–79, and 256.

• 2013:

All-sky view of the cosmic microwave background radiation, the oldest discernible echo of the Big Bang in the perceivable universe. This image, from the European Space Agency's Planck space telescope, reveals small fluctuations in the density of this ancient radiation, representing slight temperature differences. Visible in all directions, the cosmic microwave background was used by the Planck team in 2013 to date the Big Bang to 13.81 billion years ago—making it slightly older than previous estimates. Their results also confirm that most of the universe is made of a cryptic substance that physicists are calling "dark energy." The radiations represented in this image date back to the so-called era of recombination, when the universe became transparent to light for the first time. While more or less uniform in all directions, unexplained regional variations are visible. The temperature of this relic radiation has cooled with time; it is currently only 2.725 degrees Celsius above absolute zero.

CHAPTER 2

Earth

I don't know about other places,
but here on Earth there's quite a lot of everything.
Here chairs are made and sadness,
scissors, violins, tenderness, transistors,
water dams, jokes, teacups.

—WISLAWA SZYMBORSKA, *HERE*

IN THE BEGINNING, THE CHARTING OF THE UNIVERSE WAS the charting of the world, with the terms *world* and *universe* referring to the same thing. From classical antiquity to the Renaissance, the world was everything we could possibly perceive—the entirety of knowable reality. In his 1559 book *Cosmographical Glasse*, English physician John Cunningham provided a sixteenth-century perspective on the meaning of the word, simply by quoting Greek astronomer Cleomedes: "The world is an apte frame, made of heaven, and earth, & of thinges in them conteyned. This comprehendeth all thinges in it self, nether is there anything without the lymites of it visible."

Depictions of the planet as a distinct place in space are unknown from antiquity, though celestial globes containing the constellations were rendered in marble centuries before our oldest surviving copy, the Farnese Atlas. This seven-foot-tall statue of Atlas half-kneeling under the weight of a celestial sphere is a second-century Roman copy of a lost Greek original. The constellations spangling Atlas's burden probably date back to pre-Ptolemaic times, though their exact provenance is disputed. In any case, the statue is simply more evidence of a perceived unity of the terrestrial and celestial in antiquity. Such spheres always present an inversion of the constellations as seen from Earth: with the Farnese Atlas we're looking at a shell of stars from the outside, with Earth implicitly anchoring the arrangement at the center below. In other words, it isn't just a representation of the night sky but of the entire universe, with Earth contained inside under its outermost Aristotelian sphere.

Representations of the unity of Earth and sky are also seen in pre-Hellenistic images, even if they tend to illustrate mythological understandings of the relationship. For example Nut, the Egyptian goddess of the sky, was usually depicted as a star-covered nude woman arching protectively over the Earth, with her hands and feet still touching the ground. Working in Egypt much later, second-century Greek-Roman astronomer Claudius Ptolemy evidently considered his treatise on astronomy, the *Almagest*, only part of a whole. His second major work, *Geographia*, attempted to map the Earth in eight books using all the information available at the time. As in the *Almagest*, Ptolemy imposed a mathematical order on the material, devising a grid system imposing coordinates on geographical features. Latitude was measured from the equator, a system still in use today.

Ptolemy was well aware that the Earth is a sphere, and presented arguments to that effect in the *Almagest*. Contrary to misinformation about the state of knowledge in medieval times, this knowledge was transmitted to Europe, and in the High Middle Ages representations of planetary globes were seen with some regularity in illuminated manuscripts. Some depicted people as large figures standing at right angles to each other astride the globe, as in the manuscript illumination reproduced on page 39, depicting the seasons of Earth from thirteenth-century philosopher and composer Hildegard von Bingen's *Liber divinorum operum* (Book of Divine Works), in which southern figures are upside down in relation to northern ones, and the eastern and western reapers and sowers are sideways on the page—all due to a mysterious attraction exerted by Earth, seemingly in defiance of common sense. Its cause isn't understood to this day, though we now have a word for it.

Medieval and early Renaissance maps of Earth usually show it surrounded by the classical elements in the order originally decreed by Aristotle, who observed that fire leapt upward, soil and water fell toward the center of the Earth, and air simply blew around freely. As a result, thousands of maps exist in which a central globe has a deep blue ring for water, a lighter blue ring for air (sometimes itself divided into three thermal layers), and an orange or red outer ring representing fire. Directly beyond this begin the crystalline nested spheres containing the planets and stars, starting with the moon, which provided a kind of glass ceiling for the elements in the Aristotelian arrangement: Beyond it, nothing was ever supposed to change, even if movement was possible.

At the center of this design, Earth itself was usually positioned with Asia at the top (thus the term *to orient*), Europe on the lower left, and Africa on the right. These continents in turn were surrounded by a global ocean and further divided into climate zones—frigid at the poles, "torrid" at the center, and temperate in between. This arrangement can be seen in the map on pages 40–41, taken from the twelfth-century encyclopedia *Liber floridus* (Book of Flowers). A dramatic red slash cuts diagonally across the vertical band of sea dividing Africa from an assumed but then-unknown southern continent. The slash contains a diagram packed with zigzagging lines, representing the fluctuations of the planets, the moon, and the sun in relation to the zodiac. Since the zodiac—a band divided into twelve sections representing the sun's path in the sky—is in fact diagonal to Earth's equinoctial line, *Liber floridus*' oblique red slash constitutes a particularly masterful visual display combining spatial and temporal information. (For a more detailed look at the zigzagging simulacrum of planetary motion that *Liber floridus*' author positioned in that red slash, see page 177.)

Another popular way to portray the Earth was to illustrate it at the nucleus of an armillary sphere, by producing two-dimensional representations of three-dimensional objects. Such spheres, which resemble spherical birdcages with terrestrial globes at their centers, were designed to represent a projection of the Earth's key lines of latitude—the equator, the tropics of Cancer and Capricorn, and sometimes the arctic circles—onto the fixed sphere of stars. In the actual object this was usually accomplished with metal rings arranged in graduated diameters, all held together by such other important elements as that same diagonal band representing the zodiac and longitudinal, polar-orbiting bands as well.

Some graphic representations of armillary spheres got far more sophisticated than the actual mechanical devices ever did. For example, the plate from Andreas Cellarius's *Harmonia macrocosmica* (Cosmic Harmony) on page 46 has inserted all the known planets, the moon, and the sun into the mechanism. Were this in fact a three-dimensional object, that level of complexity would have required a full-scale orrery. But those mechanical devices, which are similar to a clockwork, were so complex they weren't produced until the eighteenth century, and then with the sun at their centers.

By the late Renaissance, Aristotle's and Ptolemy's neat categorization schemes were being subjected to some pretty heated questioning, with the entrenched view that the ancients had already figured everything out eroding under the weight of new evidence. In seventeenth-century Germany, omnivorous Jesuit polymath Athanasius Kircher turned his attentions to geology, with one of his forty books, the weighty 1664 *Mundus subterraneus* (Subterranean World), presenting two large-scale cross-sectional depictions of the Earth at the service of speculative theories concerning the subterranean distribution of water and fire. Although some putti faces playfully blow wind from the four corners in Kircher's prints—a concession to convention—here Aristotle's neatly ordered elements have run amok. Water is no longer relegated to an external elemental layer, but permeates the underworld, and fire now seethes through networks of tubes and lava lakes underground, rather than keeping its distance near the moon.

Other printed depictions of the Earth from the same period similarly imagine the unknown, or impatiently anticipate things to come. One presents the world as it would look drained of its oceans, by all appearances a gnawed apple with its continents riding on top of a ravaged global seabed. It seems to anticipate future arguments about plate tectonics and maps based on actual ocean soundings. Another illustrates the planet simply floating in space, its brilliant crescent offset by nighttime shadow—a premonitory vision, from 1664, of what humans would finally see with their own eyes when Apollo astronauts first flew to the moon in 1968.

A tiny eighteenth-century graphic sets out, in a matter-of-fact diagram, a series of trajectories originating on a mountaintop. Each ultimately falls to Earth, except the last and longest, which instead ends up circling the globe. This very early published illustration of the path to orbit—possibly the first such illustration—comes from an unauthorized 1728 English translation (from Latin) of a previously unpublished 1685 draft of a third section of Isaac Newton's *Philosophae naturalis principia mathematica* (Mathematical Principles of Natural Philosophy), usually simply referred to as *Principia*. Titled *A Treatise of the System of the World*, the book describes, in remarkably understandable style, the effects on the Earth, moon, tides, and solar system of Newton's theory of universal gravitation. It was a theory the physicist had otherwise rendered in impenetrably dense infinitesimal calculus in the *Principia*. For some reason, however, after writing such an accessible text, Newton thought better of conveying his breakthroughs to the lay reader, instead squirreling his potential popular science best seller away with the rest of his papers.

By the twentieth century, speculative Kirchnerian depictions of inaccessible terrestrial regions gave way to maps based on actual information. An increasing flow of data was transforming how we saw the world—both the world-as-universe in its ancient sense, and Earth as a planet. Working at Columbia University's Lamont-Doherty Earth Observatory in the early 1950s, oceanographer Marie Tharp defied an ingrained prejudice among male scientists that women were mere "calculators," and largely incapable of original research. She took sonar data provided by geologist Bruce Heezen and started producing the first detailed maps of the ocean floor. Prior to Tharp, mapmaking had largely been considered a follow-on duty to the more serious work of actual exploration. She would change that paradigm.

As she gradually assembled individual sonar-sounding data points, Tharp noticed something unusual about the mid-Atlantic ridge. This longitudinal mountain range running down the center of the Atlantic had been known in the abstract since the first transatlantic telegraph cables of the late 1870s, but she'd spotted something new emerging from all the pointillistic pings: an irregular, linked chain of rift valleys. They seemed to bisect the ridge in a north-south ribbon. And this jagged chain wasn't an isolated feature: It ran almost continuously from pole to pole.

Tharp had discovered a seam in the planet's crust—clear evidence in favor of a then widely dismissed theory that the continents of the Earth are riding on top of giant, slow-moving lithospheric plates. Tharp's rifts, which she continued to find in all of the Earth's oceans using Heezen's soundings, were in fact the exact places where subsurface material is gradually rising from inside the planet, replenishing its crust.

Tharp and Heezen went on to chart the entire ocean floor, revealing the staggering 70 percent of planetary surface that was previously unknown. They published their first map, of the North Atlantic seabed, in 1957—the year the Soviets launched Sputnik 1 into orbit, thus confirming the premise laid out in a thumb-size infographic in Isaac Newton's posthumous book. At the dawn of the space age, a cartographer named Marie Tharp had used mapmaking as a tool of planetary exploration.

• 1210–30:

In this illumination from a late work by the prolific medieval visionary writer, composer, and proto-feminist Hildegard von Bingen, the four seasons of a spherical Earth are represented. Although produced after her death in 1179, the illustration is thought to follow her original design. Knowledge of the spherical Earth dates back to the Greek philosophers of about the sixth century B.C., with Pythagoras said to have been among the first to describe it. By the eighth century A.D. and the early medieval period, the shape of the planet was well established. This is one of the most dramatic early representations of a spherical Earth, from Saint Hildegard's last masterpiece, *Liber divinorum operum* (Book of Divine Works).

• 1121:

In this foldout image from the medieval encyclopedia *Liber floridus* (Book of Flowers), Earth is depicted surrounded by the orbits of the planets, which cut into an early example of a graph depicting planetary motion through time (the red diagonal, seen at an oblique angle). At the bottom, Venus, the sun, and the moon move from their respective spheres into the graph. Compiled and written in his own hand between 1090 and 1120 by Lambert, the canon of St. Omer, in northern France, the encyclopedia encompasses astronomical, biblical, geographical, and natural history subjects. India is visible at the top of the map of the Earth, following medieval convention, in which Asia was frequently at the top, thus "Orienting" the globe. *Liber floridus* is considered the first encyclopedia of the High Middle Ages; this is from the original manuscript. For another example of a graph depicting planetary motions taken from the encyclopedia, see page 177. For other pages from *Liber floridus*, see pages 115 and 141.

• Circa 1450:

Although the shape of the planet was known since antiquity, occasionally the ancient concept of a flat, disc-shaped Earth under a domed firmament appeared in medieval and renaissance art. This section of a rectangular painting by Florentine master Francesco Pesellino, originally mounted on a cassoni wedding chest, illustrates one of Petrarch's vernacular allegorical *Triumph* poems—specifically the "eternity" part of the "Triumph of Fame, Time, and Eternity." In this painting no attempt was made to render coastal details of Earth's dry area, which is simply disc-shaped, and the ocean extends to the sky's edge. Above, the celestial spheres are stacked in layers to the border between the material and immaterial, with God and his angels above and outside the temporal realm. Between Earth and heaven, the sphere bearing the constellations of the zodiac is defined by the constellations Leo and Aquila, which look like they've been disturbed in the library. In Petrarch's *Triumphs*, fame triumphs over death, but time ultimately triumphs over fame, and in the end, eternity triumphs over time. The poem concludes with a kind of epic cosmological finality: "The time will come when every change shall cease, / This quick revolving wheel shall rest in peace: / No summer then shall glow, nor winter freeze; /Nothing shall be to come, and nothing past, / But an eternal now shall ever last." For another depiction of a Petrarch Triumph, see page 222.

• 1499:

Depiction of an armillary sphere in the hand of a disembodied deity, from a Venetian edition of monk-astronomer Johannes de Sacrobosco's *Tractatus de sphaera* (Treatise on the Sphere), originally written circa 1220. The Earth is surrounded by the zodiac and the tropics of Cancer and Capricorn. Such spheres, usually constructed from metal elements, were used to explain and defend the Ptolemaic system, particularly after the rise of Copernicanism. Sacrobosco's *Tractatus* was the most successful and reprinted work of pre-Copernican astronomy in Europe, with hand-copied volumes giving way to printed editions after the European invention of the printing press in 1455. More than a hundred editions are known, spanning two hundred years. In keeping with its depiction here, Sacrobosco referred to the universe as a *machina mundi*, or "machine of the world." (At the time, the word *world* referred to the entire universe.) *Tractatus* presented a coherent account of the complex planetary motions of the Ptolemaic universe.

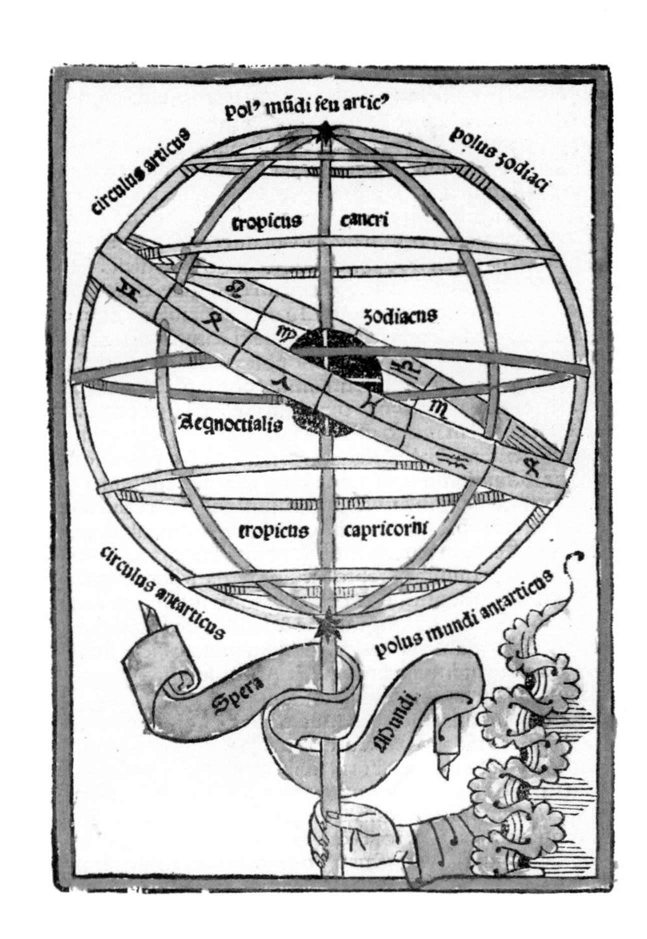

• 1410–1500:
This highly unusual fifteenth-century depiction of the Earth as a weightless sphere studded with city spires possesses a protoscience fictional quality. From a later French translation of the 1240 encyclopedia *De proprietatibus rerum* (On the Properties of Things) by Bartholomeus Anglicus (Bartholomew of England).

• 1540:
Movable paper volvelle from German printer, mathematician, and cosmographer Peter Apian's stunning *Astronomicum Caesareum* (Caesar's Astronomy; the book was dedicated to Holy Roman Emperor Charles V and his brother, King Ferdinand of Spain). Considered one of the most elaborate and beautiful scientific books ever published, this masterpiece of printing turns a text into an actual scientific instrument. This particular volvelle can be used to determine the position of the Earth's shadow at any point in time. For more from this book-instrument, see pages 77, 119, 180, 227, and 262. *Astronomicum Caesareum* was printed and hand-colored by Apian himself, at his own press in the Bavarian town of Ingolstadt.

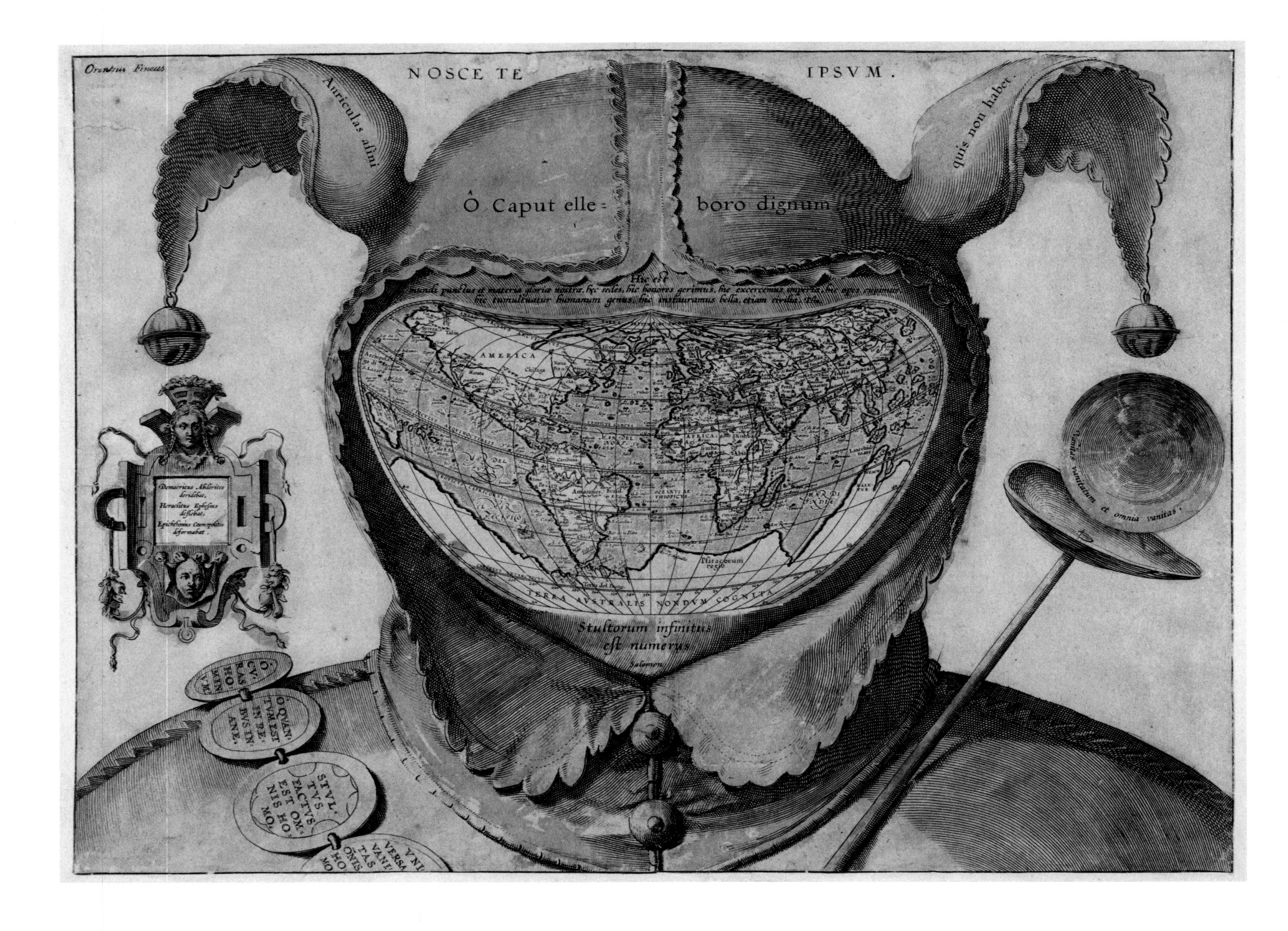

• 1580–90:

The mysterious and disturbing "Fool's Cap Map of the World" is based on French mathematician and cartographer Oronce Finé's cordiform, or heart-shaped, projection of the Earth. Whoever produced this iteration of Finé's map had an interesting worldview. By dressing the world in a jester's belled cap above a Latin inscription from Ecclesiastes, "The number of fools is infinite," the unknown author produced a "cosmigraphic" of a new kind. Other Latin quotations include the motto, in large letters across the top of the cap, "Oh head, worthy of a dose of hellebore." (Hellebore, a highly toxic herb, was used by the ancients to try to cure insanity, though it more frequently caused cardiac arrest and death.) As is clear from Shakespeare's *King Lear*, among other examples, at one time fools (jesters) were some of the few individuals permitted to speak truth to power.

• 1593:

In this depiction of the Revelation to John, who is visible on the island of Patmos to the lower right, a cross section of the heavens can be seen between the Earth and God's empyrean realm. As with the paintings by Francisco de Holanda depicting the creation on pages 26–29, an elongated, wandlike triangle connects heaven and Earth. Filling it, in this case, is the substance of John's revelations, which produced the final book of the New Testament, here appearing as a kind of broadcast directly from the Holy Trinity. The Ptolemaic arrangement of the universe is again clearly depicted, with the Earth, the elements, the planets, the constellations, and finally the angelic hosts in heaven. An inscription at the border between temporal and empyrean realms reads "Tenth Sphere, Where Movement Ends." An unknown printmaker produced this masterful engraving for Flemish engraver Nicolaas van Aelst, who dedicated it to Archbishop Ferdinand of Bavaria.

CAELVM EMPYREVM IN QVO
SANCTA TRINITAS. BEATA VIRGO
ANGELI ET SANCTI OMNES
VISIONES
APOCALYPSIS
Troni
Virtutes
Archangeli
Angeli
CAELVM CHRYSTALLINVM
IN QVO AQVAE OMNES
QVAE SVPER CAELOS
DECIMA SPHAERA QVA TERMINATVR MOTVS
NONA SPHAERA IN QVA LVCIDI CHARACTERES VT ARABES AFFIRMANT
OCTAVA SPHAERA IN QVA OMNES STELLAE FIXAE
SATVRNVS
IVPPITER
MARS
SOL
LVNA
VENVS
IGNIS
AER
MERCVRIVS
Sanctus Ioannes coelo aperto intuetur visiones quae in Apocalypsi continentur et eas ab Angelo scribere iubetur.
Ser.mo Principi Ferdinãdo Comiti Palatino Rheni Vtriusq. Bauariae Ducis Guilielmi F. picturae, et elegantiorũ artium admiratori max.o Dño ac Mecoenati Colen.mo
Nicolaus van Aelst Bruxellens. obseruantiae ergo Romae d.d. 1593.
GALLIAE PARS
ITALIA
HISPANIAE PARS
GRECIA
FRETVM HERCVLEVM
OCEANVS OCCIDENTALIS
AFRICAE PARS
MARE MEDITERRANEVM

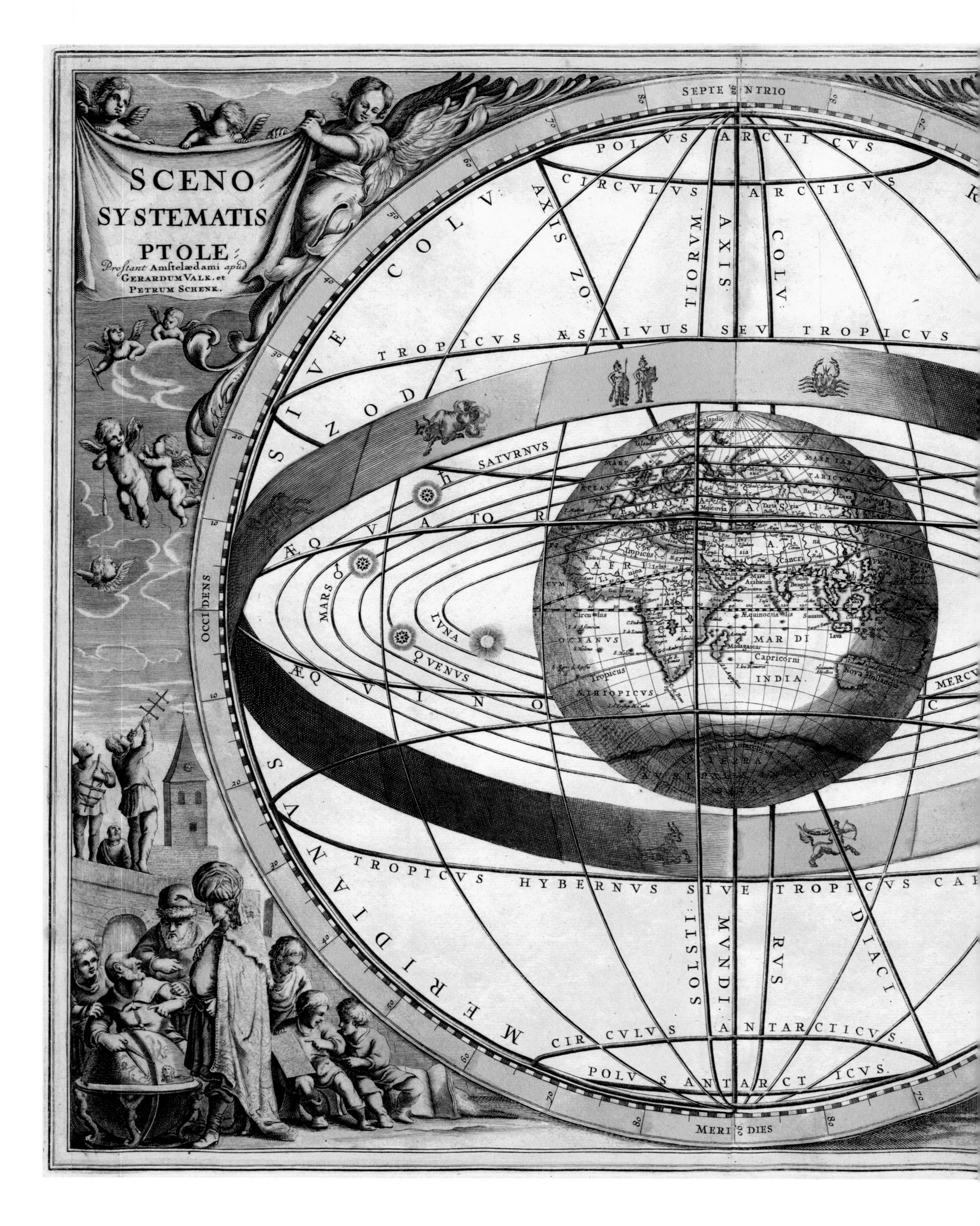

SCENO
SYSTEMATIS
PTOLE
Prostant Amstelædami apud
GERARDUM VALK. et
PETRUM SCHENK.
SEPTENTRIO
POLVS ARCTICVS
CIRCVLVS ARCTICVS
SIVE COLV
AXIS ZO
TIORVM
AXIS
COLV
TROPICVS ÆSTIVUS SEV TROPICVS
ZODI
SATVRNVS
ÆQVATOR
MARS
LVNA
VENVS
ÆQVINO
OCCIDENS
MERCV
EUROPA
ASIA
AFRICA
MARE DI
INDIA
Capricorni
Nova Hollandia
TROPICVS HYBERNVS SIVE TROPICVS CA
SOLSTI
MVNDI
RVS
DIACI
MERIDIANVS
CIRCVLVS ANTARCTICVS.
POLVS ANTARCTICVS.
MERIDIES

• 1660:

The geocentric Ptolemaic cosmos, from Andreas Cellarius's lavish Baroque *Harmonia macrocosmica* (Cosmic Harmony), one of the high points among seventeenth-century celestial atlases. Orbiting a large central Earth, the planets are depicted as starlike shapes, each identified with its traditional symbol. The zodiac, divided into twelve thirty-degree divisions of celestial longitude, defines the apparent path of the sun through the constellations as seen from Earth. The axis of the universe is defined by the terrestrial poles, and the Earth's equator is projected outward, creating a celestial equator as well (see page 41 for an earlier example). Ptolemy himself might be represented by one of the figures on the lower right, in a crumbling Alexandria, possibly also symbolic of the decline of his cosmological design, following the revolutionary findings of Copernicus. For other maps from Cellarius, see pages 84, 122–23, 147–48, 182–83, and 232–35.

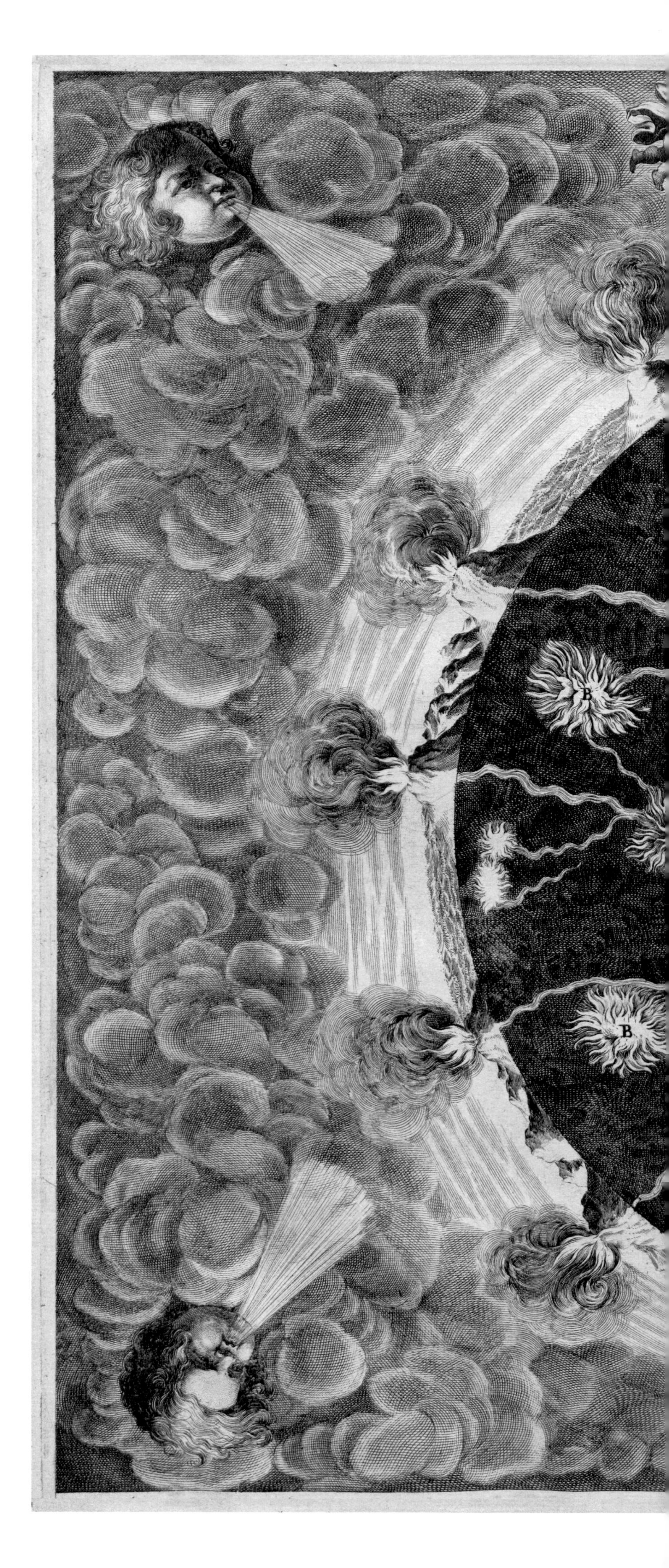

• 1664:
Depiction of a subterranean network of molten lava from German Jesuit polymath Athanasius Kircher's book *Mundus subterraneus* (Subterranean World). Kircher, who reportedly had himself lowered into the crater of a restive Mount Vesuvius in 1638, has been described as "one of the last thinkers who could rightfully claim all knowledge as his domain." He developed a theory involving intertwined ducts of water and fire reaching down to the core of the planet. By attempting to understand the subterranean structures of the planet and how they modify its surface features, Kircher could be said to have exhibited a "planetary" consciousness three centuries before the Gaia hypothesis proposed that we look at Earth as a giant self-regulating system. For Kircher's depiction of the sun, see pages 124–25.

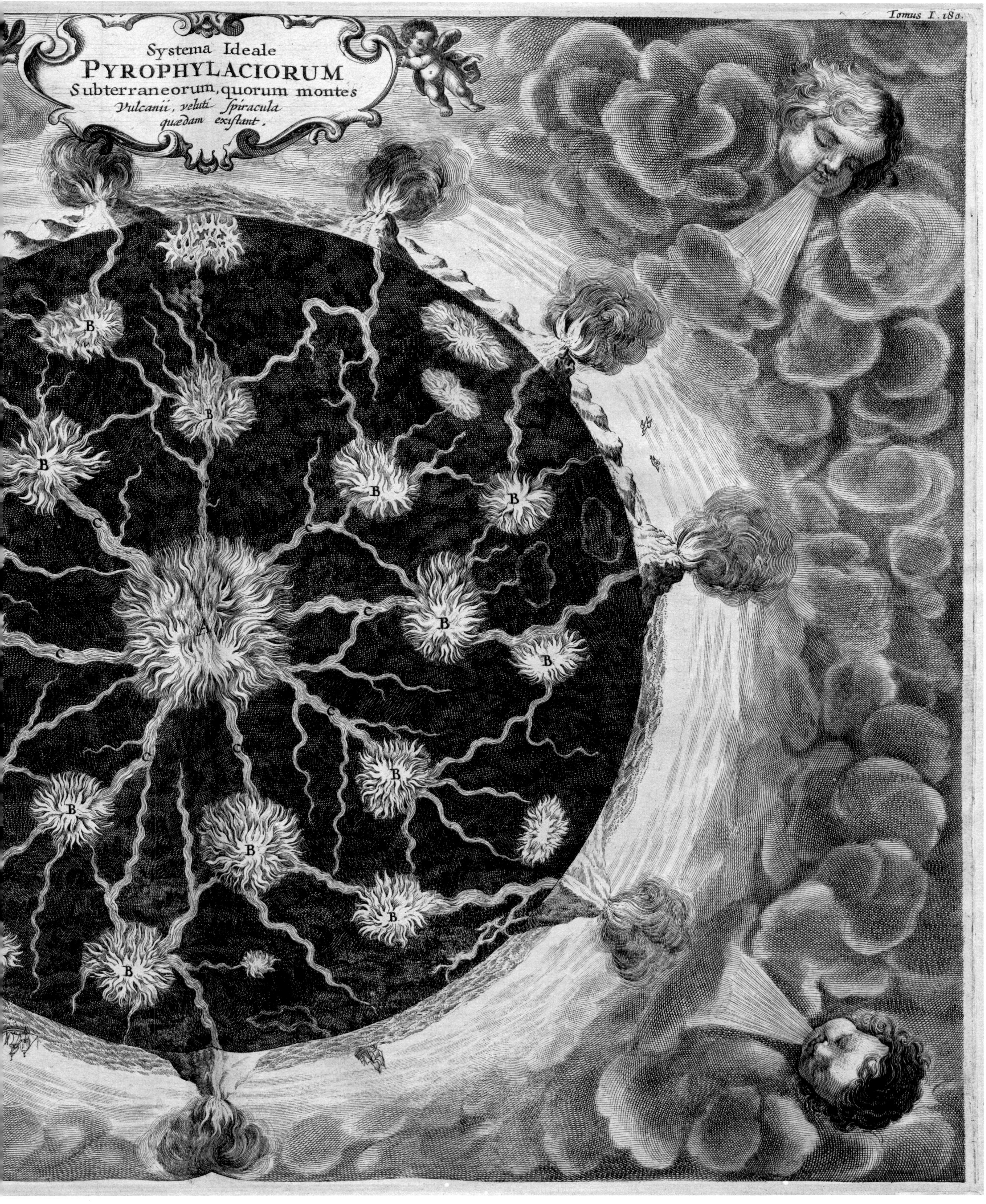

Systema Ideale
PYROPHYLACIORUM
Subterraneorum, quorum montes
Vulcanii, veluti Spiracula
quædam existant.

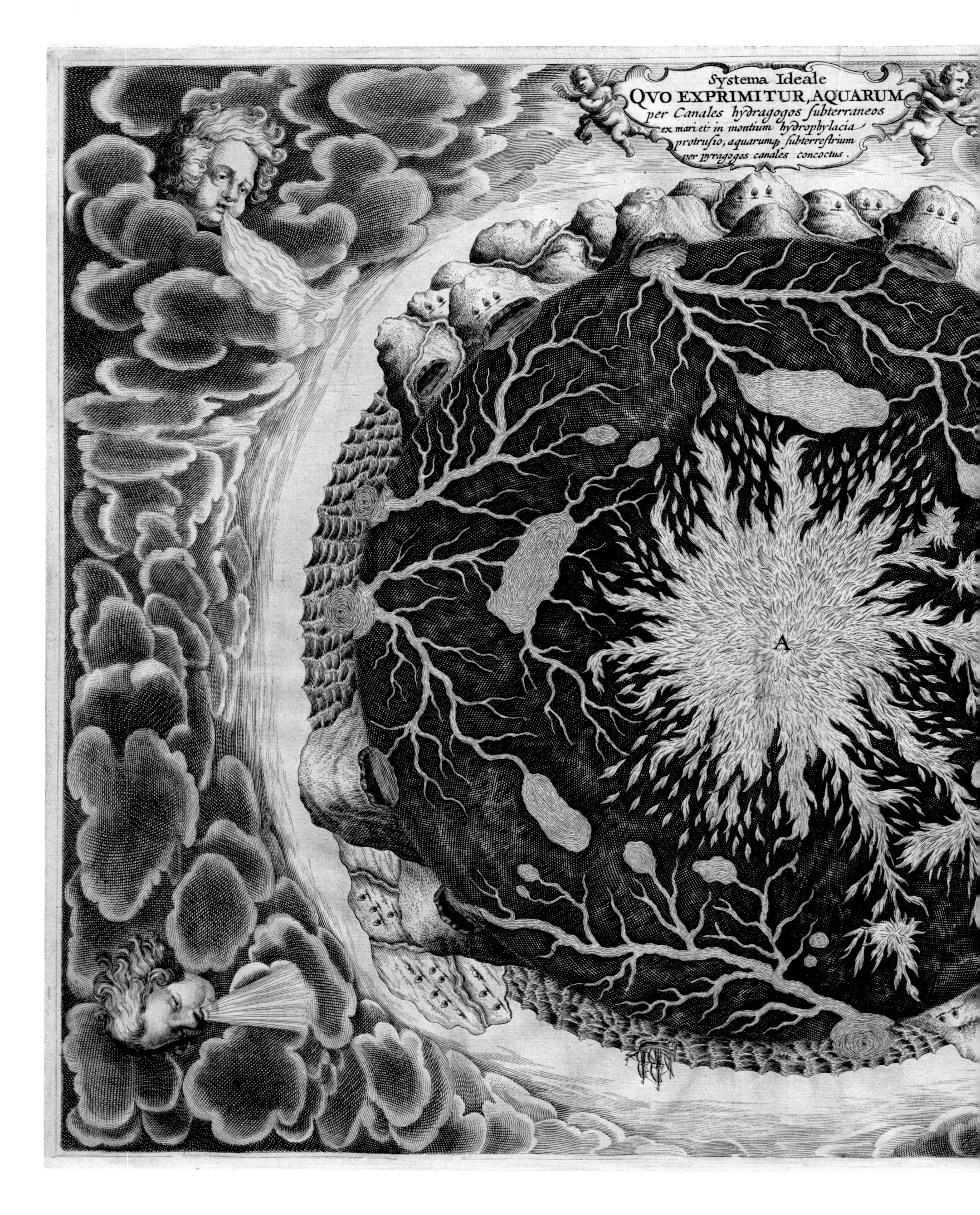
Systema Ideale
QVO EXPRIMITUR, AQUARUM
per Canales hydragogos ſubterraneos
ex mari et in montium hydrophylacia
protruſio, aquarumq; ſubterreſtrium
per pyragogos canales concoctus.
A

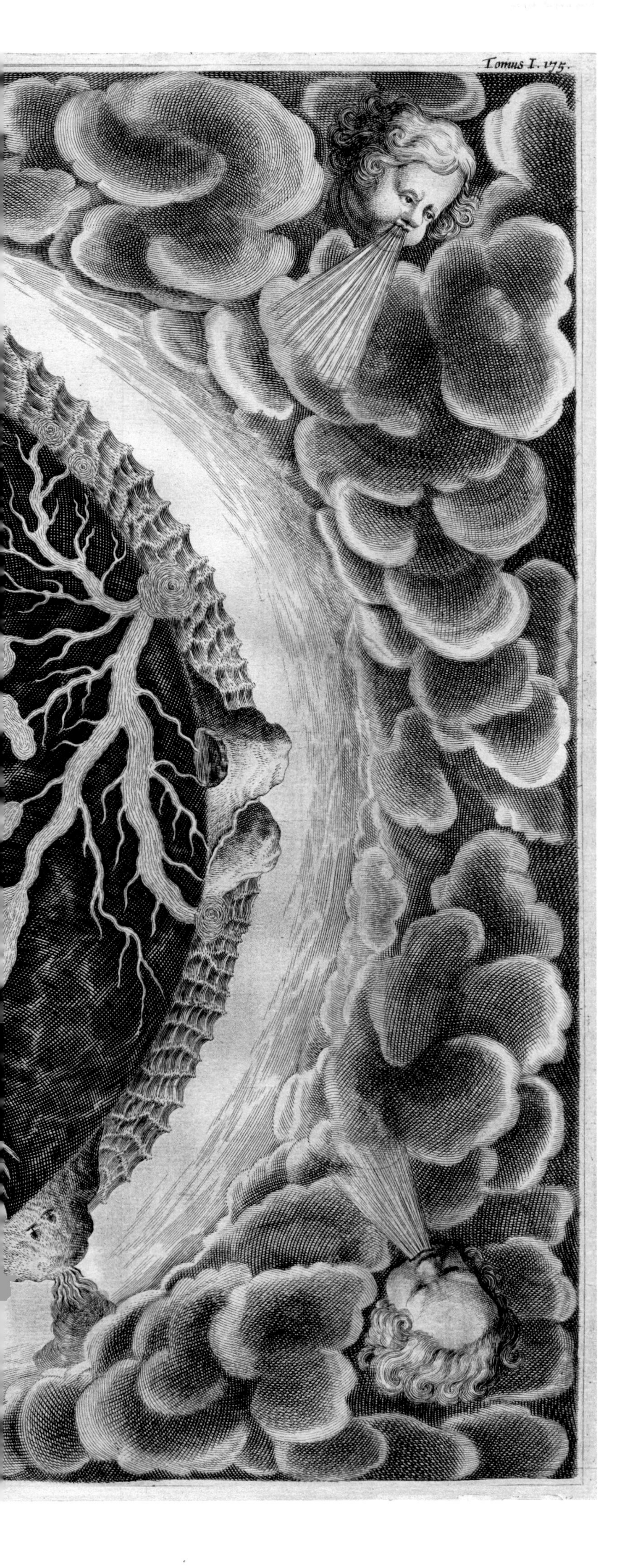

Kircher's depiction of a subsurface network of karst-style waterways, from *Mundus subterraneus*. (More than 150 years earlier, in a text almost certainly unknown to Kircher, Leonardo da Vinci had reached a similar conclusion about the subterranean world, writing "as the blood works in all animals so water does in the world, which is a living animal.") Although many of Kircher's ideas seem wildly off base today, his depictions of subsurface Earth are not entirely wrong. Research during the last two decades indicates that water penetrates farther into the planet's crust than previously supposed, with some estimates having it that, as a result, the planet possesses a considerable subsurface biosphere, mostly comprised of microorganisms hardened against the extreme pressures, temperatures, and radioactivity.

Erhardi Weigelij P.P. SPECULUM TERRÆ Cometam secundum reflectens A. 1665.
Septent.
MAR DEL NOR
Tropicus
C Verd
Æqua
Tropicus
reliquorum ab Anno 1652 hucusq. visorum vias Ostendens. I. B. Paravicin. f.

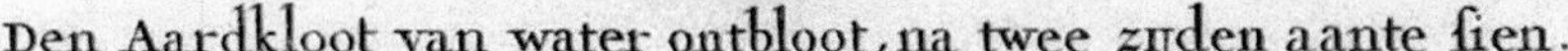

Den Aardkloot van water ontbloot, na twee zijden aante sien.

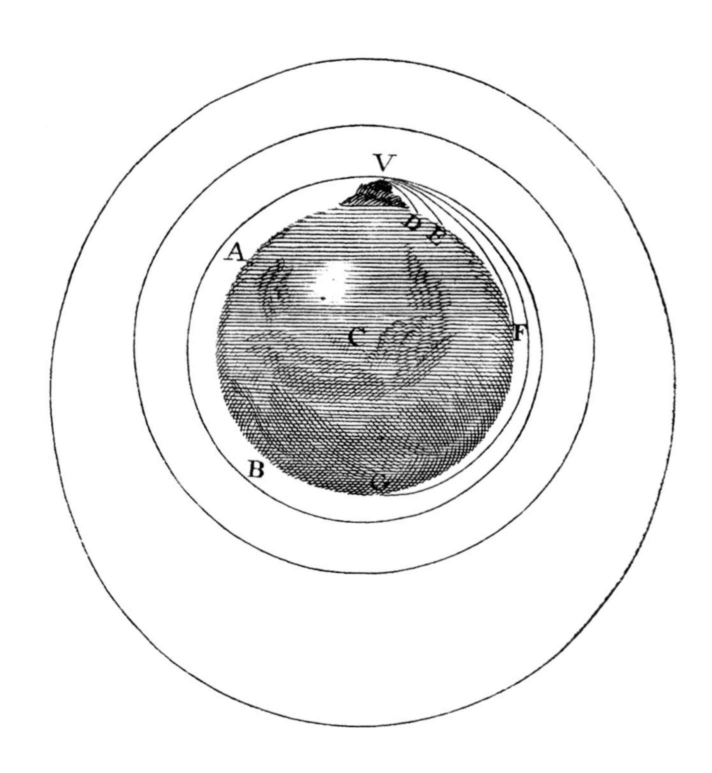

• 1728:

Above: In this illustration from a previously unpublished 1685 draft of a section of Isaac Newton's landmark *Principia*, the trajectory necessary to achieve Earth orbit is depicted—among the earliest such depictions. Titled *A Treatise of the System of the World*, the remarkably accessible volume described the effects on the Earth, moon, tides, and solar system of the principles Newton outlined in his revolutionary first book. In the end, Newton thought better of writing in such an approachable style, releasing instead a densely mathematical second volume.

• 1684:

Left: Planet Earth devoid of water, from a Dutch copy of a map by English theologian and cosmologist Thomas Burnet. In Burnet's *Telluris theoria sacra*, or "Sacred Theory of the Earth" (originally published in Latin in 1681), he wrote that a hollow Earth had likely contained most of its water in subterranean reservoirs until Noah's Flood, after which the oceans appeared. Note that California appears here as an island, albeit in this case one drained of surrounding ocean.

• 1665:

Facing page: A comet swings across the zodiac in this prescient depiction of the Earth in space from *Speculum terrae, das ist, Erd-spiegel* (Speculum Terrae, That Is, the Earth-Mirror) by Erhard Weigel. Between 1664 and 1665, two prominent comets appeared, causing widespread consternation across Europe, particularly because they came on either side of a lunar eclipse. Such a concatenation of omens was unprecedented, and was duly followed by the last major arrival of the Great Plague in London, which killed 15 percent of its population, and then the city's Great Fire in September 1666. For more on comets and eclipses, see chapters 8 and 9.

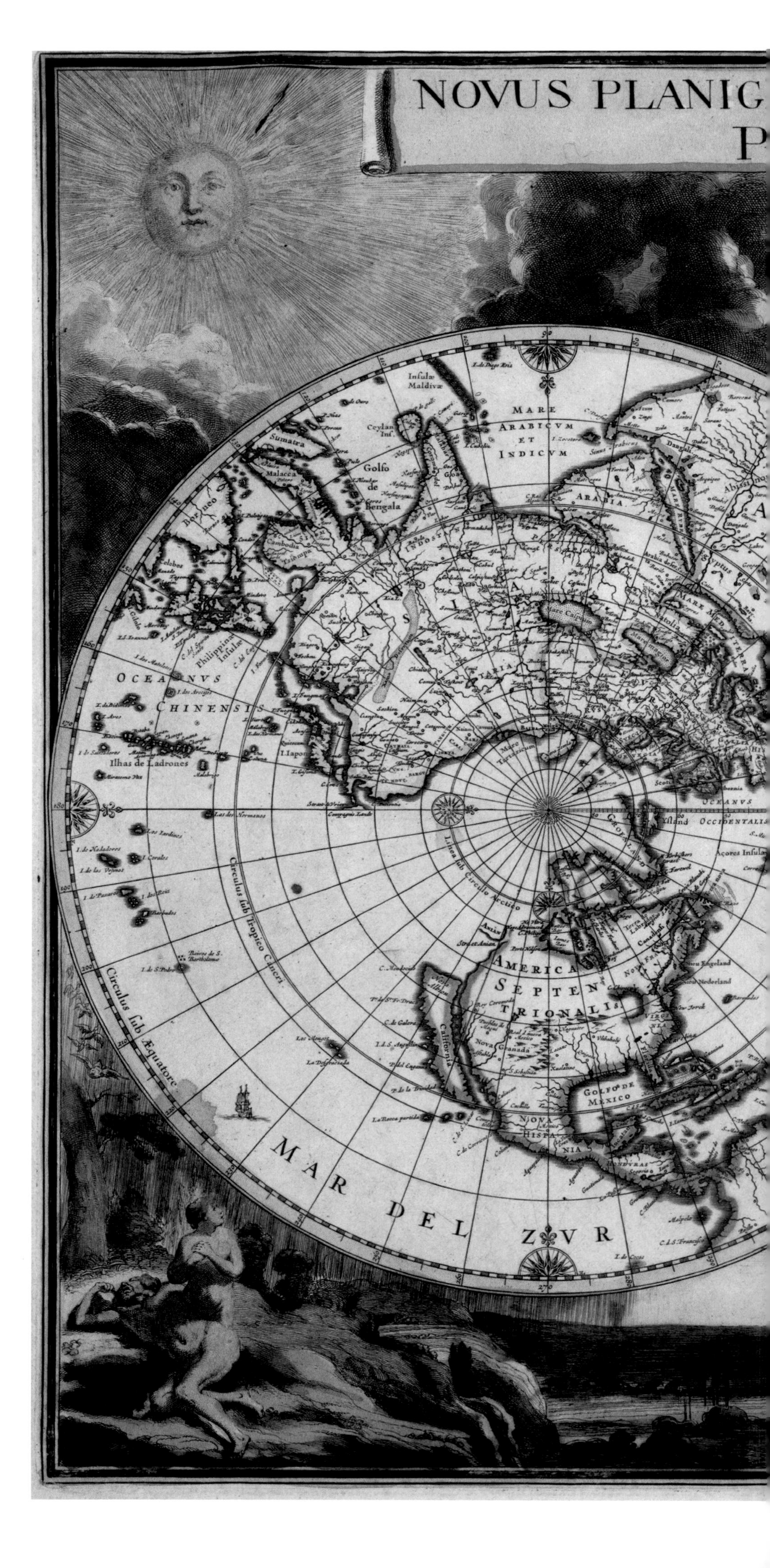

• 1695:
On this polar-oriented projection of a world map, from Dutch atlas producer Joan Blaeu, California is an island, Antarctica is still unknown, and only part of the Australian coast is delineated. In the upper corners of the map, the sun and moon emerge through clouds, with a rendition of the Earth in space surrounded by an armillary sphere given prominent placement at the top center. Although seaborne explorations were clearly the preoccupation of the era, such depictions of intricate machinery floating in a starry void seem to presage the arrival of spaceflight by 250 years. (For an earlier rendition of an armillary sphere, see page 41.)

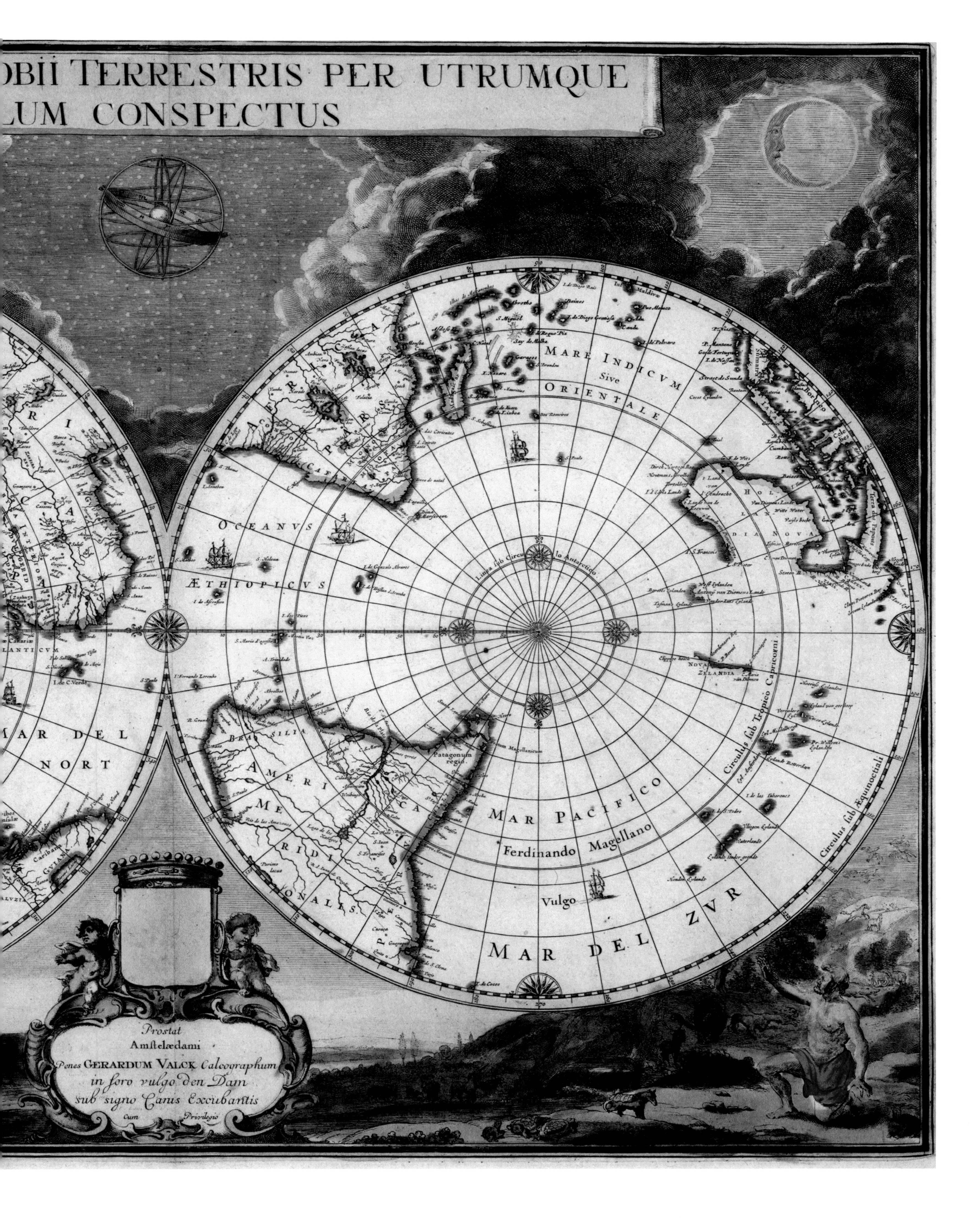
OBII TERRESTRIS PER UTRUMQUE
LUM CONSPECTUS
MARE INDICVM
Sive
ORIENTALE
OCEANVS
ÆTHIOPICVS
Linea sub Circulo Antarctico
MAR DEL
NORT
MAR PACIFICO
Ferdinando Magellano
Vulgo
MAR DEL ZVR
Circulus sub Tropico Capricorni
Circulus sub Æquinoctiali
Patagonum regio
Prostat
Amstelædami
Penes GERARDUM VALCK Calcographum
in foro vulgo den Dam
sub signo Canis Excubantis
Cum Privilegio

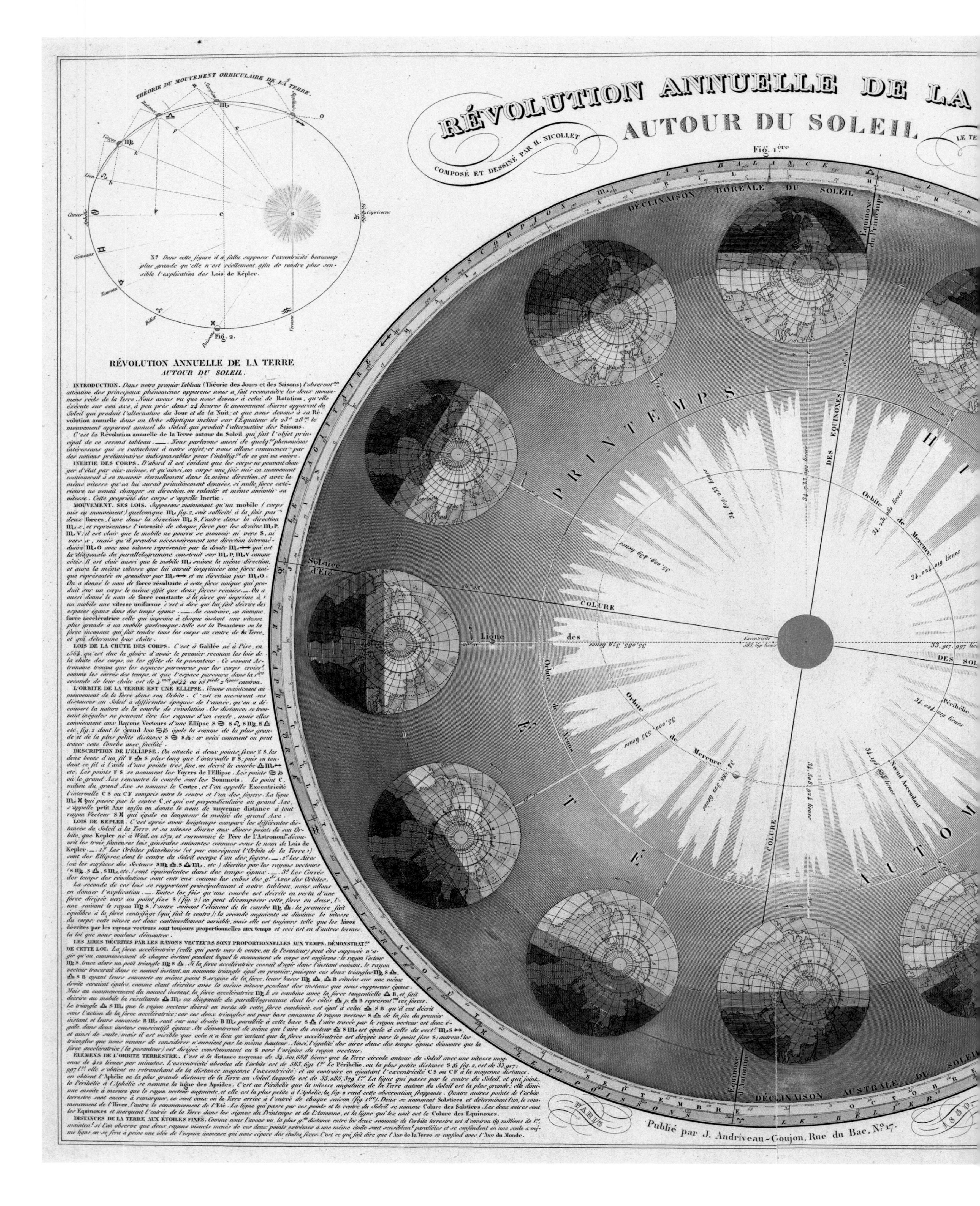
RÉVOLUTION ANNUELLE DE LA
AUTOUR DU SOLEIL
COMPOSÉ ET DESSINÉ PAR H. NICOLLET
Fig. 1.re
THÉORIE DU MOUVEMENT ORBICULAIRE DE LA TERRE.
Vierge
Balance
Scorpion
Sagittaire
Lion
Cancer
Aphélie
Gémeaux
Taureau
Bélier
Poissons
Verseau
Capricorne
Périhélie
Fig. 2.
N.B. Dans cette figure il a fallu supposer l'excentricité beaucoup plus grande qu'elle n'est réellement, afin de rendre plus sensible l'explication des Lois de Képler.
DÉCLINAISON BORÉALE DU SOLEIL
Equinoxe du Printemps
PRINTEMPS
ÉTÉ
AUTOM
Solstice d'Été
23° 28'
COLURE
Ligne des
DES SOL
DES EQUINOXES
Orbite de Mercure
Orbite de Vénus
Excentricité 583,691 lieues
Périhélie
Nœud Ascendant
Equinoxe d'Automne
DÉCLINAISON AUSTRALE DU SOLEIL
PARIS
Publié par J. Andriveau-Goujon, Rue du Bac, N°17.
RÉVOLUTION ANNUELLE DE LA TERRE
AUTOUR DU SOLEIL.
INTRODUCTION. Dans notre premier Tableau (Théorie des Jours et des Saisons) l'observat.on attentive des principaux phénomènes apparens nous a fait reconnaître les deux mouvemens réels de la Terre. Nous avons vu que nous devons à celui de Rotation, qu'elle exécute sur son axe, à peu près dans 24 heures le mouvement diurne apparent du Soleil qui produit l'alternative du Jour et de la Nuit; et que nous devons à sa Révolution annuelle dans un Orbe elliptique incliné sur l'Equateur de 23° 28' le mouvement apparent annuel du Soleil, qui produit l'alternative des Saisons.
C'est la Révolution annuelle de la Terre autour du Soleil qui fait l'objet principal de ce second tableau. —. Nous parlerons aussi de quelq.es phénomènes interessans qui se rattachent à notre sujet, et nous allons commencer par des notions préliminaires indispensables pour l'intellig.ce de ce qui va suivre.
INERTIE DES CORPS. D'abord il est évident que les corps ne peuvent changer d'état par eux-mêmes, et qu'ainsi, un corps une fois mis en mouvement continuerait à se mouvoir éternellement dans la même direction, et avec la même vitesse qu'on lui aurait primitivement donnée, si nulle force extérieure ne venait changer sa direction, ou ralentir et même anéantir sa vitesse. Cette propriété des corps s'appelle Inertie.
MOUVEMENT. SES LOIS. Supposons maintenant qu'un mobile (corps mis en mouvement) quelconque ♏ fig. 2, soit sollicité à la fois par deux forces, l'une dans la direction ♏ S, l'autre dans la direction ♏ x, et représentons l'intensité de chaque force par les droites ♏ P, ♏ V; il est clair que le mobile ne pourra se mouvoir ni vers S, ni vers x, mais qu'il prendra nécessairement une direction intermédiaire ♏ O avec une vitesse représentée par la droite ♏ ♐ qui est la diagonale du parallélogramme construit sur ♏ P, ♏ V comme côtés. Il est clair aussi que le mobile ♏ suivra la même direction, et aura la même vitesse que lui aurait imprimées une force unique représentée en grandeur par ♏ ♐ et en direction par ♏ O. On a donné le nom de force résultante à cette force unique qui produit sur un corps le même effet que deux forces réunies. — On a aussi donné le nom de force constante à la force qui imprime à un mobile une vitesse uniforme c'est à dire qui lui fait décrire des espaces égaux dans des temps égaux. — Au contraire, on nomme force accélératrice celle qui imprime à chaque instant une vitesse plus grande à un mobile quelconque: telle est la Pesanteur ou la force inconnue qui fait tendre tous les corps au centre de la Terre, et qui détermine leur chûte.
LOIS DE LA CHÛTE DES CORPS. C'est à Galilée né à Pise, en 1564, qu'est due la gloire d'avoir le premier reconnu les lois de la chûte des corps, ou les effets de la pesanteur. Ce savant Astronome trouva que les espaces parcourus par les corps graves, comme les carrés des temps, et que l'espace parcouru dans la 1.re seconde de leur chûte est de 4 mèt. 9044 ou 15 pieds 1 lignes environ.
L'ORBITE DE LA TERRE EST UNE ELLIPSE. Venons maintenant au mouvement de la Terre dans son Orbite. C'est en mesurant ses distances au Soleil à différentes époques de l'année, qu'on a découvert la nature de la courbe de révolution. Ces distances se trouvant inégales ne peuvent être les rayons d'un cercle, mais elles conviennent aux Rayons Vecteurs d'une Ellipse S♋ S♌ S♍ S♎ etc. fig. 2 dont le grand Axe ♋♑ égale la somme de la plus grande et de la plus petite distance S♋ S♑; or voici comment on peut tracer cette Courbe avec facilité.
DESCRIPTION DE L'ELLIPSE. On attache à deux points fixes F S les deux bouts d'un fil F♎S plus long que l'intervalle F S; puis en tendant ce fil à l'aide d'une pointe très fine, on décrit la courbe ♎♏♐ etc. Les points F S, se nomment les Foyers de l'Ellipse. Les points ♋ ♑ où le grand Axe rencontre la courbe sont les Sommets. Le point C, milieu du grand Axe se nomme le Centre, et l'on appelle Excentricité l'intervalle C S ou C F compris entre le centre et l'un des foyers. La ligne ♏ ♓ qui passe par le centre C, et qui est perpendiculaire au grand Axe, s'appelle petit Axe enfin on donne le nom de moyenne distance à tout rayon Vecteur S ♓ qui égale en longueur la moitié du grand Axe.
LOIS DE KEPLER. C'est après avoir longtemps comparé les différentes distances du Soleil à la Terre, et sa vitesse diurne aux divers points de son Orbite, que Kepler né à Weil, en 1571, et surnommé le Père de l'Astronom.ie découvrit les trois fameuses lois générales suivantes connues sous le nom de Lois de Kepler. — 1.° Les Orbites planétaires (et par conséquent l'Orbite de la Terre) sont des Ellipses dont le centre du Soleil occupe l'un des foyers. — 2.° Les Aires (ou les surfaces des Secteurs S♍♎, S♎♏, etc.) décrites par les rayons vecteurs (S♍, S♎, S♏, etc.) sont équivalentes dans des temps égaux. — 3.° Les Carrés des temps des révolutions sont entr'eux comme les cubes des g.ds Axes des Orbites.
La seconde de ces lois se rapportant principalement à notre tableau, nous allons en donner l'explication. — Toutes les fois qu'une courbe est décrite en vertu d'une force dirigée vers un point fixe S (fig. 2) on peut décomposer cette force en deux, l'une suivant le rayon ♍ S, l'autre suivant l'élément de la courbe ♍♎: la première fait équilibre à la force centrifuge (qui fuit le centre); la seconde augmente ou diminue la vitesse du corps; cette vitesse est donc continuellement variable, mais elle est toujours telle que les Aires décrites par les rayons vecteurs sont toujours proportionnelles aux temps et ceci est en d'autres termes la loi que nous voulons démontrer.
LES AIRES DÉCRITES PAR LES RAYONS VECTEURS SONT PROPORTIONNELLES AUX TEMPS. DÉMONSTRAT.on DE CETTE LOI. La force accélératrice (celle qui porte vers le centre, ou la Pesanteur) peut être supposée n'agir qu'au commencement de chaque instant pendant lequel le mouvement du corps est uniforme: le rayon Vecteur ♍ S, trace alors un petit triangle ♍ S ♎. Si la force accélératrice cessait d'agir dans l'instant suivant, le rayon vecteur tracerait dans ce nouvel instant, un nouveau triangle égal au premier; puisque ces deux triangles ♍ S ♎, ♎ S B ayant leurs sommets au même point S, origine de la force, leurs bases ♍♎, ♎B situées sur une même droite seraient égales, comme étant décrites avec la même vitesse pendant des instans que nous supposons égaux. Mais au commencement du nouvel instant, la force accélératrice ♍ k se combine avec la force tangentielle ♎ B, et fait décrire au mobile la résultante ♎♏ ou diagonale du parallélogramme dont les côtés ♎ p, ♎ B représent.nt ces forces. Le triangle ♎ S ♏ que le rayon vecteur décrit en vertu de cette force combinée, est égal à celui ♎ S B qu'il eut décrit sans l'action de la force accélératrice; car ces deux triangles ont pour base commune le rayon vecteur S ♎ de la fin du premier instant, et leurs sommets B ♏ sont sur une droite B ♏ parallèle à cette base S ♎ l'aire tracée par le rayon vecteur est donc égale, dans deux instans consécutifs égaux. On démontrerait de même que l'aire du secteur ♎ S ♏ est égale à celle du sect.r ♏ S ♐, et ainsi de suite; mais il est visible que cela n'a lieu qu'autant que la force accélératrice est dirigée vers le point fixe S; autrem.t les triangles que nous venons de considérer n'auraient pas la même hauteur. Ainsi, l'égalité des aires dans des temps égaux démontre que la force accélératrice (la pesanteur) est dirigée constamment en S vers l'origine du rayon vecteur.
ÉLÉMENS DE L'ORBITE TERRESTRE. C'est à la distance moyenne de 34,501,688 lieues que la Terre circule autour du Soleil avec une vitesse moyenne de 412 lieues par minutes. L'excentricité absolue de l'orbite est de 583,691 l.es Le Périhélie, ou la plus petite distance S ♑ fig. 2, est de 33,917,997 l.es elle s'obtient en retranchant de la distance moyenne l'excentricité; et au contraire en ajoutant l'excentricité C S ou C F à la moyenne distance, on obtient l'Aphélie ou la plus grande distance de la Terre au Soleil, laquelle est de 35,085,379 l.es La ligne qui passe par le centre du Soleil, et qui joint le Périhélie à l'Aphélie se nomme la ligne des Apsides. C'est au Périhélie que la vitesse angulaire de la Terre autour du Soleil est la plus grande; elle diminue ensuite à mesure que le rayon vecteur augmente, et elle est la plus petite à l'Aphélie; la fig. 2 rend cette observation frappante. Quatre autres points de l'orbite terrestre sont encore à remarquer, ce sont ceux où la Terre arrive à l'entrée de chaque saison (fig. 1.re). Deux se nomment Solstices et déterminent l'un, le commencement de l'Hiver, l'autre le commencement de l'Eté. La ligne qui passe par ces points et le centre du Soleil se nomme Colure des Solstices. Les deux autres sont les Equinoxes et marquent l'entrée de la Terre dans les signes du Printemps et de l'Automne, et la ligne qui les unit est le Colure des Equinoxes.
DISTANCES DE LA TERRE AUX ÉTOILES FIXES. Comme nous l'avons vu, la plus g.de distance entre les deux sommets de l'orbite terrestre est d'environ 69 millions de l.es, mainten.t si l'on observe que deux rayons visuels menés de ces deux points extrêmes à une même étoile sont sensiblem.t parallèles et se confondent en une seule & même ligne, on se fera à peine une idée de l'espace immense qui nous sépare des étoiles fixes. C'est ce qui fait dire que l'Axe de la Terre se confond avec l'Axe du Monde.

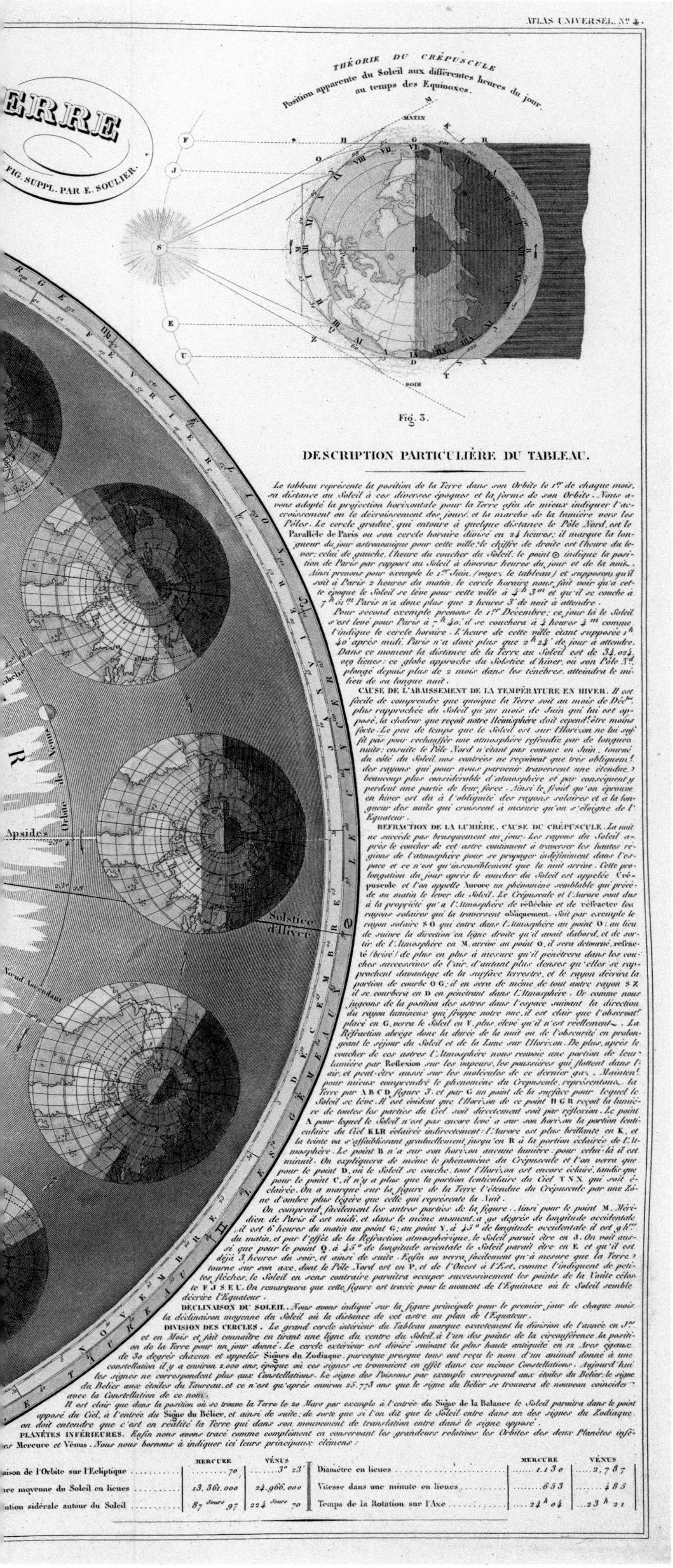

• 1850:

Depiction of the annual revolution of the Earth around the sun, by H. Nicollet in the *Atlas classique et universel de geographie ancienne et moderne* (Classic and Universal Atlas of Ancient and Modern Geography). Note that seasonal changes can clearly be seen, with the northern polar region exposed to sunlight in summer but shaded in winter. Here the Ptolemaic system has long been replaced by the Copernican scheme, in which the Earth revolves around the sun. For more on this transition, see chapter 5.

• 1881:

Illustration from French astronomer and prolific author Camille Flammarion's best-selling *L'Astronomie populaire* (Popular Astronomy). The caption reads "Carried away by time, vaulting towards a vanishing goal, Earth rolls speedily through space." Apart from his science books, he also wrote science fiction.

Un missionnaire du moyen âge raconte qu'il avait trouvé le point où le ciel et la Terre se touchent...

• 1888:

One of the most famous visual depictions ever created of the human situation within the machinery of a vast and complex cosmos. Captioned "A missionary of the Middle Ages recounts that he had found the point where heaven and Earth meet," the engraving first appeared in Camille Flammarion's *L'Atmosphère: météorologie populaire* (The Atmosphere: Popular Meteorology). Flammarion was apprenticed to an engraver in his teens, and it is believed that most of the many illustrations in his prolific output of more than fifty books were based on his own drawings. Text on the facing page helps clarify that the illustration was purpose-made for this book. Contrary to the impression left by this engraving, which constitutes a nineteenth-century idea of the medieval worldview, by the Middle Ages it was well accepted that the Earth was spherical.

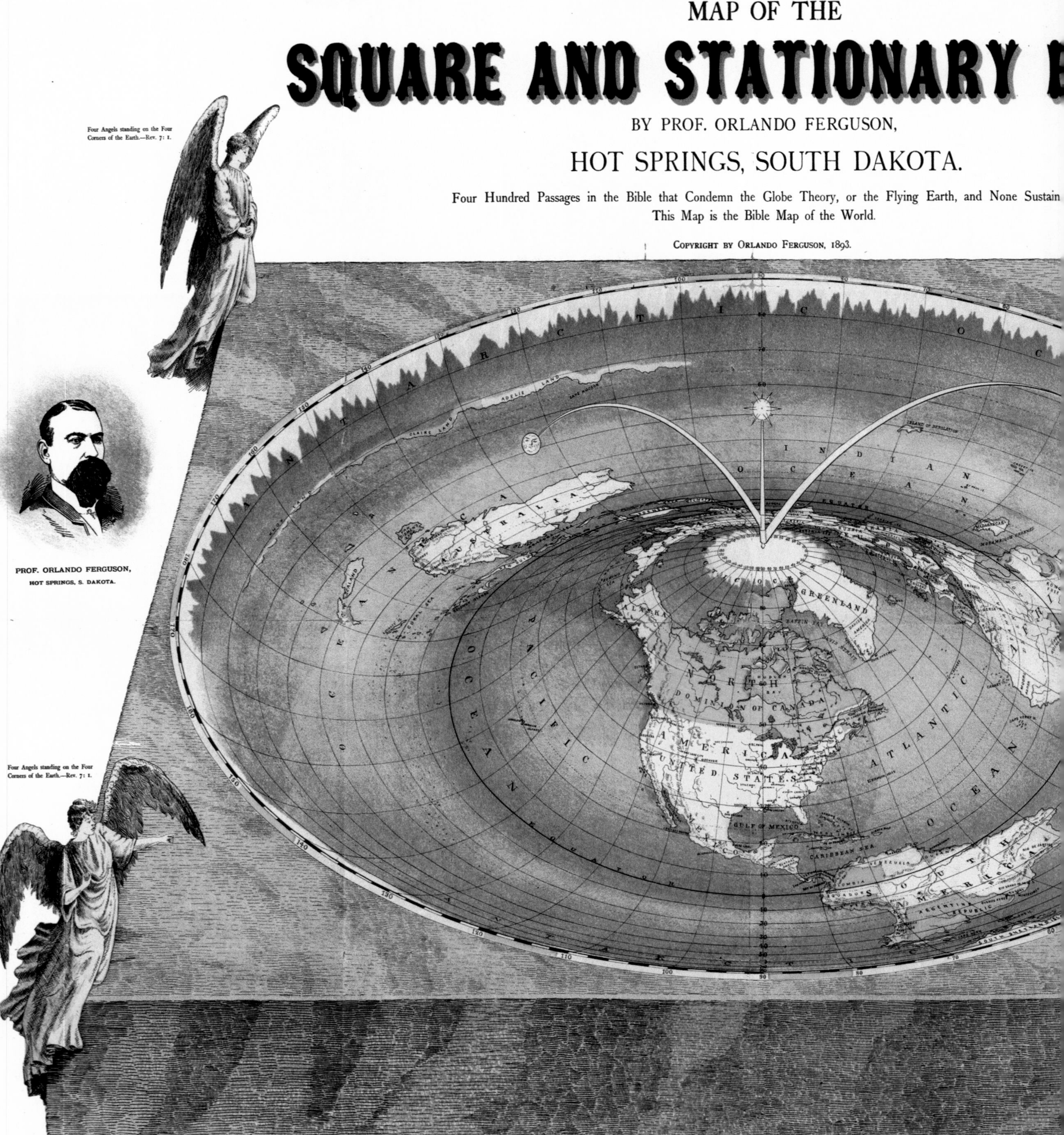

SCRIPTURE THAT CONDEMNS THE GLOBE THEORY.

And his hands were steady until the going down of the sun.—Ex. 17: 12. And the sun stood still, and the moon stayed.—Joshua 10: 12–13. The world also shall be stable that it be not moved.—Chron. 16: 30. To him that stretched out the earth, and made great lights (not worlds).—Ps. 136: 6–7. The sun shall be darkened in his going forth.—Isaiah 12: 10. The four corners of the earth.—Isaiah 11: 12. The whole earth is at rest.—Isaiah 14: 7. The prophecy concerning the globe theory.—Isaiah: 29th chapter. Woe to the rebellious children, sayeth the Lord, that take counsel, but not of me.—Isaiah 30: 1. So the sun returned ten degrees.—Isaiah 38: 8–9. It is he that sitteth upon the circle of the earth.—Isaiah 40: 22. He that spread forth the earth.—Isaiah 52: 5. That spreadeth abroad the earth by myself.—Isaiah 54: 24. My hand also hath laid the foundation of the earth.—Isaiah 58: 13. Thus sayeth the Lord, which giveth the sun for a light by day, and the moon and stars for a light by night (not worlds).—Jer. 31: 35–36. The sun shall be turned into darkness, and the moon into blood.—Acts 2: 20.

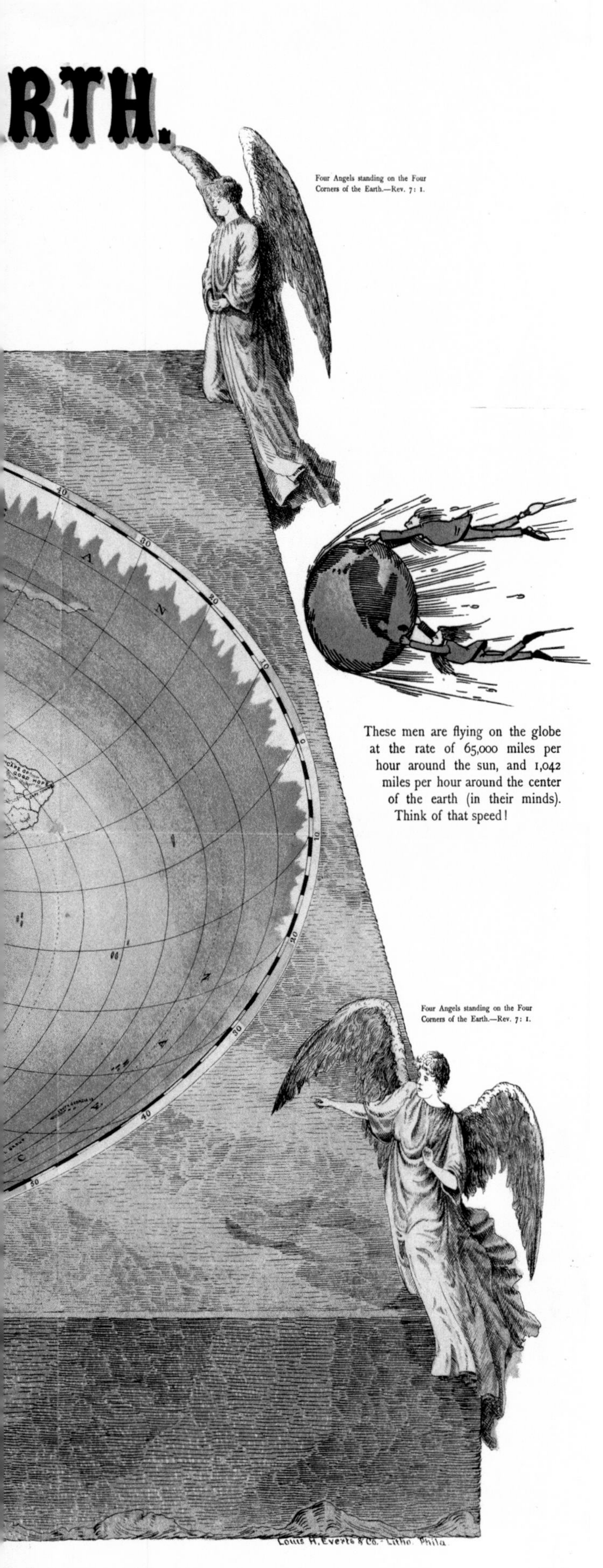

RTH.
Four Angels standing on the Four Corners of the Earth.—Rev. 7: 1.
These men are flying on the globe at the rate of 65,000 miles per hour around the sun, and 1,042 miles per hour around the center of the earth (in their minds). Think of that speed!
Four Angels standing on the Four Corners of the Earth.—Rev. 7: 1.
d 25 Cents to the Author, Prof. Orlando Ferguson, for a book
xplaining this Square and Stationary Earth. It Knocks the Globe Theory
Clean Out. It will Teach You How to Foretell Eclipses. It is Worth
Its Weight in Gold.

• 1893:

Left: Professor Orlando Ferguson refutes the "globe theory" in this broadsheet bulletin from Hot Springs, South Dakota, proposing instead a kind of four-cornered, roulette-wheel world. Note the sun, moon, and north star all suspended on wands projecting from the pole. Ferguson's cosmology didn't catch on.

• 1944:

Overleaf: Starting in the modern era with English geologist William Smith in the early nineteenth century, the idea of mapping geology augmented the communication of spatial information in cartography. Because geology deals with history and processes that unfold in time, this introduced the element of time into mapping. In the early 1940s, geologist Harold N. Fisk, of Louisiana State University, conducted a comprehensive geological survey of the alluvial plain of the lower Mississippi, producing a stunning series of maps for the Army Corps of Engineers. By charting the multiple overlapping meanders of the waterway through time, he produced a beautiful temporal tracery. For geographic maps of the moon and the other planets and moons of the solar system, see chapters 3 and 6.

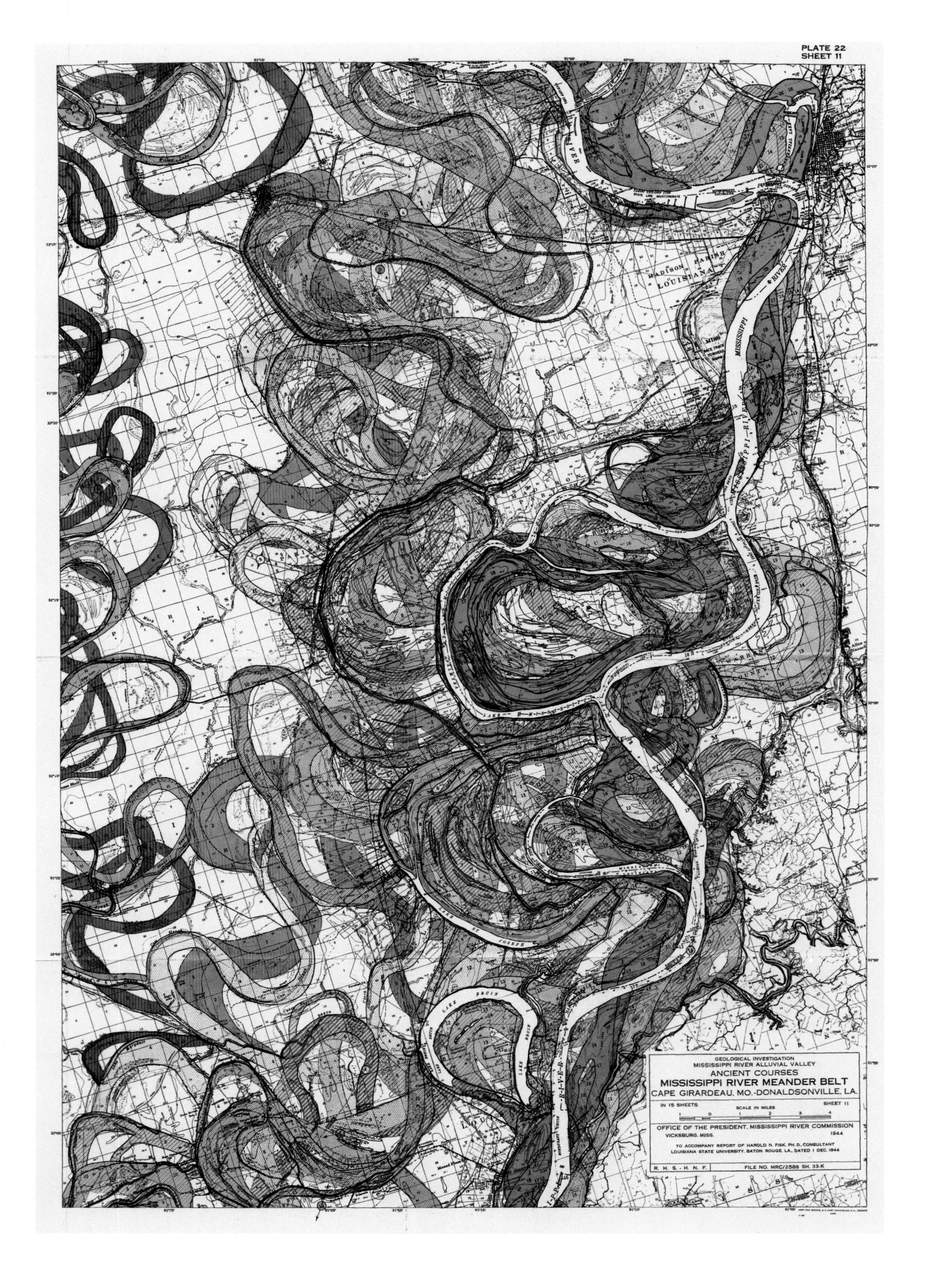
PLATE 22
SHEET 11
MADISON PARISH
LOUISIANA
MISSISSIPPI RIVER
LAKE BRUIN
LAKE ST JOSEPH
GEOLOGICAL INVESTIGATION
MISSISSIPPI RIVER ALLUVIAL VALLEY
ANCIENT COURSES
MISSISSIPPI RIVER MEANDER BELT
CAPE GIRARDEAU, MO.-DONALDSONVILLE, LA.
IN 15 SHEETS
SHEET 11
SCALE IN MILES
1 0 1 2 3 4
OFFICE OF THE PRESIDENT, MISSISSIPPI RIVER COMMISSION
VICKSBURG, MISS.
1944
TO ACCOMPANY REPORT OF HAROLD N. FISK, PH. D., CONSULTANT
LOUISIANA STATE UNIVERSITY, BATON ROUGE, LA., DATED 1 DEC. 1944
R. H. S. - H. N. F.
FILE NO. MRC/2588 SH. 33-K

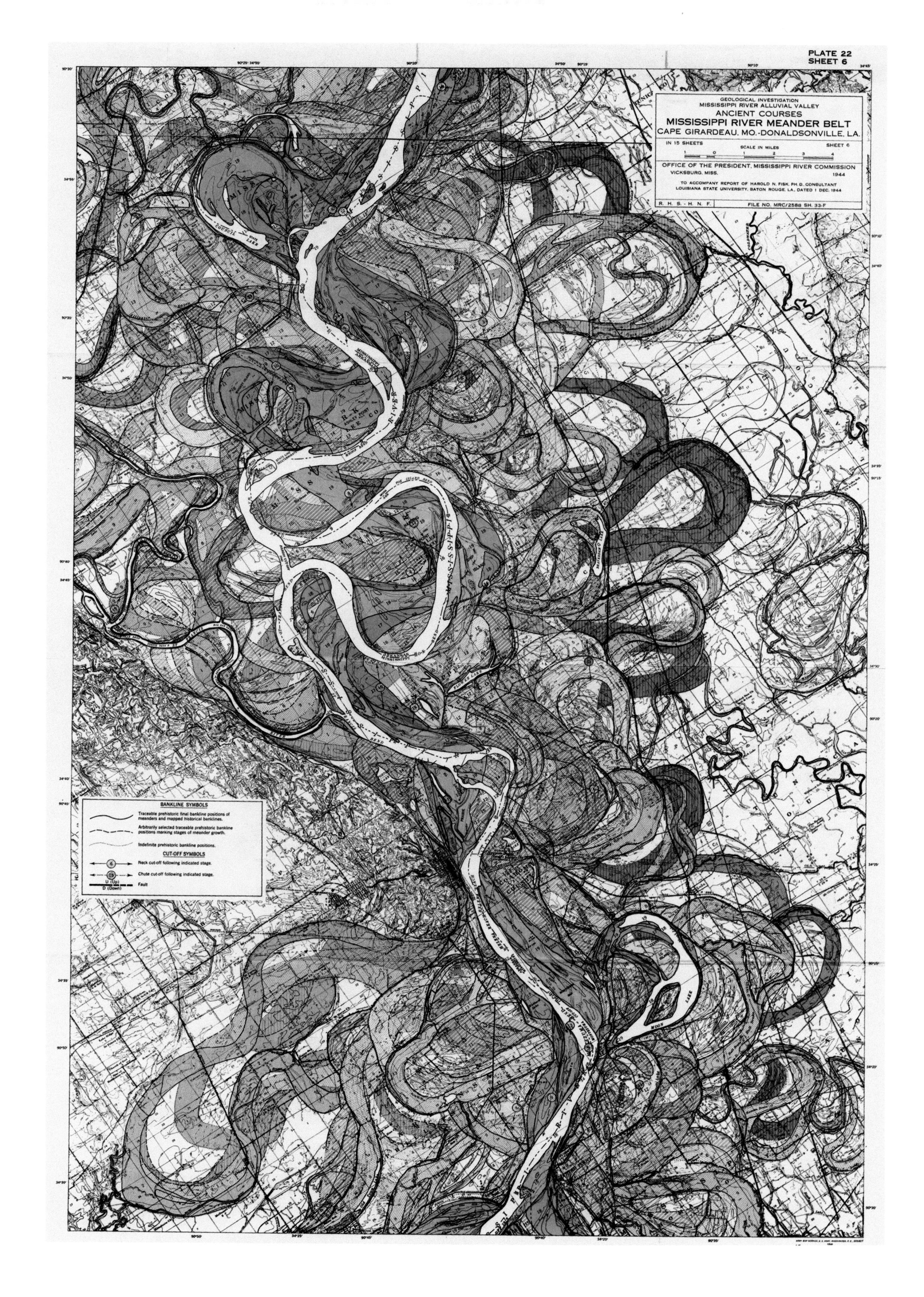

PLATE 22
SHEET 6
GEOLOGICAL INVESTIGATION
MISSISSIPPI RIVER ALLUVIAL VALLEY
ANCIENT COURSES
MISSISSIPPI RIVER MEANDER BELT
CAPE GIRARDEAU, MO.-DONALDSONVILLE, LA.
IN 15 SHEETS
SHEET 6
SCALE IN MILES
1 0 1 2 3 4
OFFICE OF THE PRESIDENT, MISSISSIPPI RIVER COMMISSION
VICKSBURG, MISS.
1944
TO ACCOMPANY REPORT OF HAROLD N. FISK, PH. D. CONSULTANT
LOUISIANA STATE UNIVERSITY, BATON ROUGE, LA., DATED 1 DEC. 1944
R. H. S. - H. N. F.
FILE NO. MRC/2588 SH. 33-F
BANKLINE SYMBOLS
Traceable prehistoric final bankline positions of meanders and mapped historical banklines.
Arbitrarily selected traceable prehistoric bankline positions marking stages of meander growth.
Indefinite prehistoric bankline positions.
CUT-OFF SYMBOLS
6
Neck cut-off following indicated stage.
15
Chute cut-off following indicated stage.
U (Up)
D (Down)
Fault

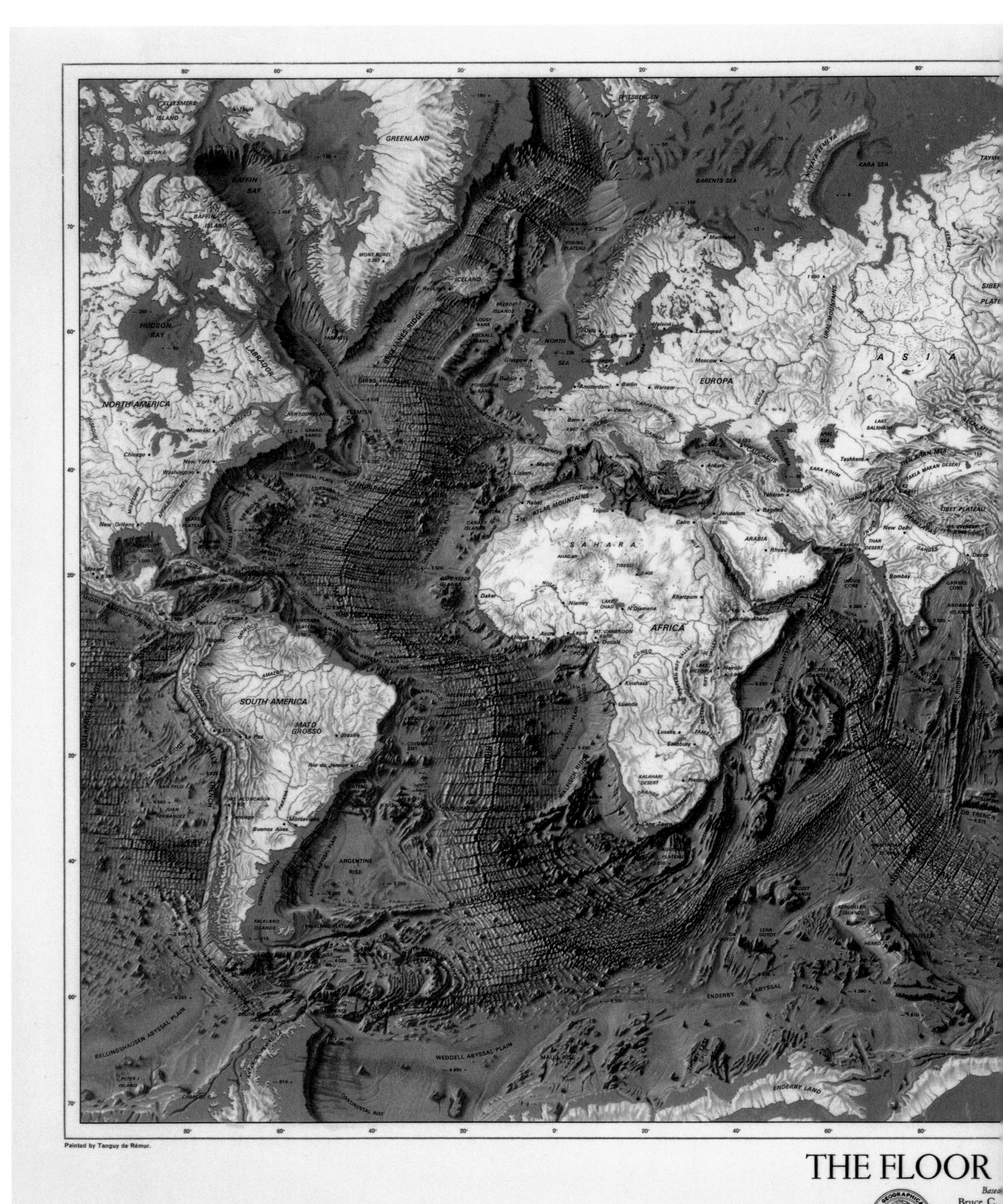

THE FLOOR
SUPPORTED B
OFFICE
Painted by Tanguy de Rémur.
GREENLAND
BAFFIN BAY
BAFFIN ISLAND
ELLESMERE ISLAND
HUDSON BAY
LABRADOR
NORTH AMERICA
ICELAND
SPITSBERGEN
BARENTS SEA
KARA SEA
NORTH SEA
EUROPA
URAL MOUNTAINS
A S I A
ATLAS MOUNTAINS
S A H A R A
ARABIA
AFRICA
SOUTH AMERICA
MATO GROSSO
ARGENTINE RISE
KALAHARI DESERT
MADAGASCAR
WEDDELL ABYSSAL PLAIN
BELLINGSHAUSEN ABYSSAL PLAIN
ENDERBY LAND
ENDERBY ABYSSAL PLAIN
New Orleans
Chicago
Montréal
New York
Washington
London
Berlin
Warsaw
Moscow
Cairo
Jerusalem
Lagos
Kinshasa
Dakar
New Delhi
Bombay
Tashkent
Buenos Aires
Santiago
Montevideo
Rio de Janeiro
Brasília
La Paz
Quito
Bogotá

THE OCEANS

tudies by
Marie Tharp
ical Observatory
New York, 10964
D STATES NAVY
ESEARCH

work exposed the existence of a continuous rift running down the middle of the mid-Atlantic ridge, clear evidence that the then widely dismissed theory of continental drift must be true. She and Heezen went on to map the entire ocean floor of Earth, in 1976 producing their magnum opus—the complete map of all the world's oceans seen here. Tharp's initial discovery of a tectonic plate boundary seam on the Atlantic seabed led to the finding of similarly irregular but continuous rifts in the Indian and Pacific Oceans. With this map, the entire surface of the Earth had finally been charted to a high degree of accuracy.

Pacific Ocean - Plate Tectonics

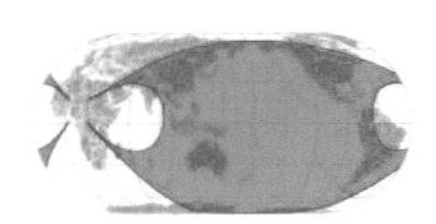

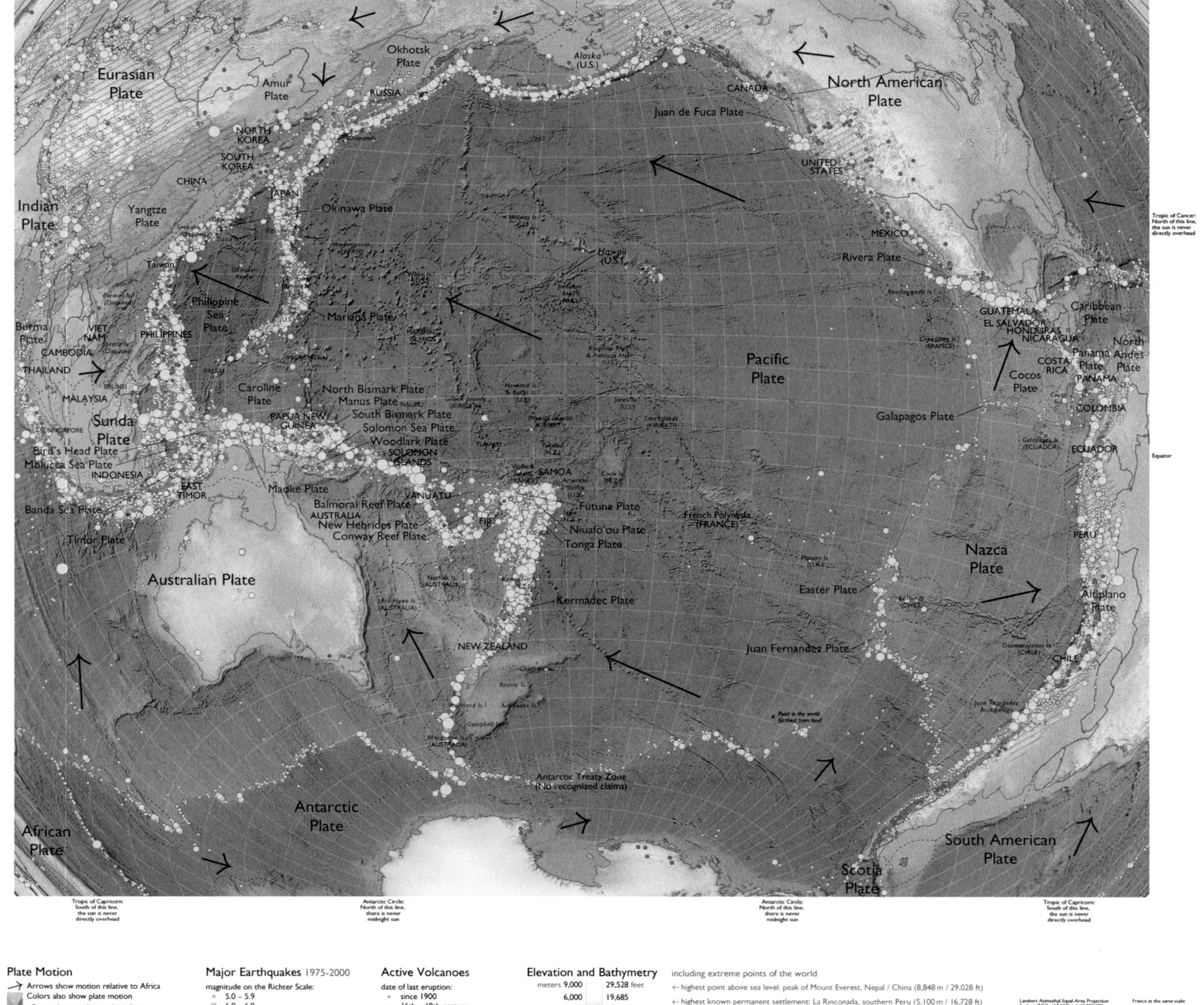

• 2006:

In this map of Pacific Ocean plate tectonics by Bill Rankin, recorded earthquake data is overlaid as yellow discs, illuminating the clear correlation between plate borders and quake frequency and intensity. Volcanoes, visible as purple dots, also line the plate boundaries. Arrows indicate plate motion. Such sophisticated multilayered mapping has produced new ways to visualize data and conceptualize natural processes.

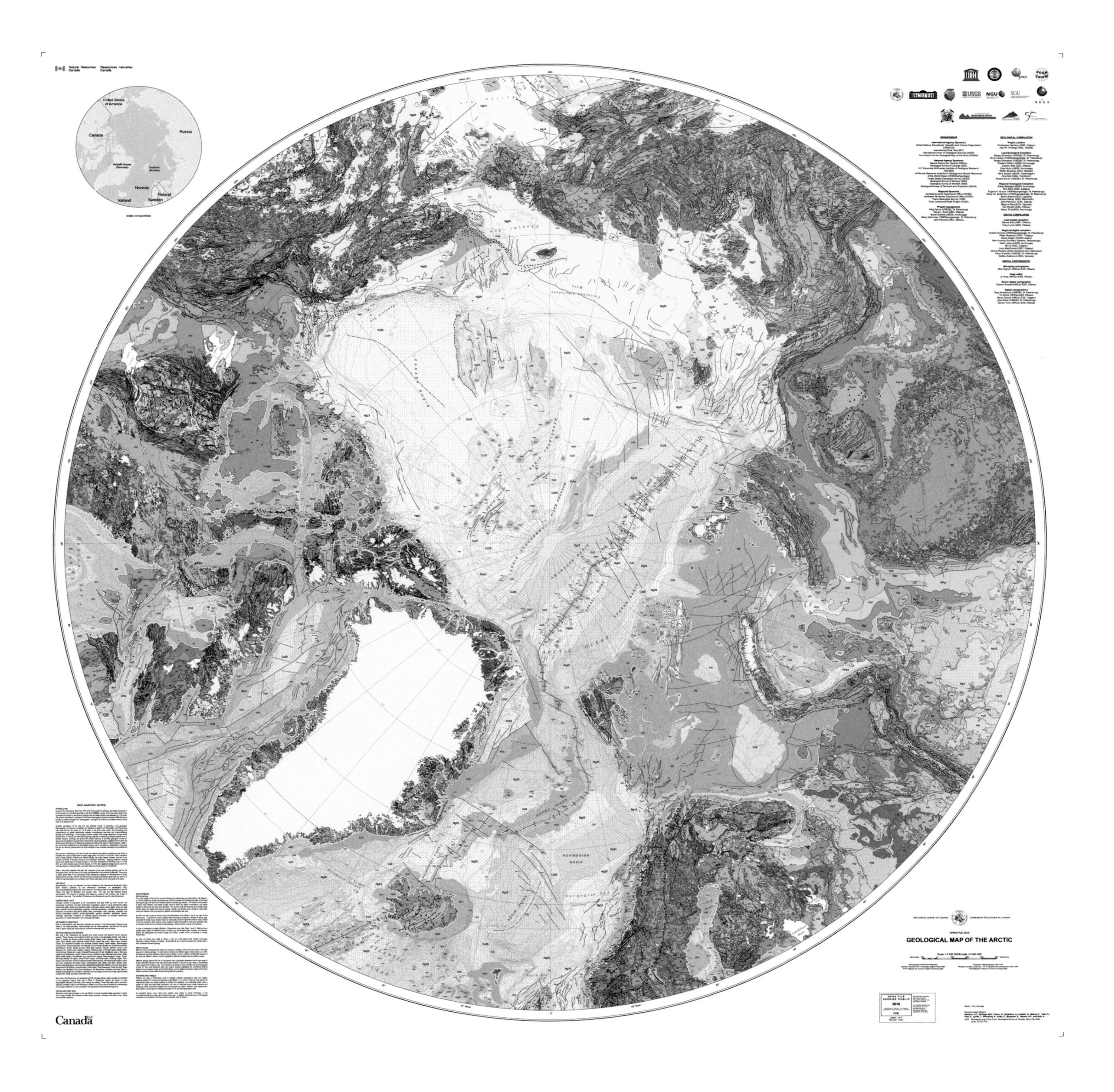

• 2008:
Geological map of the Arctic by the Canadian Geological Survey, featuring complete coverage of all inshore and offshore bedrock down to sixty degrees north. The color shades are used to define the types of rock, with volcanic rocks in shades of green; yellow, gray, and brown for clastic rocks (comprised of fragments of preexisting minerals); blue for carbonate rocks (those containing the carbonate ion); etc. Such maps show the great distance cartography has traveled since entire continents had yet to be discovered, let alone charted. For geographic maps of the moon and the other planets and moons of the solar system, see chapters 3 and 6.

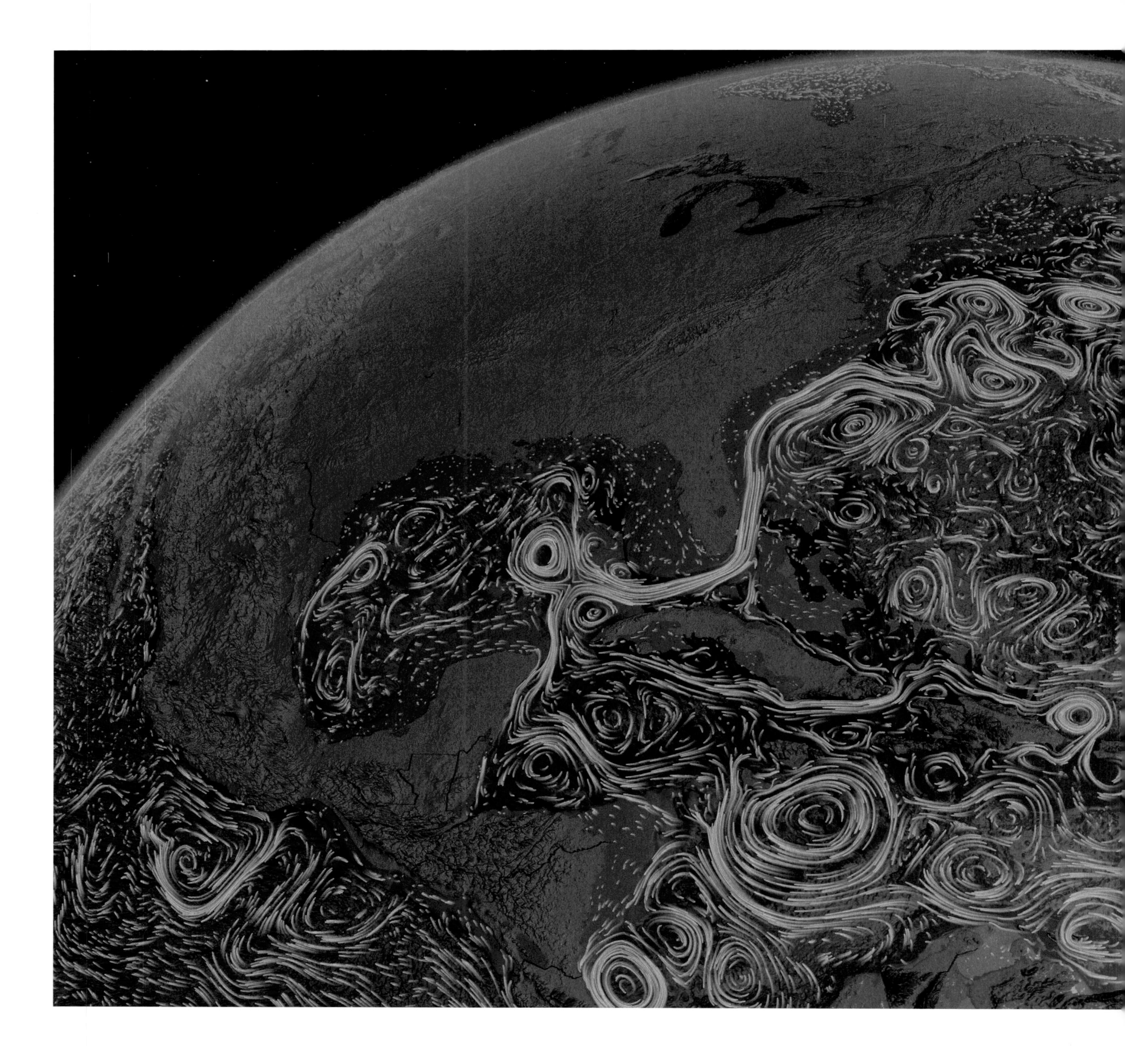

• 2011:

In the early twenty-first century, attempts to visualize such complex ephemeral phenomena as ocean currents, wind direction, and speed grew increasingly sophisticated as the volume of real-time data increased and supercomputers proved capable of processing it. This ocean surface current visualization was produced by NASA's Goddard Space Flight Center's Scientific Visualization Studio and based on data collected from 2005–7.

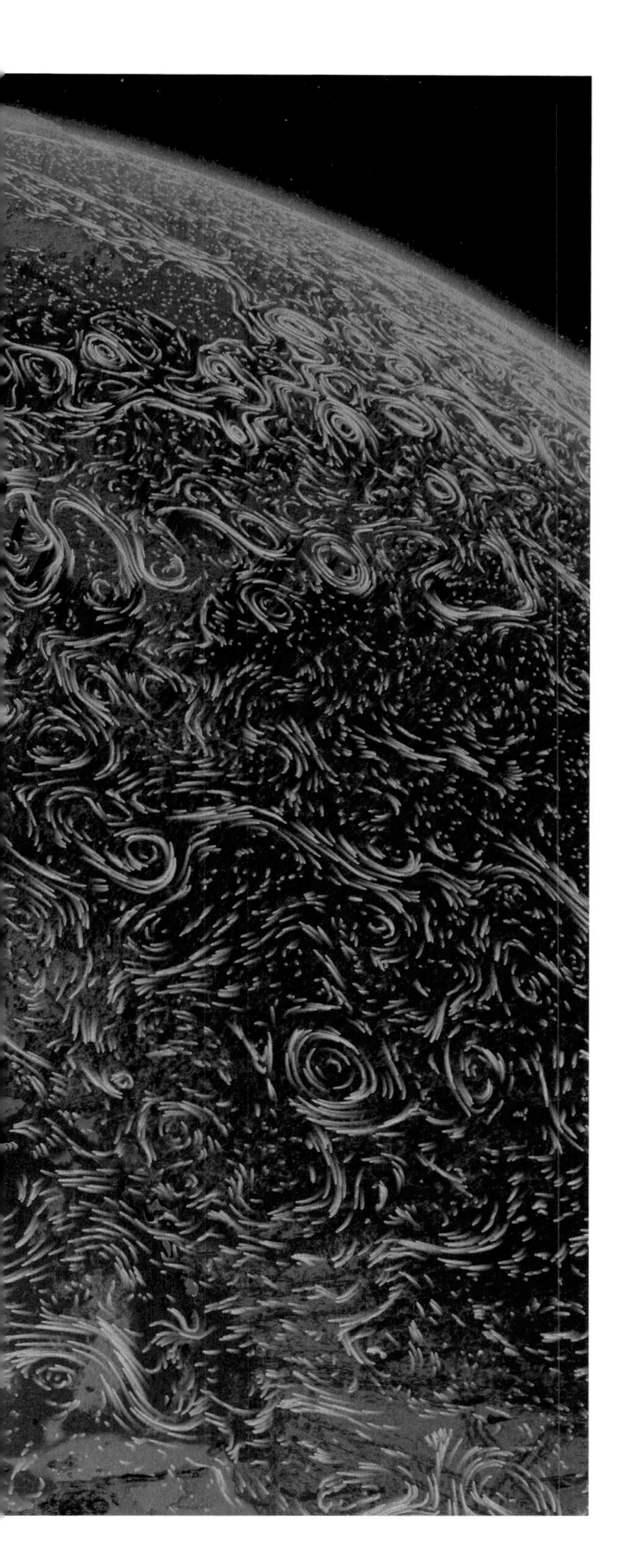

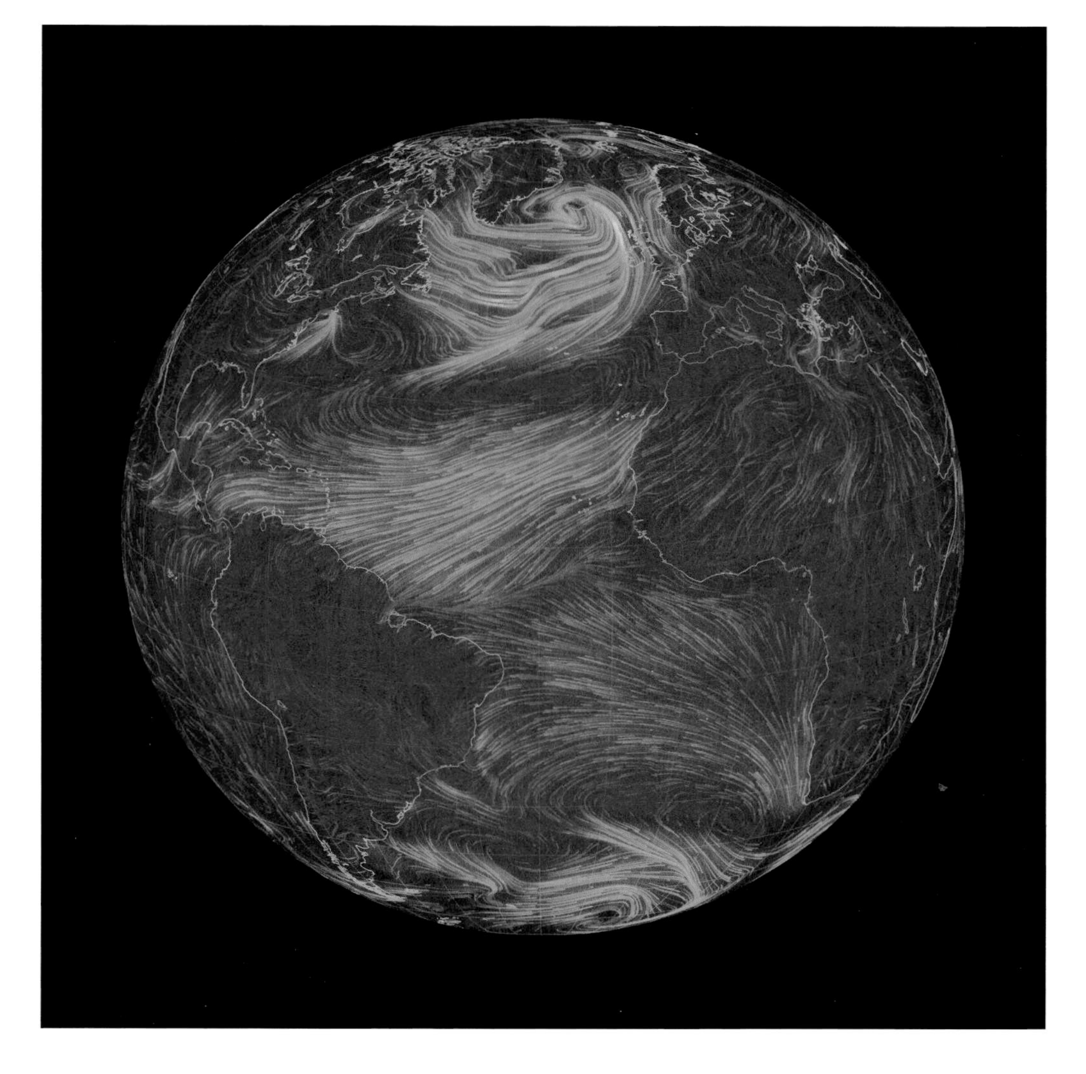

• 2013:

Right, top and bottom: Late-December wind-flow charts over the Indian and Atlantic Oceans. From the global interactive wind-map site earth.nullschool.net, developed by web designer Cameron Beccario in 2013.

CHAPTER 3

The Moon

The moonsheep, dreaming, does with himself converse:
"I am the dark space of the universe."
The moonsheep.

—Christian Morgenstern, *The Moonsheep*

More than any other object in the sky with the exception of the sun, the moon influenced the imaginations of early humans, stirring both a fascination and a need to understand. While the sun, with incontrovertible power, rules the daily rhythm of almost every living creature on Earth, its sheer clout makes it less questionable, less subject to study, even less inspirational. It is of course a celestial object par excellence, but it has so much blinding authority that it immediately wipes all other visible objects from the sky on rising. Except for the moon.

Earth's giant natural satellite—the largest in the solar system, relative to the size of its parent planet—can be seen quite clearly in daylight, even when only a thin crescent. The moon's phases are a monthly metronome, the temporal backbone of any calendar divided into twelve, with the word *month* itself a cognate of "moon." These calendrical facts are already one key unlocking secrets of the moon's long-term, history-spanning influence on civilization. The changing geometries of sunlight on the moon's face, its phases, are the reason we perceive a monthly cycle in the first place. Apart from the sun's daily light switch—a binary, on-or-off, zero-or-one signaling system less obviously in need of critical analysis, for all its gradations—the moon's phases, which of course also reflect the power of the sun, are the leading means by which the celestial clockwork became known on Earth. How did those changes work, and by what principle, and how was that useful in human affairs?

The first known gear wheels, which were discovered by sponge divers off the Greek island of Antikythera in 1900, comprised the internal mechanism of an ancient Greek analog computer superbly designed to track lunar phases, monitor solar, lunar, and planetary positions, and predict eclipses. Although no predecessor devices are known, it's hard to imagine that the 2,100-year-old

Antikythera Mechanism was a one-off device. The sophistication evident in its thirty surviving bronze gears and the workmanship of their intricate assembly suggest a developmental chain of antecedents. For us, that chain broke about twenty centuries ago, for whatever reason. Technological devices approaching this level of complexity didn't appear again until the astronomical clocks of the early Renaissance. Circumstantial evidence exists that the mechanism is the work of one of the great geniuses of antiquity, Greek mathematician and astronomer Archimedes.

The intricacy of the Antikythera Mechanism suggests that the foundation stone of our technologies ultimately rests on a civilizational need to track celestial movements. The foremost signifier of those movements was the changing play of light on the moon—that and eclipses, which result when the sun's light on our natural satellite is cut off due to the intervening presence of the Earth, or the sun itself is occulted by the moon. Predicting the periodicities of these eclipses was one of the leading challenges to early astronomers. The largest gear wheel in the Antikythera Mechanism has 223 teeth—one for every month of the eighteen-year saros cycle that ancient Babylonian astronomers discovered could be used to predict lunar and solar eclipses. Devising a mechanical device capable of operating at such time frames was a staggering achievement for the ancient world, demanding a thorough reconsideration of our understanding of the history of technology. In any case, it's clear that the drive to construct devices capable of simulating celestial movements in all their complexity is at the fountainhead of successor technologies.

Although other devices comparable to the Antikythera Mechanism are unknown in antiquity, evidence exists that at least some Greek gearing technologies devolved to the Arab world, which played a role in preserving them, just as Arab translations of Greek scientific treatises preserved them for posterity. In any case, it is clear that all of today's mechanical timekeepers have descended from European astronomical clocks. The need to miniaturize, the necessity to pack timekeeping devices into ever-smaller packages, and the skills developed during that effort, were critical to the development of almost every technology in use today, be it a laptop, cellular phone, or GPS device. The gravitational effects of the moon, in other words, raised more than tides through time.

And the Antikythera Mechanism only takes us partway back in a very long story. It and its antecedents were long preceded by the large-scale architectonic technologies necessary for the construction of such static archaeoastronomical sites as Stonehenge and other prehistoric, stone- and earthwork artifacts. These functioned as ancient observatories with calendric functions—stationary astronomical clocks of a certain kind. Here, too, the light of the moon—its monthly phases and the mysteriously predictable process behind its eclipses—reached into human imagination, triggering a need to comprehend, and also a drive to build.

Stonehenge dates to between four and five thousand years ago, to the Stone Age's last chapter—the so-called Neolithic Revolution, a period when agriculture and domesticated animals first appeared. But the earliest direct evidence of the influence of the moon on human ingenuity dates back further still, to thirty-five thousand years ago and the mnemonic devices known as tally sticks. The Lebombo Bone, a baboon fibula discovered in a rock shelter known as Border Cave in the western Lebombo Mountains, near the contemporary frontier between South Africa and Swaziland, contains twenty-nine notches. It's assumed to be a lunar-phase counter, and is considered the oldest known prehistoric mathematical tool. (In a 365-day solar year, the average month is 30.4 days, while the moon's actual synodic cycle averages out to 29.5 days.) Given the evidence of human presence in the Border Cave dating back two hundred thousand years, the Lebombo Bone is likely representative of a line of similar tally sticks reaching much further back in time.

Three of the images in this chapter can be understood as calendrical or lunar-phase counting devices in their own right, including one by Athanasius Kircher on page 85, and a volvelle, or rotating paper instrument, from German mathematician and printer Peter Apian's *Astronomicum Caesareum* (Caesar's Astronomy) on page 77. The oldest by far—in fact the oldest representative image of the cosmos to be found anywhere on Earth—is the exceptional Nebra Sky Disc, visible on page 73. It has been reliably dated to between 1600 and 2000 B.C.—within a few hundred years of Stonehenge—and is considered both the first-known portable astronomical instrument and the oldest-known graphic depiction of celestial objects in human history.

Excavated illegally in 1999 in Saxony-Anhalt, Germany, and confiscated by the police in 2002, the Nebra disc is linked to the pre-Celtic Bronze Age Unetice culture of central Europe. Although it's a unique object—like the Antikythera Mechanism, nothing similar exists in the historical record—it has been exhaustively tested and found authentic. Made of a blue-green copper inset with lustrous gold, the twelve-inch-wide disc contains an arrangement of seven stars that has been understood to represent the Pleiades, or Seven Sisters. They're in between a crescent moon on the right and full moon (or possibly sun) in the center. Two golden bands at the disc's edge (one is missing) span eighty-two degrees, seemingly corresponding to the angle between sunset at the winter and summer solstices at the latitude where it was found. This would already make the disc a calendrical tool.

Further research has reinforced this conclusion and added new potential functionality. Three thousand six hundred years ago, the juxtaposition of the Pleiades and the new or crescent moon took place in March and with the full moon in October—always key months for agriculture. The case for the stars being the Pleiades is reinforced by the cluster playing a role in the records of subsequent agricultural civilizations, although astronomer Owen Gingerich has pointed out that in its representation on the disc, the cluster has one star too many; the contemporary Pleiades has only five visible stars. Still, evidence exists for a "Lost Pleiad" in the records of numerous ancient civilizations. Based on those records, some astronomers have argued that it is possible that a seventh easily visible Pleiades star dimmed subsequent to the production of the Nebra Sky Disc.

The other stars on the disc are distributed in an artfully random manner—as though the maker of the disc wanted to avoid the appearance of a pattern. Studies at Ruhr University Bochum revealed that a truly random arrangement would have created more seeming asterisms, or congregation of stars.

From there, problems of interpretation exist, because the Nebra Sky Disc is literally unprecedented. One group of German scholars looked into the potential role of the Pleiades in relation to the apparent phase of the crescent moon, which in its Nebra representation appears to be three to five days old, and suggested that this specific conjunction of celestial forms may have been a cue used by the Unetice to insert an intercalary month, thus harmonizing the lunar and solar calendars. However, we don't even know if they used both calendars in the first place. (As previously mentioned, with the synodic or lunar month being 29.5 days, a lunar year is only 354 days; the discrepancy between it and the solar calendar that actually governs the seasons needed to be accounted for, evidently by adding a periodic leap month.)

Attempts to buttress this conclusion have floundered due to the absence of contemporaneous corroborative sources. One German researcher wrote that he had combed through the ancient Babylonian astronomical records known as the Mul.Apin and discovered a rule, dating back to around the seventh century B.C., concerning a days-old crescent moon and the Pleiades. The cuneiform text indicated that a thirteenth leap month should be added to the lunar calendar, exactly when such a configuration occurs. Other astronomers have cautioned that such theories are highly speculative. For example Austrian Assyriologist Hermann Hunger, one of the leading experts on the Mul.Apin tablets, confirms the existence of that cuneiform instruction, even commenting that researchers believe the text itself dates back to the fourteenth century B.C.—making it roughly contemporaneous with the Nebra disc. But he questions a connection between the two.

Because the Nebra Sky Disc comes from a period before recorded history, theories concerning it will always be provisional. As Canadian medievalist Randall Rosenfeld has suggested, in some ways its contemporary role is as a device capable of revealing the limits of our knowledge, or even a mirror reflecting back our own imaginings of ancient astronomical practices.

Apart from its primal role in helping define time, and thus initiating the history of human efforts to monitor and eventually mechanically "keep" time, the moon fostered a creative cosmic consciousness in other ways. The sun, as previously suggested, is too blindingly bright to even seem tangible, let alone accessible in any meaningful way. But with its indecipherable markings, its ever-shifting phases, and its silvery perpetual cruise among the stars, the moon seemed mysterious, out of reach, and yet somehow also potentially approachable—all the more so after the invention of the telescope.

Without the moon, the solar system would have seemed unattainably distant and abstract. Without the moon—with just the sun and the impossibly far-flung planets—we wouldn't have had a plausible destination. Space travel would probably never have occurred. *With* the moon, on the other hand—with a potential port of call floating relatively nearby—such a trip seemed (just possibly) attainable. If only a way could be found to travel so high.

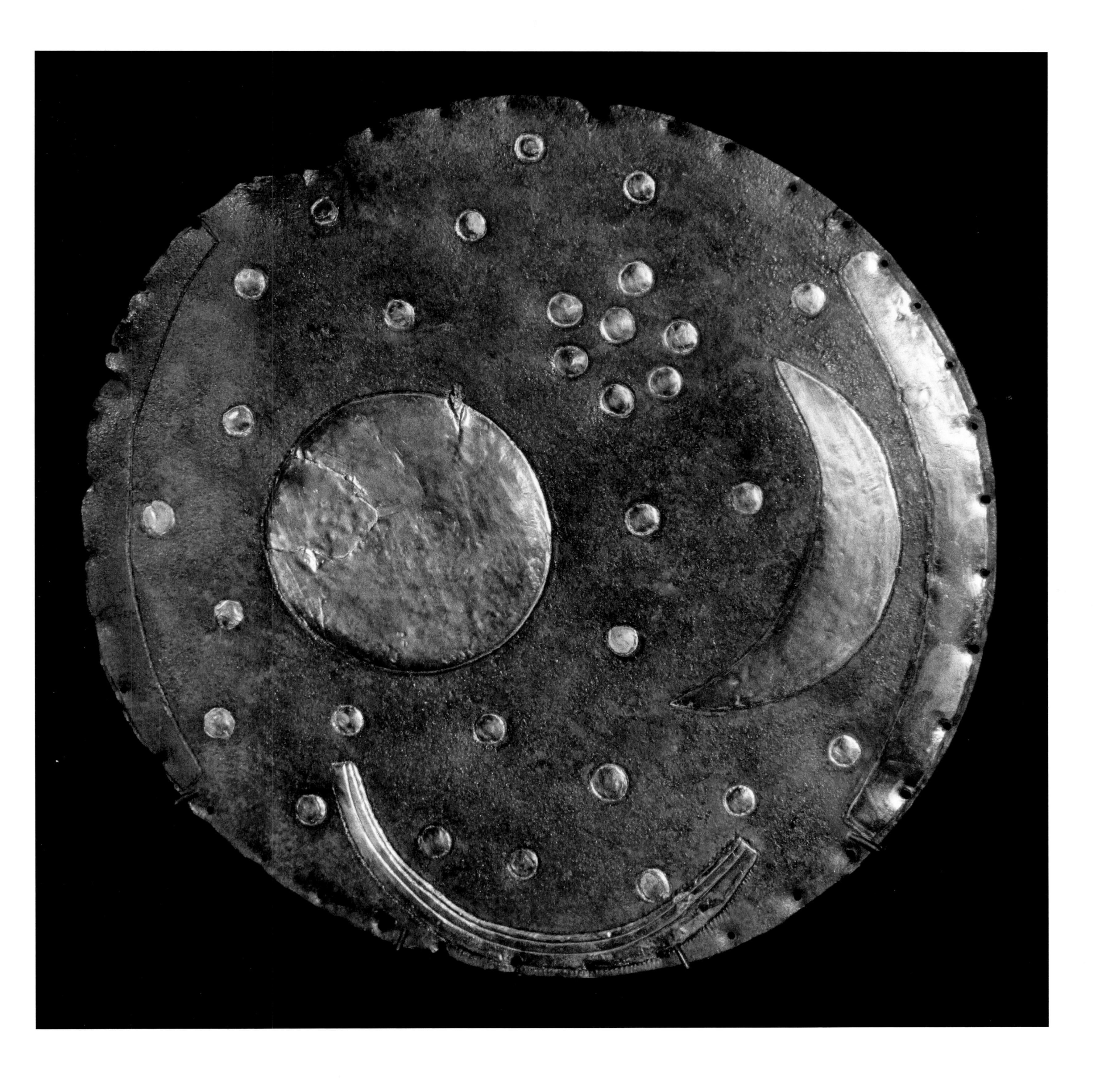

• 2000–1600 B.C.: Excavated illegally in 1999 in Saxony-Anhalt, Germany, the extraordinary Nebra Sky Disc is considered both the first-known portable astronomical instrument and the oldest-known graphic depiction of celestial objects in human history. Made of a blue-green copper inset with lustrous gold, the twelve-inch-wide disc contains an arrangement of seven stars probably representing the Pleiades. They're in between a crescent moon on the right and either the full moon or sun in the center. Two golden bands at the disc's edge (one is missing) span eighty-two degrees, corresponding to the angle between sunset at the winter and summer solstices at the latitude where it was found.

• After 1277:

Diagram of the phases of the moon. This illumination by an unknown Franco-Flemish illustrator demonstrates a clear understanding of the relationship between the moon's phases and its orientation vis-à-vis the sun.

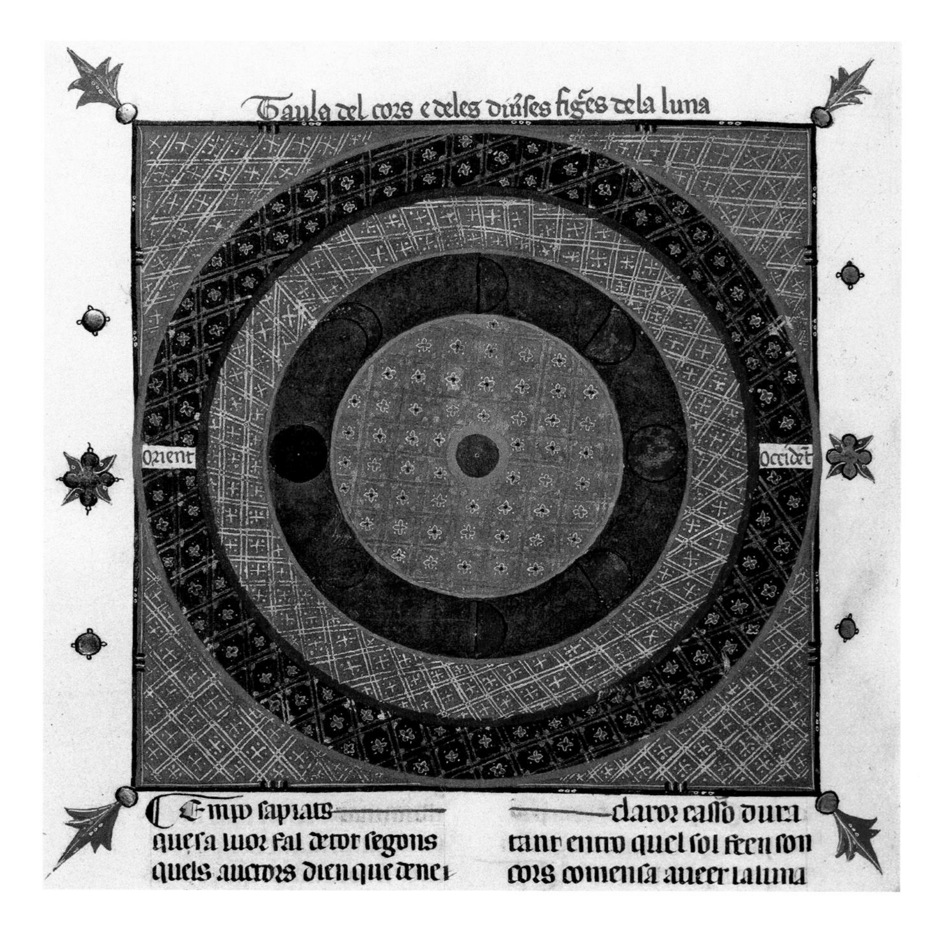

• 1375–1400:

Another diagram of the phases of the moon, this time in a manuscript originating in eastern Spain. From Matfré Ermengau of Béziers's *Breviari d'Amor* (Abstract of Love).

• 1444–50:

Dante's cosmology adhered closely to medieval conventions, with nine spheres of heaven containing the five visible planets, the sun and the moon, plus a sphere for the fixed stars and the primum mobile. Considered a symbol of purity, the moon was the "first planet," and all of the elements subject to change in time—fire, water, earth, and air—only existed in the sublunar sphere, or within its orbit. In this manuscript illumination from *The Divine Comedy* by Sienese master Giovanni di Paolo, Dante (dressed in flowing blue) and his beautiful guide, Beatrice, visit the "heaven of the moon." The canto illustrated by this image begins with a description by Dante of a miraculous, levitational form of space-flight. "I saw myself arriving, in the space of time perhaps it takes an arrow to be drawn, released, and leave the notch, there, where a marvellous thing engaged my sight: and therefore She, from whom nothing I did was hidden, turning towards me, as joyful as she was lovely, said: 'Turn your mind towards God in gratitude, who has joined us with the first planet.'" For other works by di Paolo, see pages 33, 116, 144, 178–79, and 256.

• 1540:

Lunar volvelle, a kind of analog computer, from German printer and cosmographer Peter Apian's *Astronomicum Caesareum* (Caesar's Astronomy), a book that doubled as a scientific instrument used to predict celestial motion. The word *volvelle* derives from the Latin word for "to turn." All of the volvelles were hand-colored and assembled in Apian's own printing plant. This one allowed the user to track lunar phases, as visible in the blue ring, and eclipses. Timing of eclipses could be determined by turning the wheels with the green dragons on them. The association between dragons and eclipses dates back to when it was believed a solar eclipse was caused by a dragon eating the sun. This was one of the most complicated volvelles to use in Apian's book; the instructions go on for several pages, and it's the only example where explanatory text spills over directly onto the page holding the volvelle itself. For more from Apian, see pages 43, 119, 180, 227, and 262.

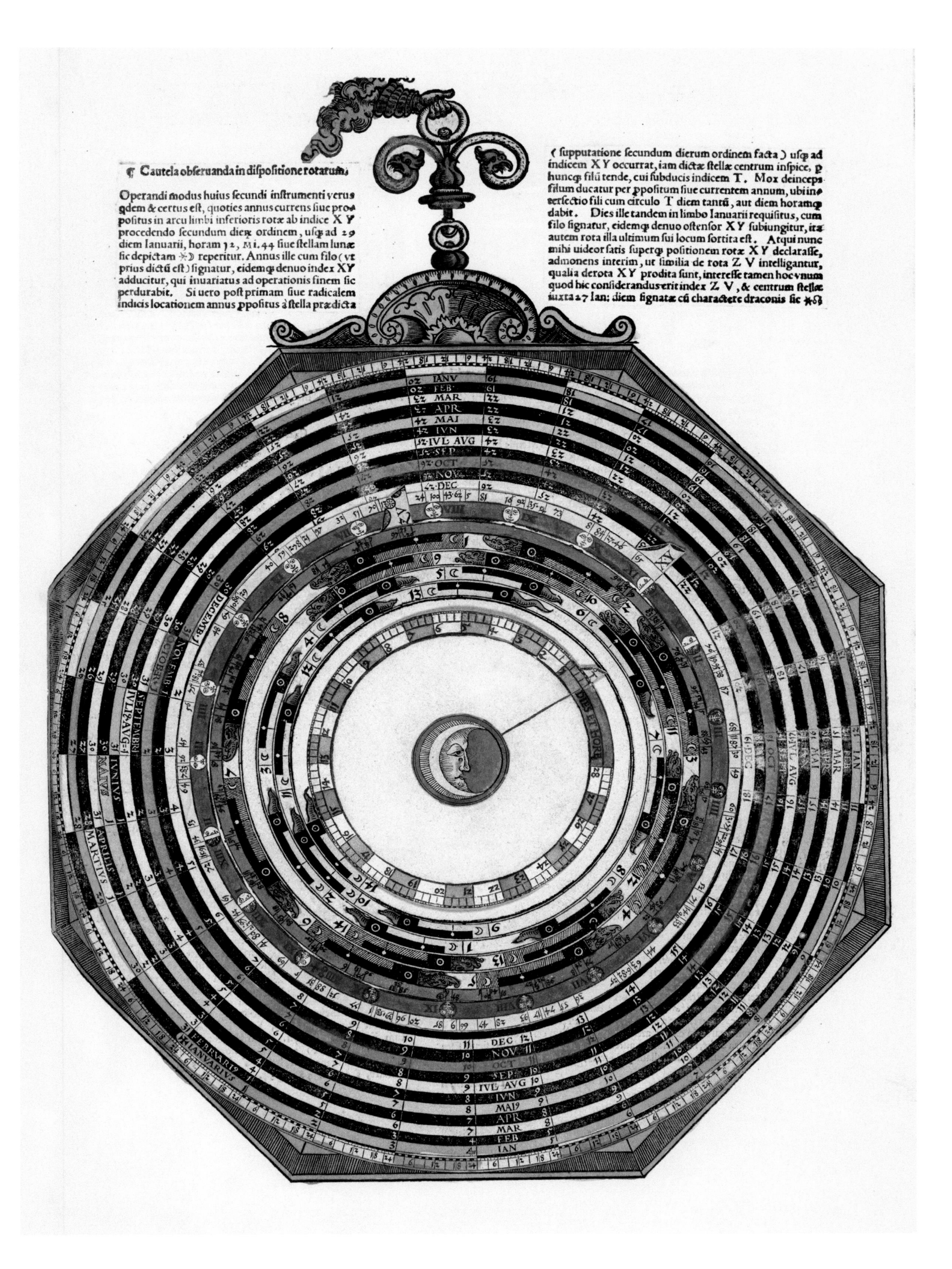

¶ Cautela obseruanda in dispositione rotarum.

Operandi modus huius secundi instrumenti verus quidem & certus est, quoties annus currens siue propositus in arcu limbi inferioris rotæ ab indice X Y procedendo secundum dierum ordinem, usque ad 29 diem Ianuarii, horam 12, Mi. 44 siue stellam lunæ sic depictam ☽ reperitur. Annus ille cum filo (ut prius dictũ est) signatur, eidemque denuo index XY adducitur, qui inuariatus ad operationis finem sic perdurabit. Si uero post primam siue radicalem indicis locationem annus ꝓpositus à stella prædicta (supputatione secundum dierum ordinem facta) usque ad indicem X Y occurrat, iam dictæ stellæ centrum inspice, ꝑ huncque filũ tende, cui subducis indicem T. Mox deinceps filum ducatur per ꝓpositum siue currentem annum, ubi intersectio fili cum circulo T diem tantũ, aut diem horamque dabit. Dies ille tandem in limbo Ianuarii requisitus, cum filo signatur, eidemque denuo ostensor X Y subiungitur, ita autem rota illa ultimum sui locum sortita est. Atqui nunc mihi uideor satis superque positionem rotæ X Y declarasse, admonens interim, ut similia de rota Z V intelligantur, qualia de rota X Y prodita sunt, interesse tamen hoc vnum quod hic considerandus erit index Z V, & centrum stellæ iuxta 27 Ian: diem signatæ cũ charactere draconis sic ☊

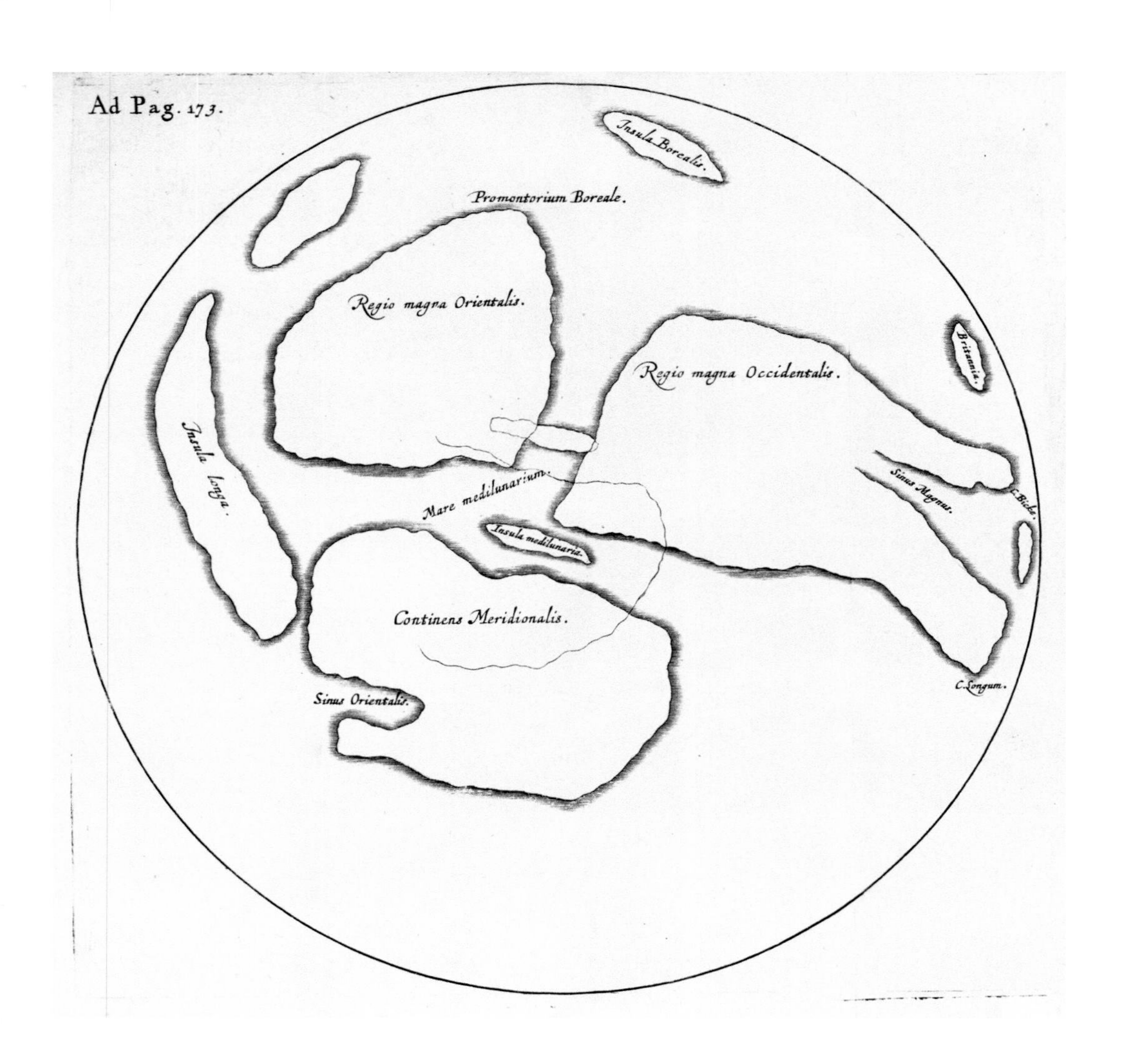

• Circa 1600:

The oldest-known map of the moon from naked eye observations, drawn by English physician and physicist William Gilbert and not published until 1651 in his *De mundo nostro sublunari philosophia nova* (New Sublunary Philosophy of the World). Gilbert believed that the lighter areas of the moon were water, and the darker, land—the exact opposite of the prevailing views of the time. Here, what we still call the lunar *mare*, or "seas," are depicted as islands. The moon is of course utterly dry.

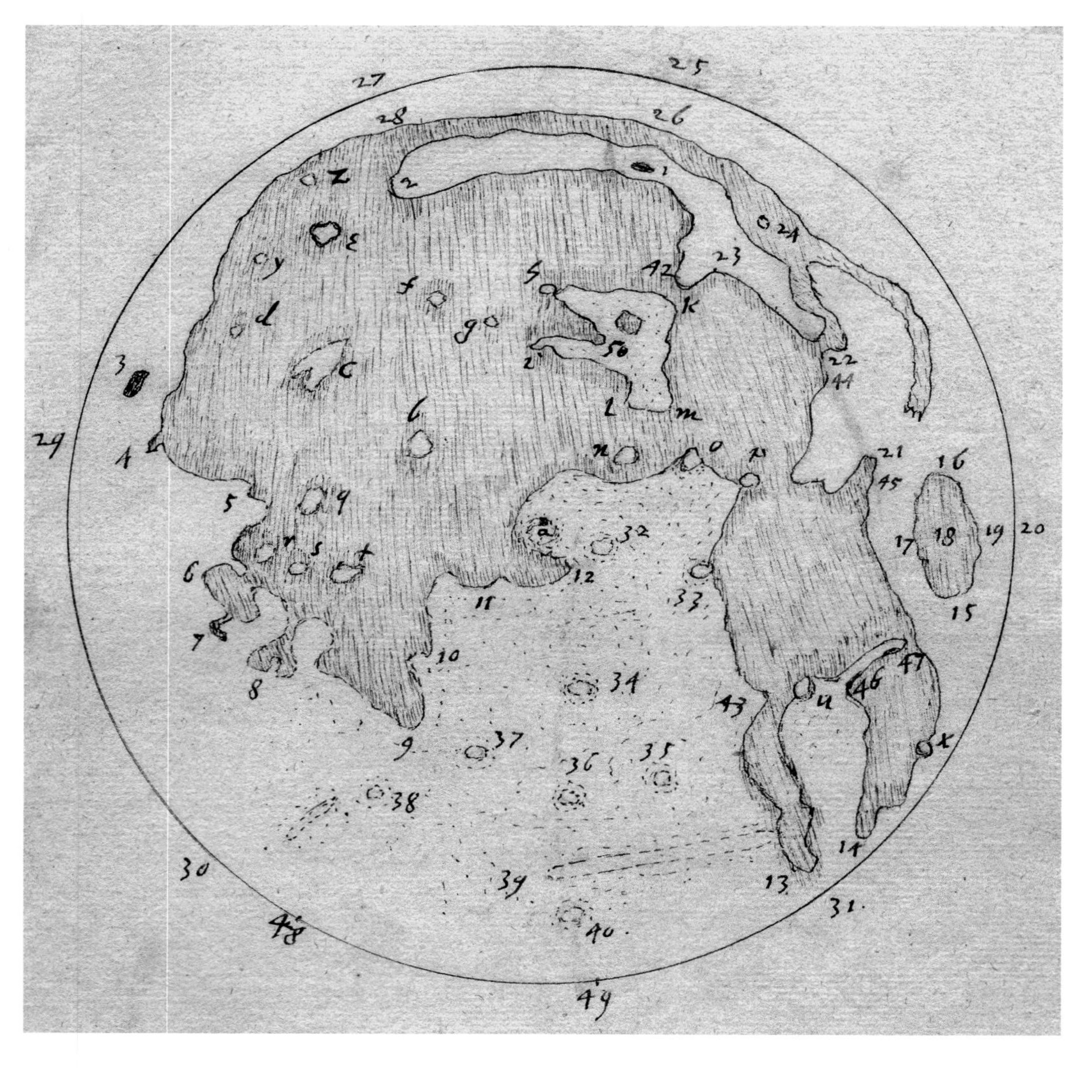

• 1613:

Detailed depiction of the moon made using a telescope by British astronomer Thomas Harriot, who was the first to use the new instrument to attempt to map the moon on July 26, 1609. Unlike Galileo, who observed our natural satellite the following November, Harriot didn't publish his results. This later map of a full moon by Harriot is better than any ever attempted by Galileo, who found the full moon uninteresting, in part because he was fascinated with its mountains and craters, which are best observed in a raking light (as on the facing page).

• 1610:

Facing page: Depiction of the moon from Galileo's *Sidereus nuncius* (Starry Messenger). The large crater visible at the terminator (line between the day and night sides) doesn't really exist; rather Galileo was attempting to convey the essence of the moon's uneven surface and the nature of craters as a phenomenon, and used this superb engraving, which may be by his own hand, to do so. Note that the image impinges on the text, evidence that the plate containing the engraving was used during a second pass through the printing press.

Hæc eadem macula ante ſecundam quadraturam nigrioribus quibuſdam terminis circumuallata conſpicitur; qui tanquam altiſſima montium iuga ex parte Soli auerſa obſcuriores apparent, quà verò Solem reſpiciunt lucidiores extant; cuius oppoſitum in cauitatibus accidit, quarum pars Soli auerſa ſplendens apparet, obſcura verò, ac vmbroſa, quæ ex parte Solis ſita eſt. Imminuta deinde luminoſa ſuperficie, cum primum tota fermè dicta macula tenebris eſt obducta, clariora mõtium dorſa eminenter tenebras ſcandunt. Hanc duplicem apparentiam ſequentes figuræ commoſtrant.

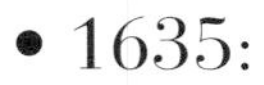

• 1635:

Depictions of the moon in its final quarter **(right)** and first quarter **(facing page)** by virtuoso French engraver Claude Mellan. Prepared at the request and under the direction of astronomer Pierre Gassendi, these representations of the moon are among the finest ever made. Mellan's evocative etching technique rejected cross-hatching in favor of simply thickening or thinning evenly spaced horizontal lines. Interestingly, Galileo found these depictions of the moon irritating when Gassendi sent him copies in the fall of 1637, responding peevishly that they couldn't have been made by anyone who had seen the moon with his own eyes.

Cl. Mellan Gal. ping. et sculp.
Phasis Aquis sextijs An. 1635. Octob. 7. a claro adhuc crepusculo in occasū usq.

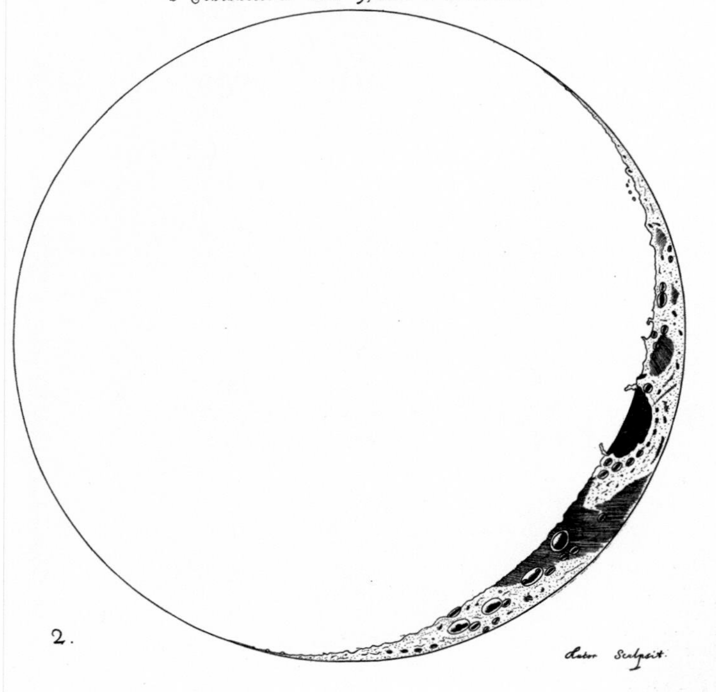
Phasis Lunæ Corniculatæ Crescentis
Observata in 14 Gradu ♈. circa Limit. Austr. et Apogeum.
GEDANI.
Anno Christi 1645. Die 28 Februar. horâ 7. à meridie numerata,
à Coniunctione verò 9. Diei 3 Currentis.
2.
Autor Sculpsit.

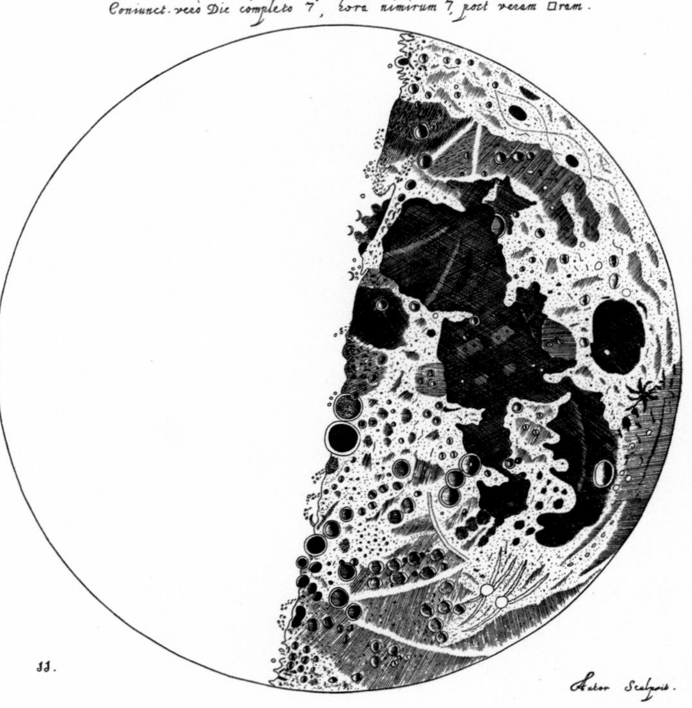
Luna Bisecta Crescens.
Observata in 7 Gradu II circa limit. A.
GEDANI
Anno Christi 1644. Die 15 Martii, hora 7 à merid. numer. à
Coniunct. verò Die completo 7, hora nimirum 7 post veram □ram.
11.
Autor Sculpsit.

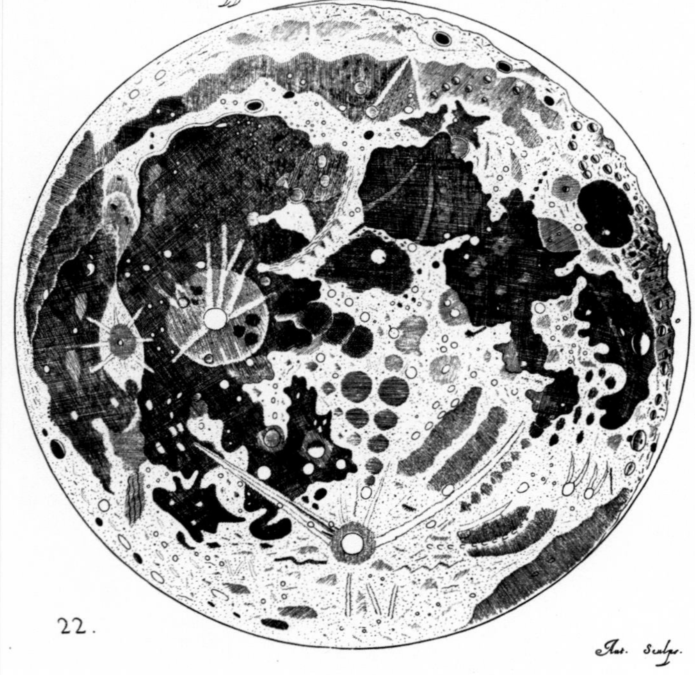
Phasis Lunæ ab Oppositione recentis.
Observata in 21 gradu II circa limit. A.
GEDANI.
Anno Christi 1643. Die 26 Novemb. hora 11 à merid. num. ab
Oppositione vero 6. Diei 2 Current.
22.
Aut. Sculps.

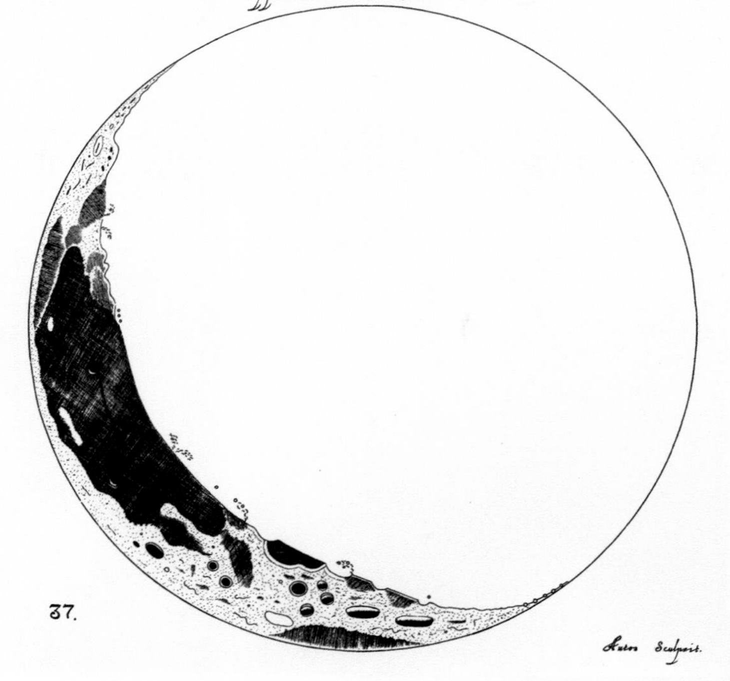
Phasis Lunæ Cornigeræ Decrescentis.
Observata in 26 Gradu ♏. propè ☊ et Apog.
GEDANI.
Anno Christi 1643. Die 7 Novemb. hora 7 à med. nocte num. ab
Oppositione verò 2. Diei 12 Curr.
37.
Autor Sculpsit.

• 1647:

Phases of the moon (**facing page**) and lunar map (**above**) from the first lunar atlas, *Selenographia, sive lunae descriptio* (Selenographia, or a Description of the Moon). Just over a decade after Mellan's engravings, German-Polish astronomer Johannes Hevelius followed four years of extensive observations by producing this massive book. Packed with illustrations, it became the definitive such atlas for over a century. Although Hevelius introduced a system of lunar nomenclature based largely on names from ancient Rome and Greece, it was supplanted by a different system, still in use today, introduced by Italian astronomer and Jesuit priest Giovanni Battista Riccioli in 1651. A lunar crater is named after Hevelius, however, and he is now generally considered the founder of lunar topography (a term that in planetary science means the study of the surface features and shape of a planet or moon). The double rings in the map above encompass two projections of the lunar surface that share most of its features. They are the result of a slow rocking back and forth of the moon relative to the Earth over time. As a result of this oscillation, about 59 percent of the lunar surface can be observed from Earth. By using this double projection technique, Hevelius could present close to the full 59 percent.

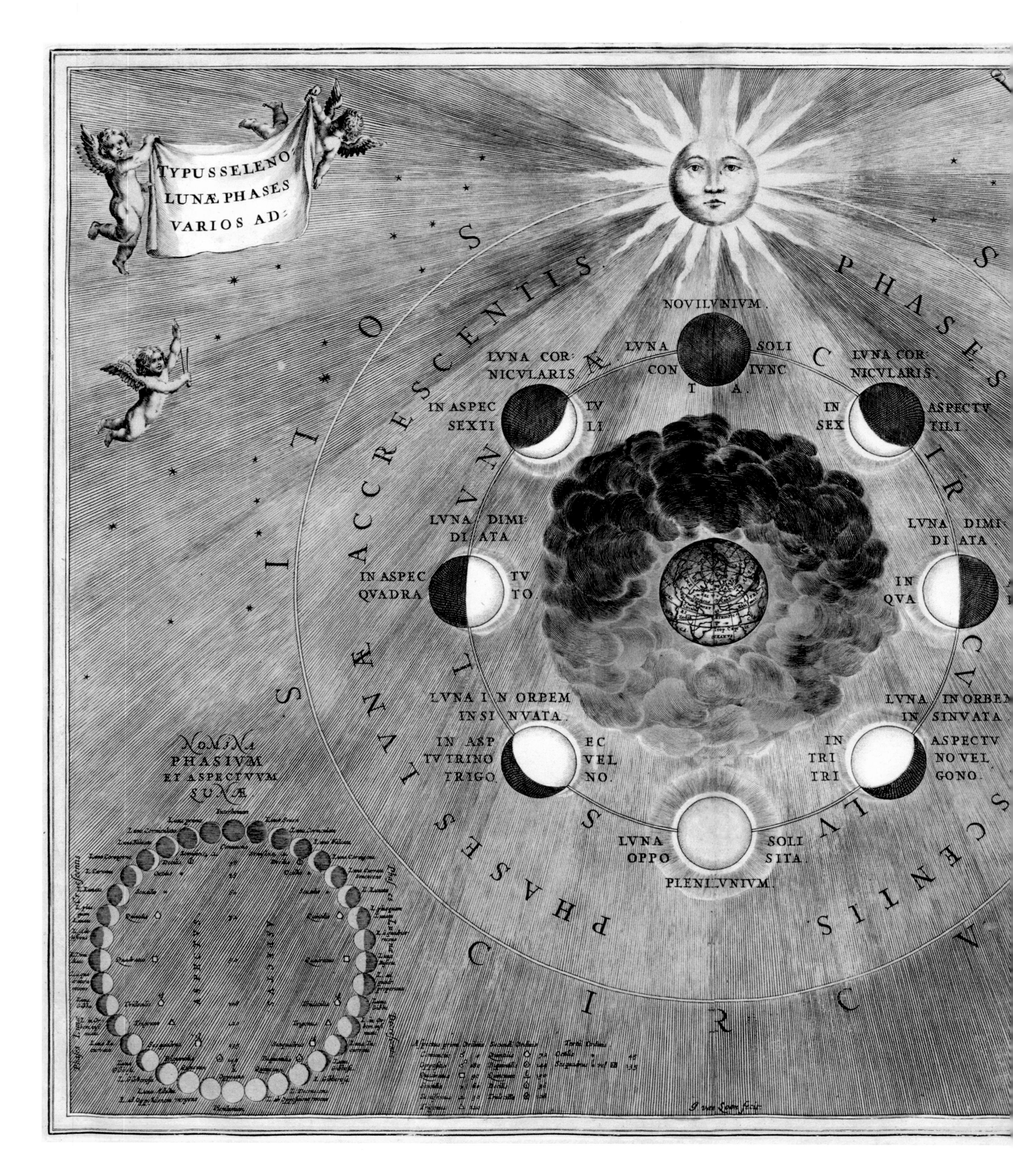

TYPUS SELENO:
LUNÆ PHASES
VARIOS AD:
CIRCULUS PHASIUM LUNÆ ACCRESCENTIS.
PHASES CIRCULI LUNÆ DECRESCENTIS.
NOVILVNIVM.
LVNA SOLI CONIVNCTA.
LVNA COR: NICVLARIS.
IN ASPECTV SEXTILI.
LVNA COR: NICVLARIS.
IN SEX ASPECTV TILI.
LVNA DIMI: DIATA.
IN ASPECTV QVADRATO.
LVNA DIMI DIATA
IN QVA
LVNA IN ORBEM INSINVATA.
IN ASPECTV TRINO VEL TRIGONO.
LVNA IN ORBEM INSINVATA
IN TRI ASPECTV NO VEL TRIGONO.
LVNA SOLI OPPOSITA.
PLENILVNIVM.
NOMINA PHASIVM ET ASPECTVVM LUNÆ

• 1660:

Left: A plate from Andreas Cellarius's *Harmonia macrocosmica*, one of the greatest celestial atlases. The moon has been used to mark time and determine when to plant and harvest since before recorded history. Here Cellarius depicts lunar phases, which are due to its varying orientation in relation to the sun when viewed from Earth. The sun's apparent path is defined by a ring around what appears to be a smog-shrouded Earth. That cloud represents the Aristotelian idea that all the elements are circumscribed by the moon's sphere, with everything beyond uncorruptible and unchanging, even in movement. The two smaller flanking diagrams are copied almost unchanged from Hevelius's *Selenographia*, a common practice of the time. For other maps from Cellarius, see pages 47, 122–23, 147–48, 182–83, and 232–35.

• 1671:

Above: Under the heading "The Selenic Shadowdial or the Process of the Lunation," German polymath Athanasius Kircher presented an entirely different graphic depiction of lunar phases only a year after Cellarius's atlas appeared. In this innovative depiction by engraver Pierre Miotte, the moon's phases are depicted as two equal-opposite spirals, with the waning moon on the top contracting as it spirals to the center, and the waxing moon on the bottom doing the opposite. The phases also create an oval surrounding this mirroring. From Kircher's typically gargantuan tome *Ars magna lucis et umbrae*, or "The Great Art of Light and Shadows."

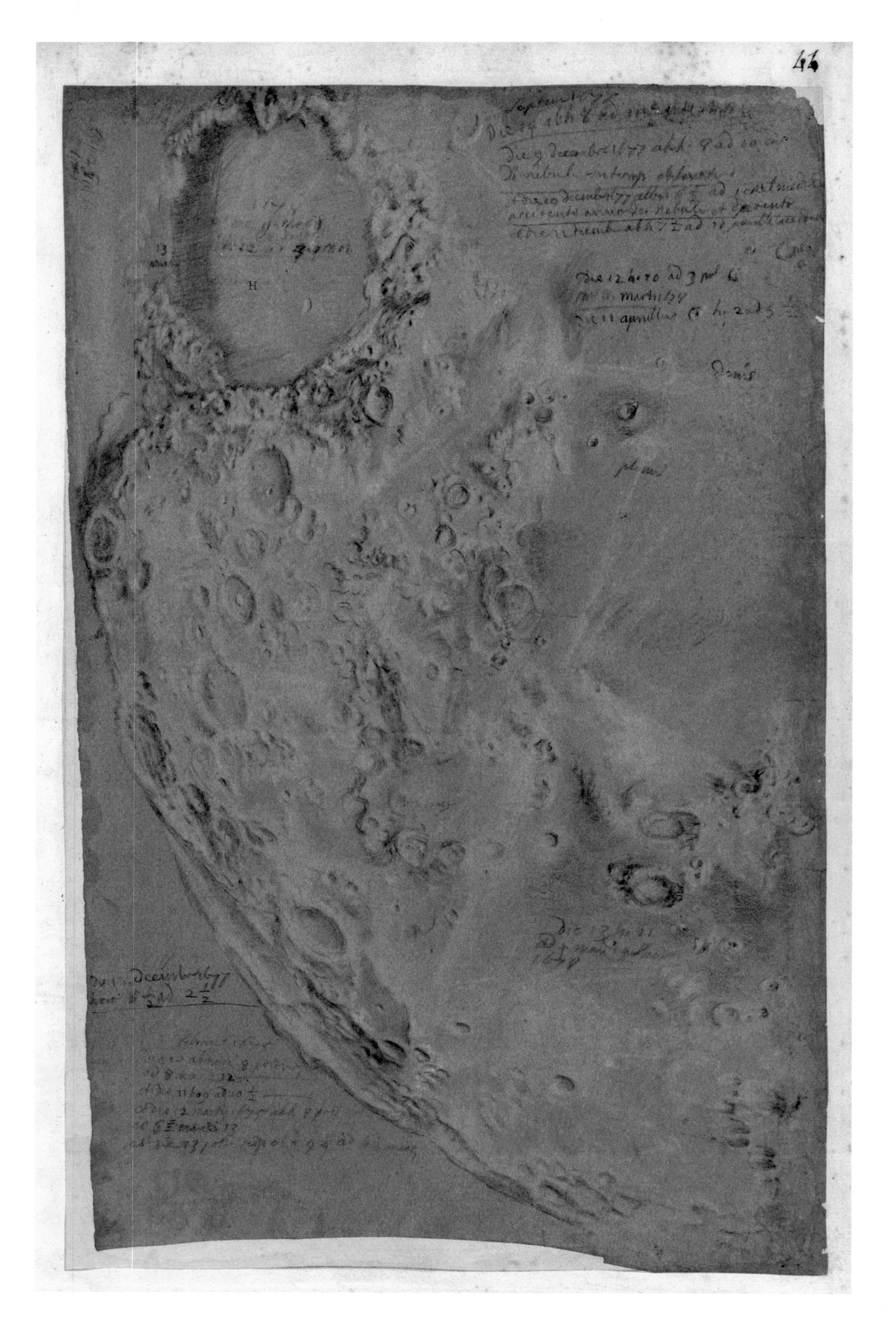

• 1679:

Between 1671 and 1679 Italian-French astronomer Gian Domenico Cassini worked with the artists Jean Patigny and Sébastien Leclerc to produce an atlas of the moon. Comprising more than fifty detailed drawings, which he annotated in black pencil, they enabled him to make a large map, which was also completed in 1679 and presented to the Académie royale des sciences. This particular drawing depicts a section of the northeastern limb, with the vast crater-shaped basin of Mare Crisium (the Sea of Crises) visible on the top left. (As was typical of the period, in this depiction south is up, inverting our standard view in the same way a Keplerian refractor telescope does.) In the middle on the right is Mare Serenitatis, or the Sea of Serenity. Under the handwritten notes on the top right is the Sea of Tranquility, where astronauts would first land on the moon three hundred years later.

• 1693–98:

Facing page: At the end of the seventeenth century, German astronomer and artist Maria Clara Eimmart made more than 250 pastel renderings of the moon on blue paper, including this superb study of a full moon. The daughter of a Nuremberg artist and amateur astronomer, Eimmart produced the images for use in her father's book, *Micrographia stellarum phases lunae ultra 300* (A Detailed Illustration of Over 300 Lunar Phases). North is up in this view; as a result, Mare Crisium is here visible as a circular patch near the upper right limb. For another painting by Eimmart, see page 184.

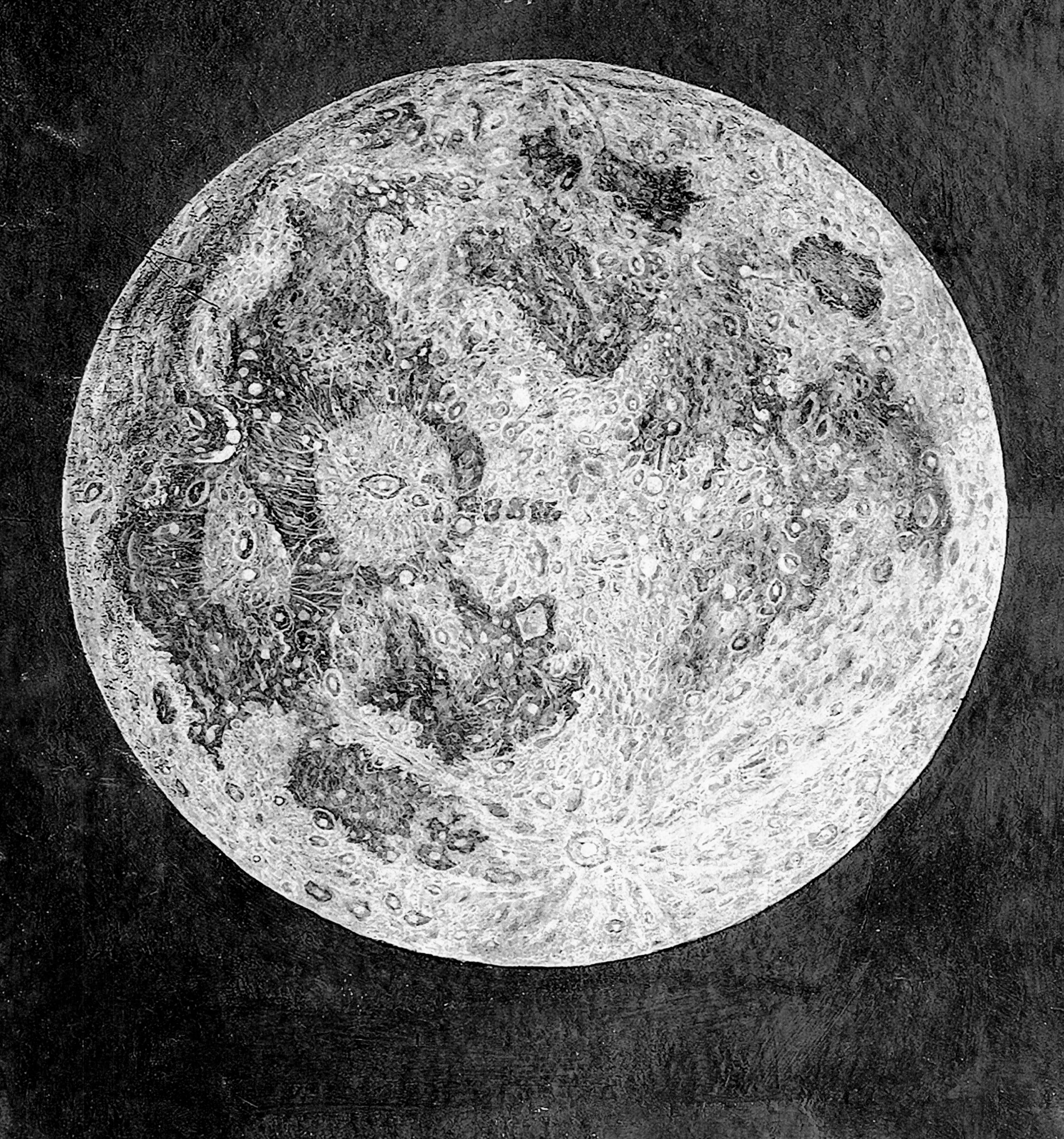
PLENILUNIUM.
pinxit ad Archetypum M.C. Eimmarta Norimb:

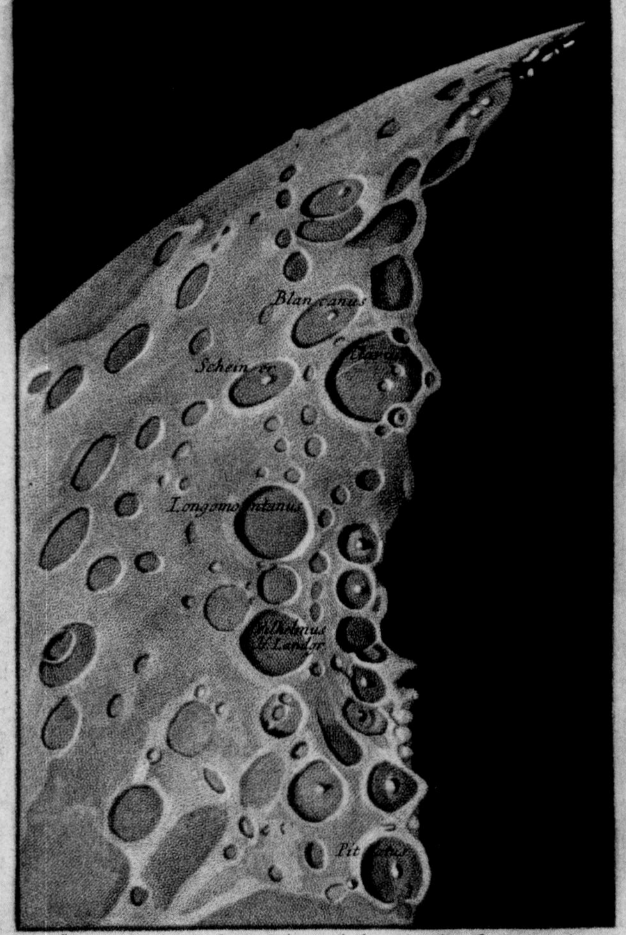

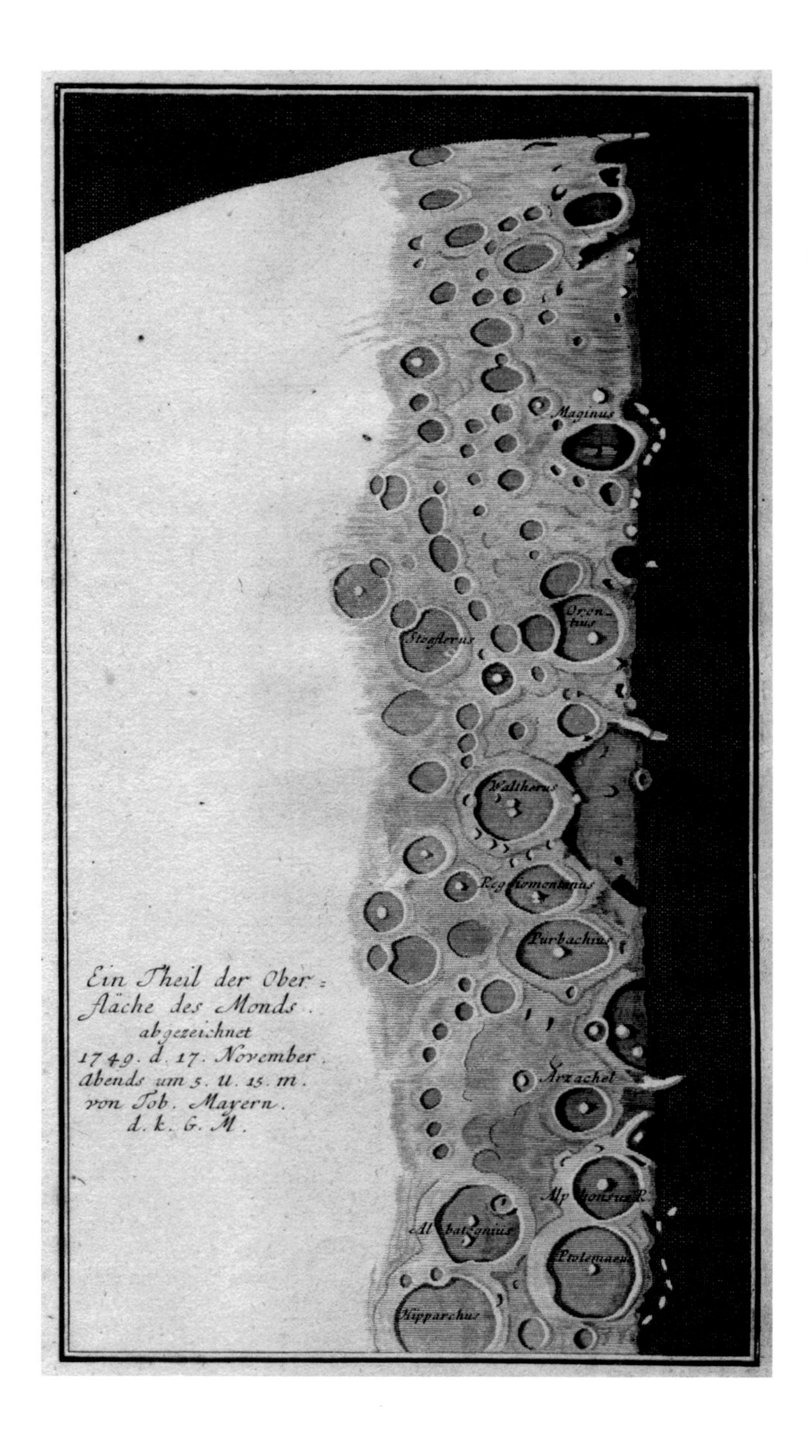

• 1750:

Above left: Lunar highlands observed on the evening of November 17, 1749, with Regiomontanus and Hipparchus craters visible. **Above right:** Southern highlands of the moon, with Clavius and Scheiner craters at the lunar terminator. German astronomer and cartographer Tobias Mayer made a name for himself in the mid-eighteenth century as a careful observer and tabulator of lunar motion, which he used in determining accurate calculations of longitude in order to make accurate terrestrial maps. He also depicted the surface features of the moon and their relationships with each other with unprecedented accuracy by using a micrometer in his telescope, a tool first developed in the late seventeenth century to enable precise measurement of celestial position, or in this case, latitude and longitude on the moon. Like Cassini and Eimmart before him, Mayer made numerous drawings as preparation for a large-scale lunar map and globe; these mezzotint engravings came from that effort. (The globe was never produced, and the map wasn't published until 1879.) An embarrassing error by the engraver resulted in these mezzotints producing reversed images; the text below the print on the left instructs the viewer to use a mirror for an accurate depiction of the subject.

• 1842:

Early calotype photograph, ostensibly of the Copernicus crater, probably made by the English astronomer and photography pioneer John Herschel. While this seems to be a photograph of a crater on the moon, in fact the emulsions of the mid-1850s, some of which Herschel was instrumental in developing, were far too slow to permit such high-resolution lunar photography. On the evidence of this photograph, Herschel was the first to solve the problem by making highly detailed plaster models of the lunar surface, which he then photographed in controlled studio conditions. (For a later example of such a technique, see pages 93–95. For prints based on drawings by Herschel of comets, see page 292.) The calotype photographic process used paper coated with a layer of light-sensitive silver iodide. The paper itself was then exposed to light within the camera.

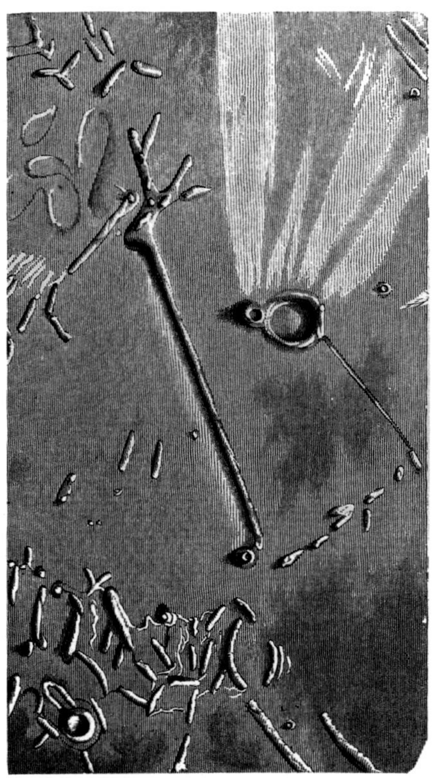

Chaîne de montagnes dans le Mare Nubium avant le coucher du soleil.

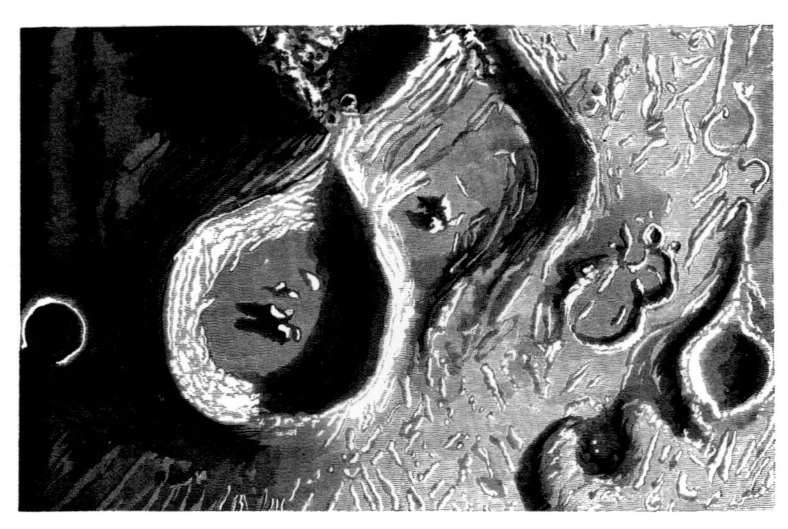

Theophilus cinq jours après la pleine lune, d'après M. Bulard (p. 120).

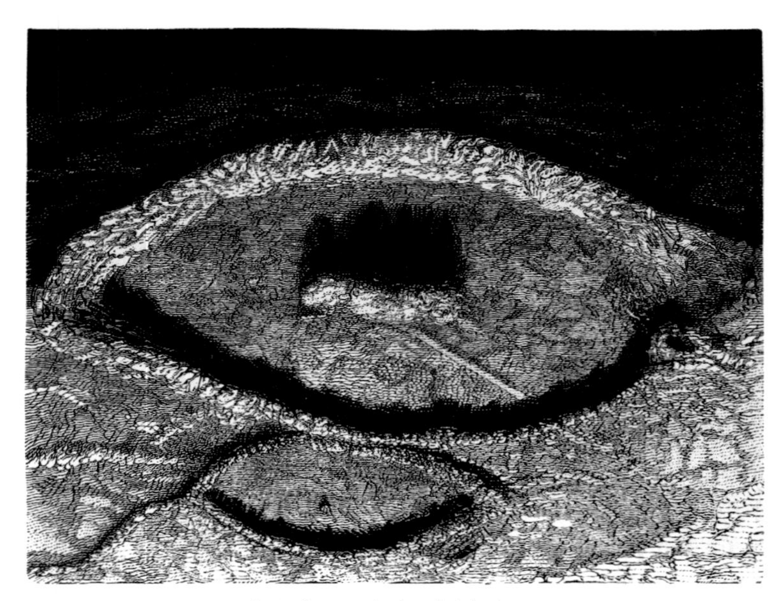

Petavius après la pleine lune.

Chaîne de montagnes dans le Mare Nubium au coucher du soleil.

• 1866:

Engravings of lunar surface features from the book *L'Espace selestial* (Celestial Space) by French astronomer and botanist Emmanuel Liais. Liais, who did much of his work in Brazil, served as director of the National Observatory in Rio de Janeiro from 1874 to 1881. **Top left:** Mountain range in Mare Nubium before sunset. **Bottom right:** Same area at sunset. **Top right:** Theophilus crater, five days after the full moon. **Bottom left:** Petavius crater, after the full moon. **Facing page:** Fragment of the surface of the moon, near the Bayer mountain, southeast of the Tycho crater, seen with high magnification.

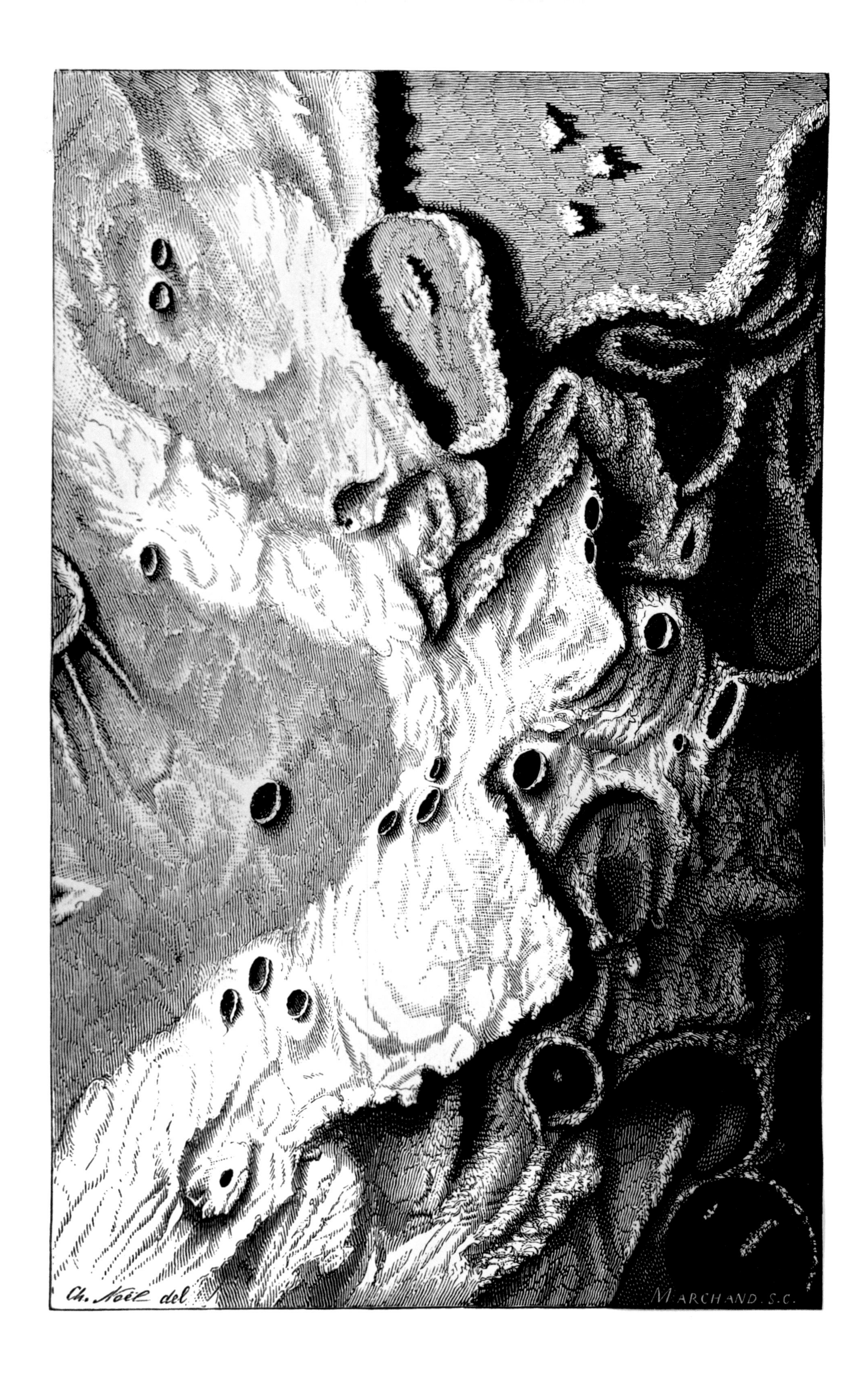
Ch. Noël del
MARCHAND S.C.

PLATE XXII.

J. Nasmyth, del. Vincent Brooks Day & Son Lith

ASPECT OF AN ECLIPSE OF THE SUN BY THE EARTH, AS IT WOULD APPEAR AS SEEN FROM THE MOON.

• 1874:

One of the most interesting attempts to visualize the moon came with the 1874 publication of a heavily illustrated book titled *The Moon: Considered as a Planet, a World, and a Satellite*, by Scottish inventor-engineer and amateur astronomer James Nasmyth and British astronomer James Carpenter. Nasmyth, who had made his fortune with the invention of the steam hammer and the hydraulic press, spared no expense, and the lavishly produced book featured several kinds of illustrations, including the superb "lossless" woodburytype photographic reproduction process. Like John Herschel thirty years earlier, and in an effort to improve on the relatively blurry lunar pictures that could be made with the photographic emulsions of the period, Nasmyth built elaborate plaster models of sections of the lunar surface, which he photographed in raking light in controlled studio conditions. **Above:** Nasmyth also made his own hand-drawn illustrations, including a depiction of an eclipse of the sun by the Earth as seen from the moon, which was converted to a lithograph for the book. **Facing page:** An excellent example of a woodburytype reproduction, in this case of Nasmyth's photograph of a plaster model of the Lunar Appenines.

J. Nasmyth

Brooks Day & Son.

THE LUNAR APENNINES, ARCHEMEDES &c., &c.

10 5 0 10 20 30 40 50 60 70 80 90 100 110 120 130 140 150
MILES

SCALE.

Published by John Murray, Albemarle Street, Piccadilly

PLATE XXI

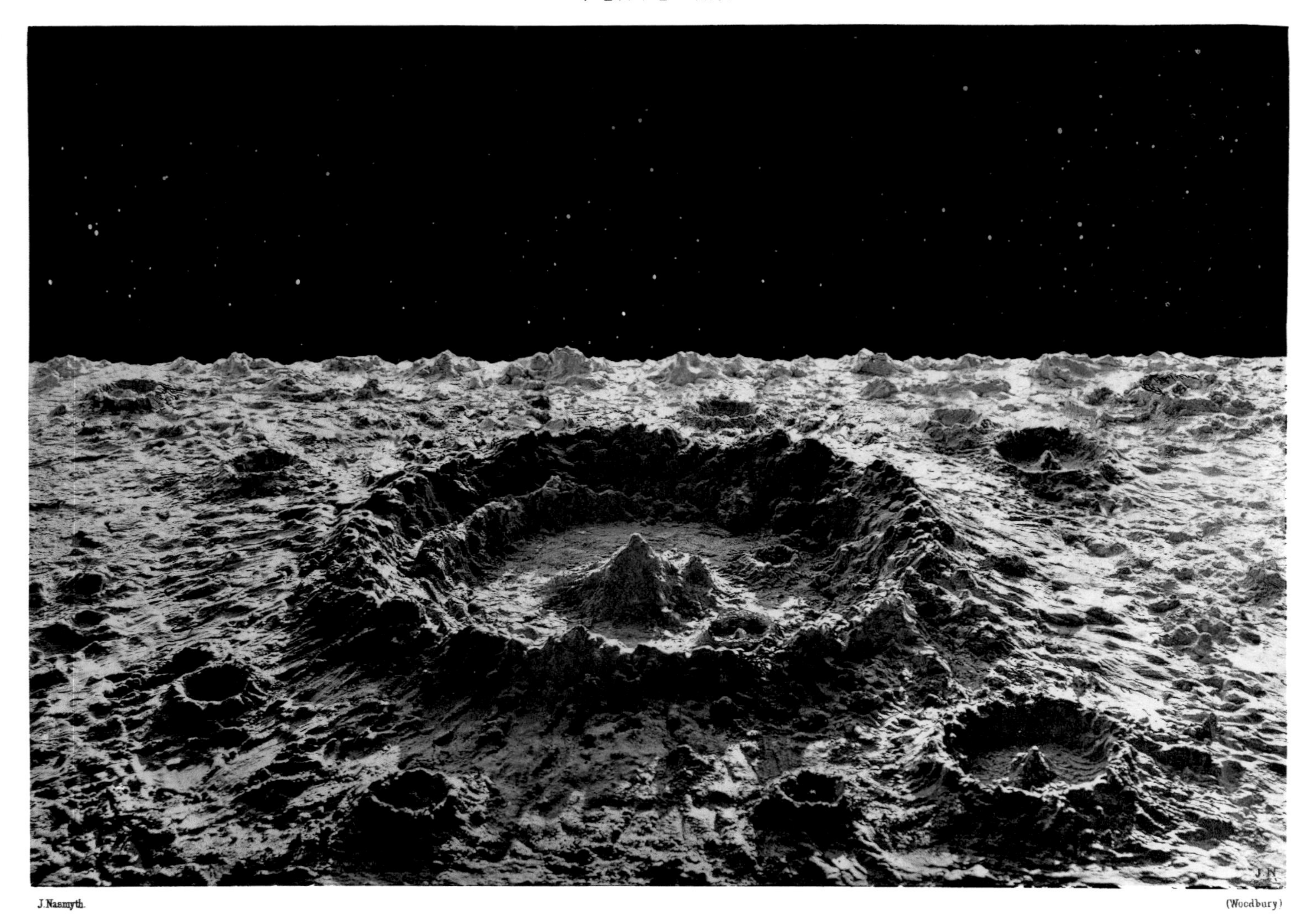

NORMAL LUNAR CRATER.

An oblique view of a lunar crater with the horizon and a black sky beyond is depicted in this photograph of a plaster model from Nasmyth and Carpenter's *The Moon: Considered as a Planet, a World, and a Satellite.* Much of Nasmyth and Carpenter's book is occupied by arguments that the lunar craters are a result of volcanic activity—a theory that was widely promulgated in the nineteenth and early twentieth centuries, but was discredited when actual lunar exploration arrived in the late 1960s. (Lunar craters are due to asteroid impacts.)

PLATE XXIII.

J.Nasmyth. (Woodbury)

GROUP OF LUNAR MOUNTAINS. IDEAL LUNAR LANDSCAPE.

Lunar mountains, from Nasmyth and Carpenter's book. Before spaceflight allowed actual reconnaissance of the moon, it was widely assumed that the surface was extremely rugged in appearance, as in these models. This was due to an optical illusion created by the extremes of dark and light that play across the moon, a body without an atmosphere, during each lunar day as observed by telescope from the Earth. In fact, a steady rain of micrometeorite impacts across more than four billion years has softened and rounded all the lunar mountains. There are no peaks like this on the moon.

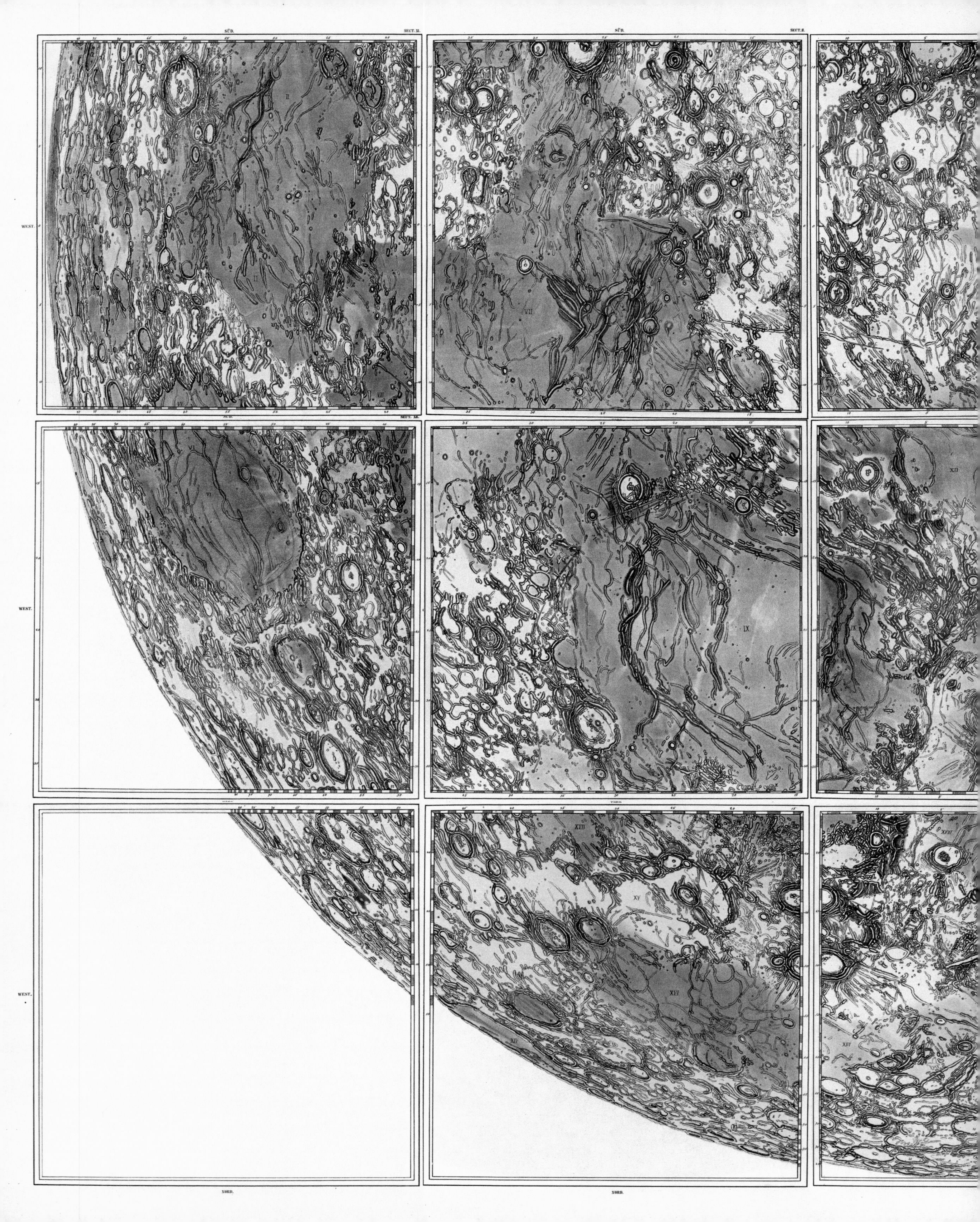
SÜD.
SECT. XI.
SÜD.
SECT. II.
WEST.
WEST.
WEST.
NORD.
NORD.

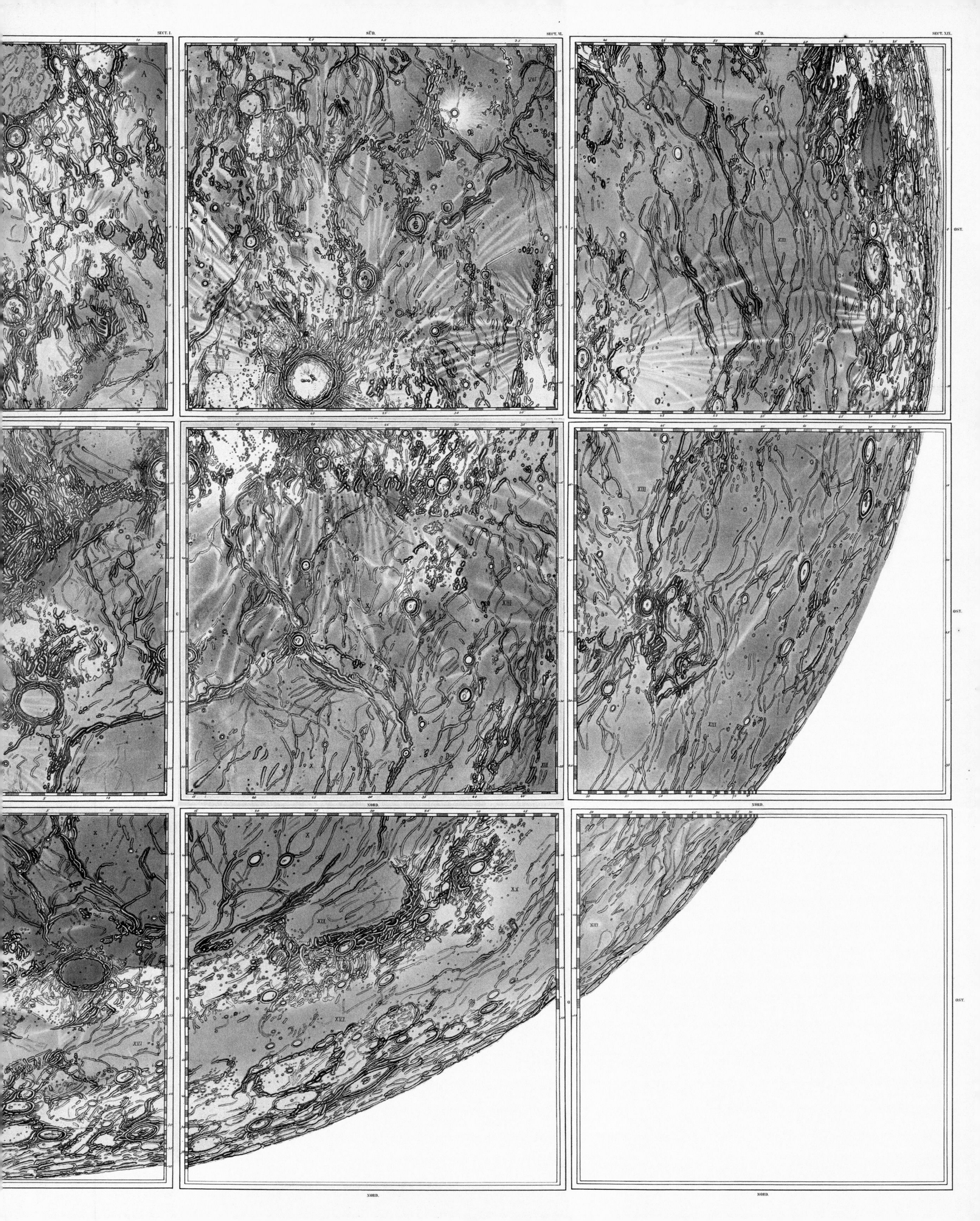
SECT. I.
SÜD.
SECT. VI.
SÜD.
SECT. XIX.
ØST.
ØST.
NORD.
NORD.
ØST.
NORD.
NORD.

• 1878:

Previous spread: During the nineteenth century, maps of the moon became increasingly sophisticated. From 1821 to 1836, German cartographer and astronomer Wilhelm Gotthelf Lohrmann conducted a campaign of lunar observations from Dresden, producing a series of map segments, twenty-five in all, that remained unpublished at his death in 1840. In 1878, another German astronomer and lunar cartographer, Johann Friedrich Julius Schmidt, edited and published Lohrmann's work as *Charte der gebirge des mondes* ("Chart of the Mountains of the Moon"—although as can be seen here, the maps delineated far more than mountains). Fifteen of the map segments are presented here, depicting the northern hemisphere of the moon. (Lohrmann and Schmidt's "upside down" presentation is an artifact of the inversion produced by reflector telescopes, and was common in lunar maps of the period.)

• 1881:

From 1870 to around 1880, artist-astronomer Étienne Trouvelot produced a series of superb illustrations of the moon, the planets, nebulae, comets, and the Milky Way. In 1872 the French expatriate was invited onto the staff of the Harvard College Observatory, where he could use powerful telescopes to assist his renderings. By 1881, Charles Scribner's Sons had released a limited-edition collection of his best pastel illustrations—one of the first serious attempts to popularize the results of observations using technology developed for scientific research, but essentially presented as belonging to the arts. This chromolithograph of Mare Humorum, a vast impact basin on the southwest side of the Earth-facing hemisphere of the moon, is from the Scribner's collection and based on a preparatory study made in 1875. Mare Humorum means the "Sea of Moisture," but is, of course, entirely dry. For more by Trouvelot, see pages 127, 128–29, 158, 188–91, 241, 270–71, 295, and 297–99.

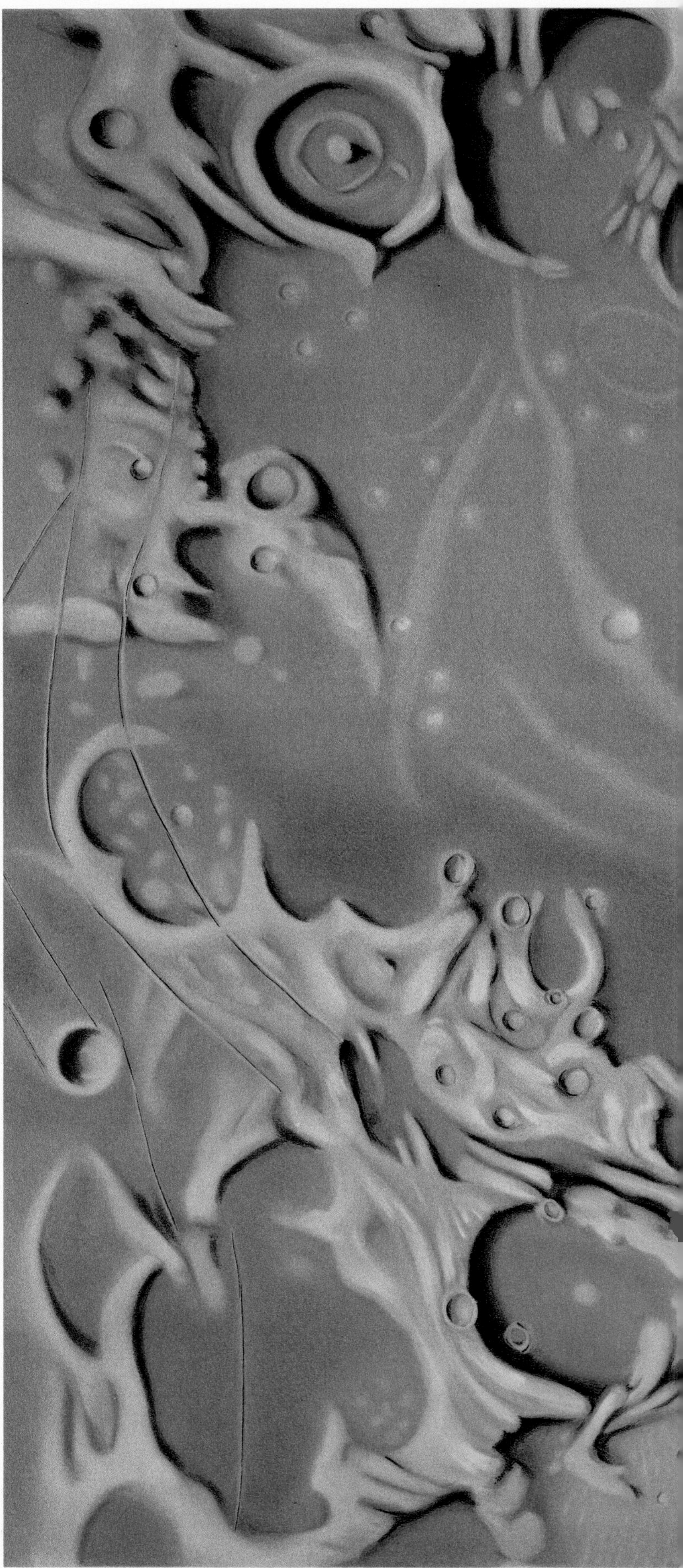

PLATE VI.

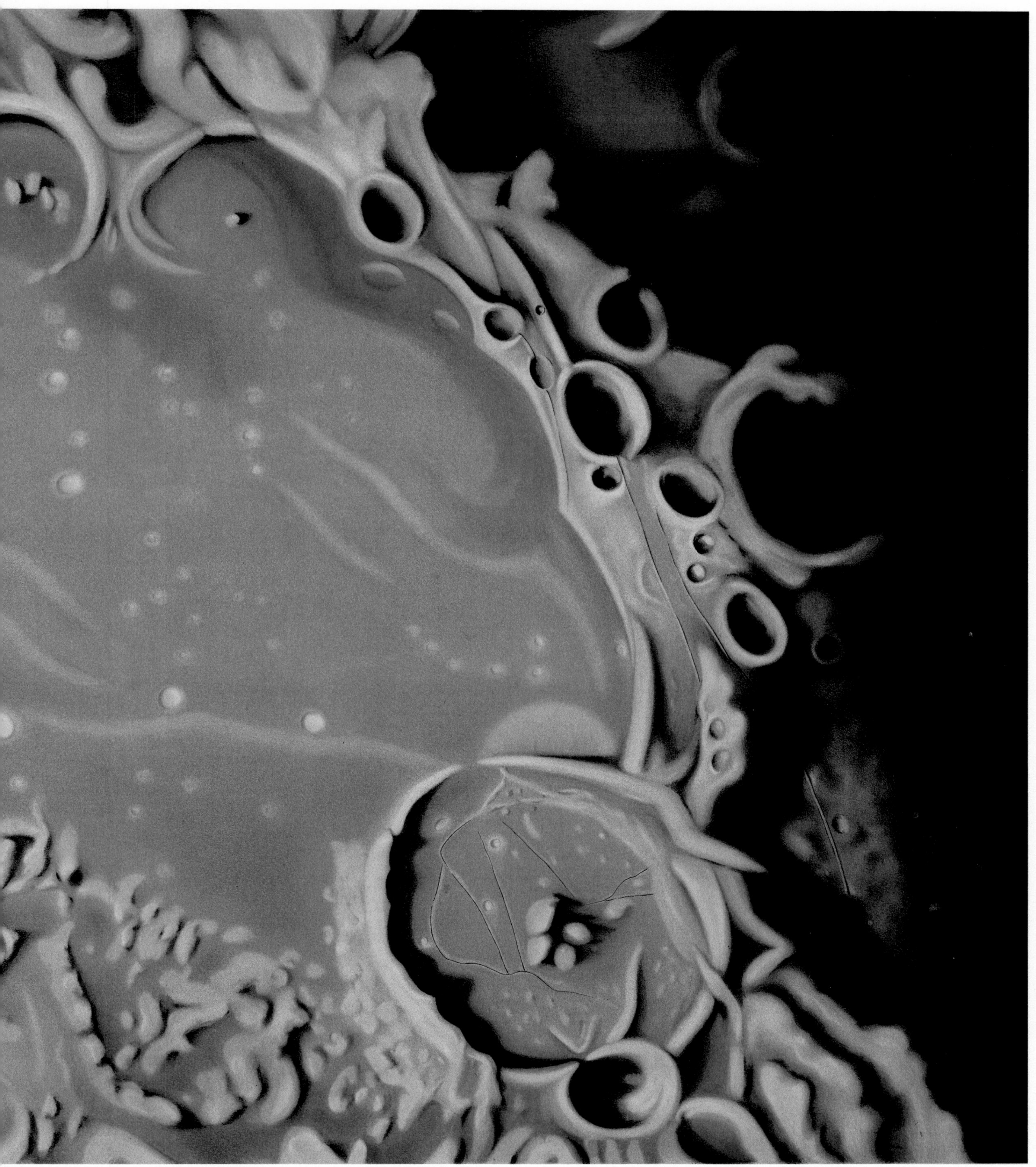

E. L. Trouvelot

MARE HUMORUM.

From a Study made in 1875.

• 1888:

In 1887, Julius Grimm, an amateur astronomer and photographer in the court of Frederick I, the Grand Duke of Baden, showed the duke his photographs of the moon. Convinced by Frederick's interest, he set about creating a detailed oil painting of our natural satellite, which he presented to the duke the following year. Grimm's painting, which features a highly textured surface, is unusual. Although it depicts a full moon—which in reality is always seen in a flat light that doesn't accentuate details—Grimm's lunar portrait was designed to be illuminated by a raking light from the left. The effect is to produce a highly variegated surface with strong relief, despite the moon being nominally "full." (As with Lohrmann's map in the previous pages, here lunar north is at the bottom.)

• Late 1930s:

Facing page: French "slide holder"-style presentation of scenes featuring the moon, including solar and lunar eclipses as seen from space, of unknown provenance and authorship. The bottom two images on the left are clearly taken from French space illustrator Lucien Rudaux's 1937 book *Sur les autres mondes* (On Other Worlds). The bottom right frame compares the relative sizes of the moon seen from Earth, and vice versa.

N° 1 - Partie du parcours lunaire

N° 2 - Premier quartier

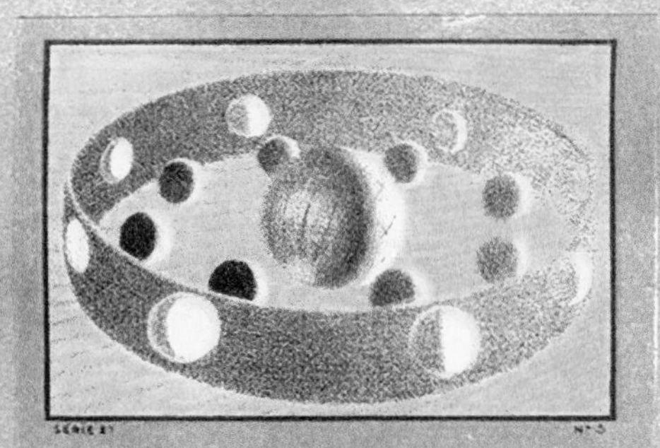
N° 3 - Explication des phases lunaires

N° 4 - Eclipse de Lune

N° 5 - Eclipse de Soleil

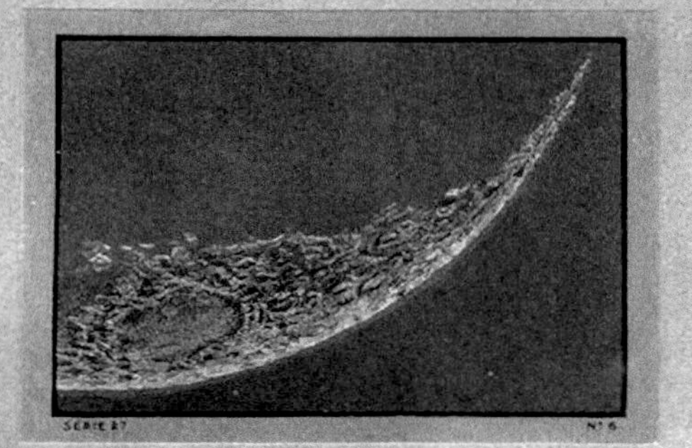
N° 6 - Corne de croissant lunaire

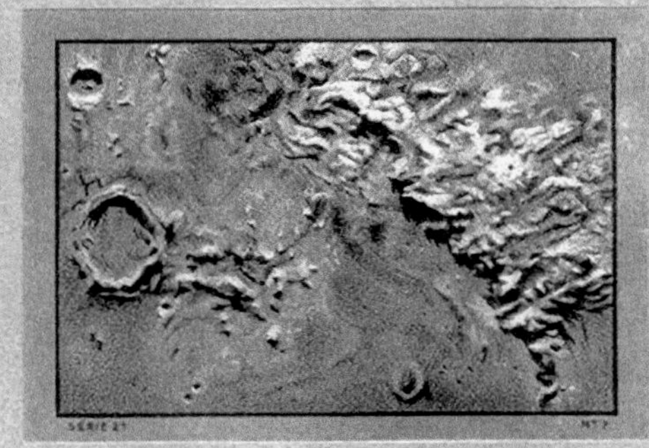
N° 7 - Cratères, cirques montagneux et plateaux

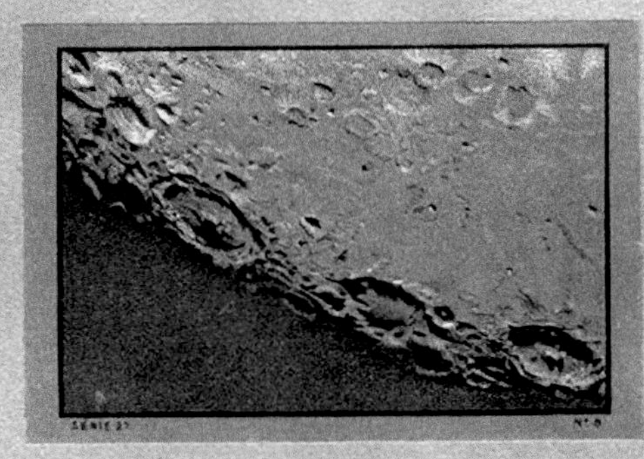
N° 8 - Cirque montagneux lunaire

N° 9 - Bords d'un cratère avec ravins et terrasses

N° 10 - Protubérances et couronne solaires

N° 11 - Coucher de Soleil, vu de la Lune

N° 12 - Pleine Terre, vue de la Lune

• 1963:
At the dawn of the space age, and before any meaningful exploration of space had really commenced, Czech illustrator Ludek Pesek depicted the solar system in multiple paintings made for a book titled *The Moon and Planets*, by Josef Sadil. This painting, of a craggy lunar landscape with distant Earth low on the horizon, comes from the book. Like his contemporary, American space illustrator Chesley Bonestell, Pesek was influenced by pioneering French space illustrator Lucien Rudaux (see previous caption). The rugged lunar mountains of pre-spaceflight popular imagination were soon to vanish as photos of the moon's softly rounded topography came to Earth.

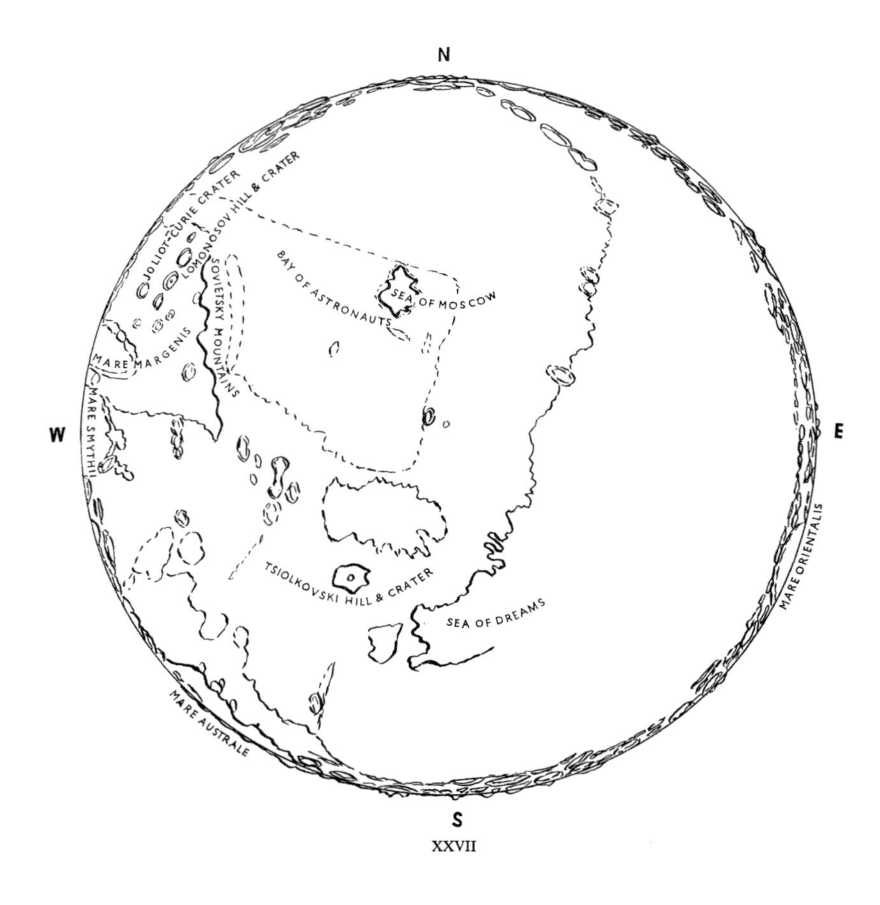

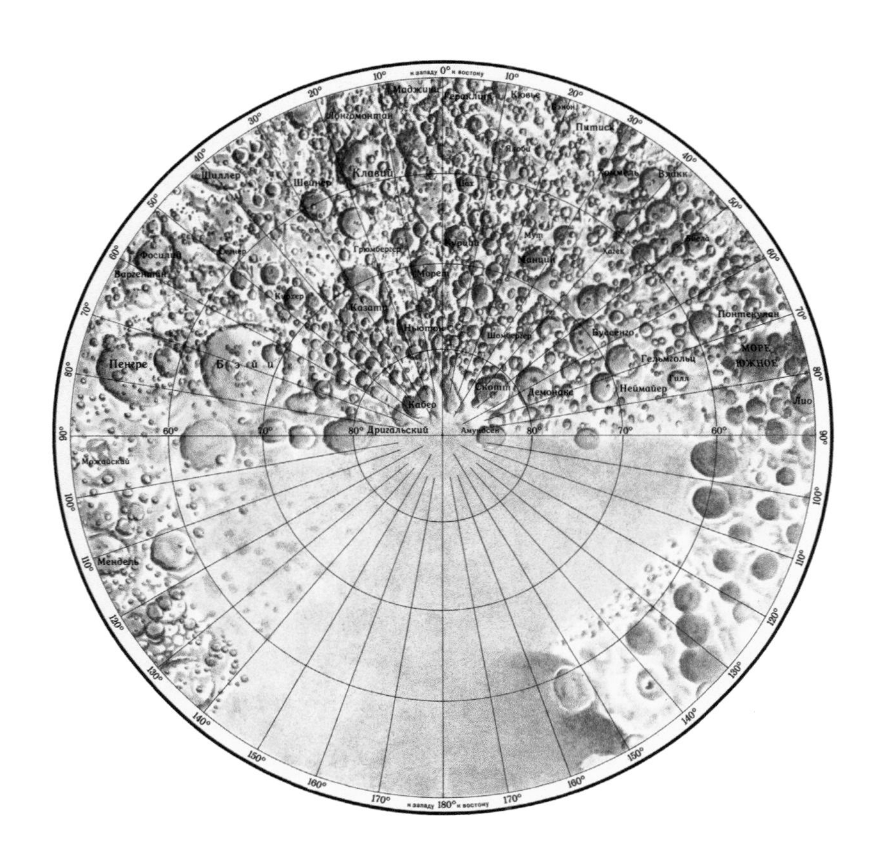

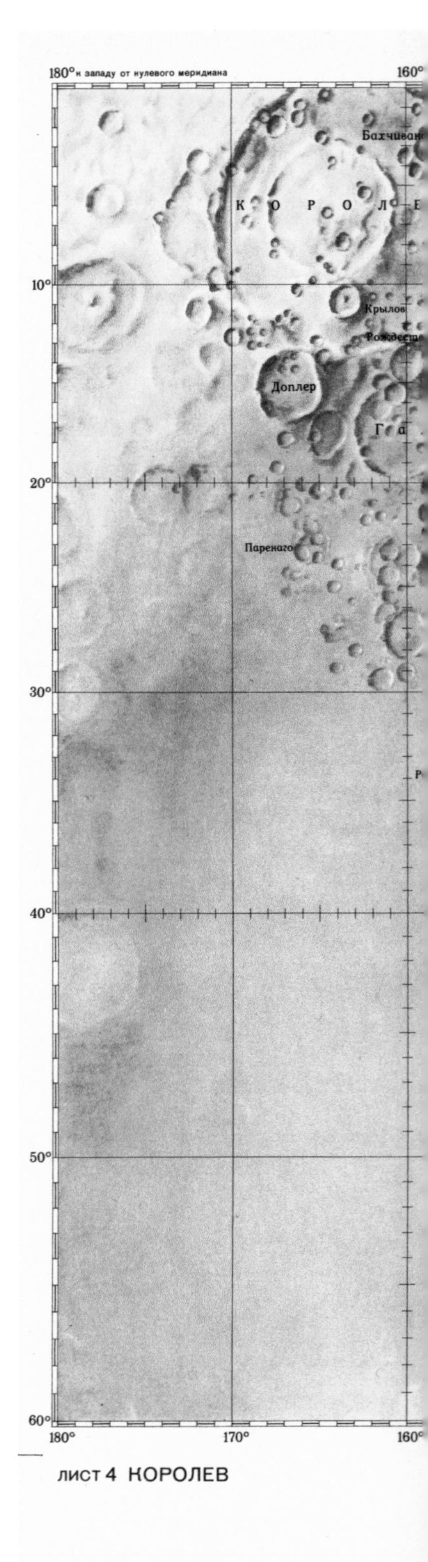

• 1960:

Top: In October of 1959, the Soviet Union launched its third robotic moon mission, Luna 3, providing humanity with its first glimpse of the far side. Although its images were blurry, they were enough for preliminary mapping. This far-side map is from a lunar atlas by English engineer and amateur astronomer Hugh Percival Wilkins. Note the newly named "Sea of Moscow" (Mare Moscoviense), Tsiolkovski crater, and the "Sea of Dreams" (Mare Desiderii). The latter name didn't stick, but the others are still in use. **Above:** See caption on facing page.

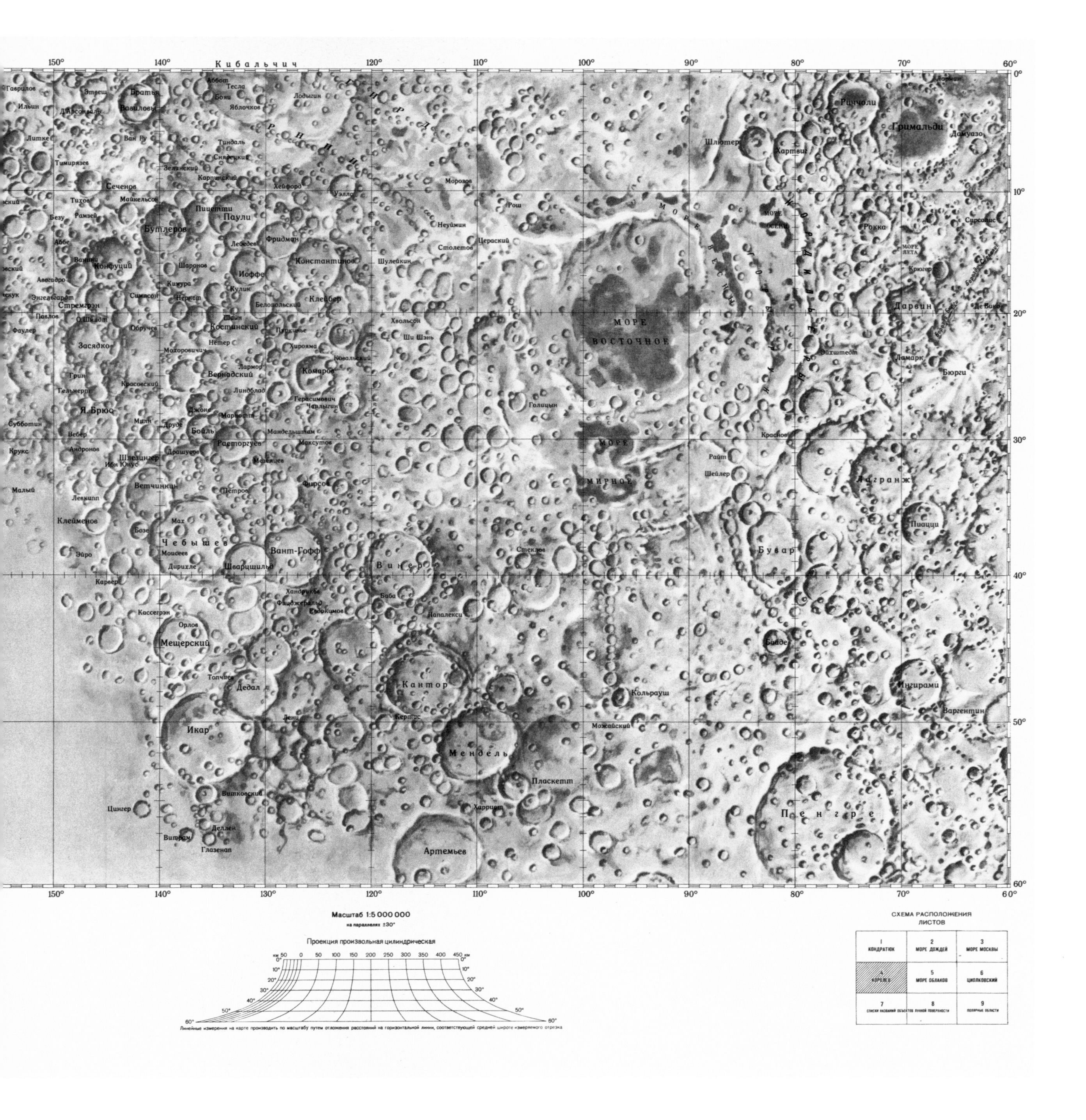

• 1967:

Above: By the time this map was made, the USSR had sent fourteen robotic missions to the moon, though as with early U.S. robotic missions, many had failed. In 1967, Moscow's Sternberg Astronomical Institute released a "full map of the moon," based on data from the Luna 3 and Zond 3 missions. Some "Luna incognita" gaps in coverage by the Soviet missions still existed on the far side, however, as is visible in this section of the southern hemisphere (constituting one-sixth of the full lunar map). As a result, in contemporary maps of the moon, a kind of irregular border exists between Russian and American place-names in this area. For example the Chebyshev crater, named after an influential Russian mathematician, is visible at the edge of the mapped area here—the last major crater before the giant, 333-mile-in-diameter Apollo crater, here invisible, which was named after the U.S. lunar missions. **Facing page, bottom:** A section of the same Soviet map with a projection of the South Pole.

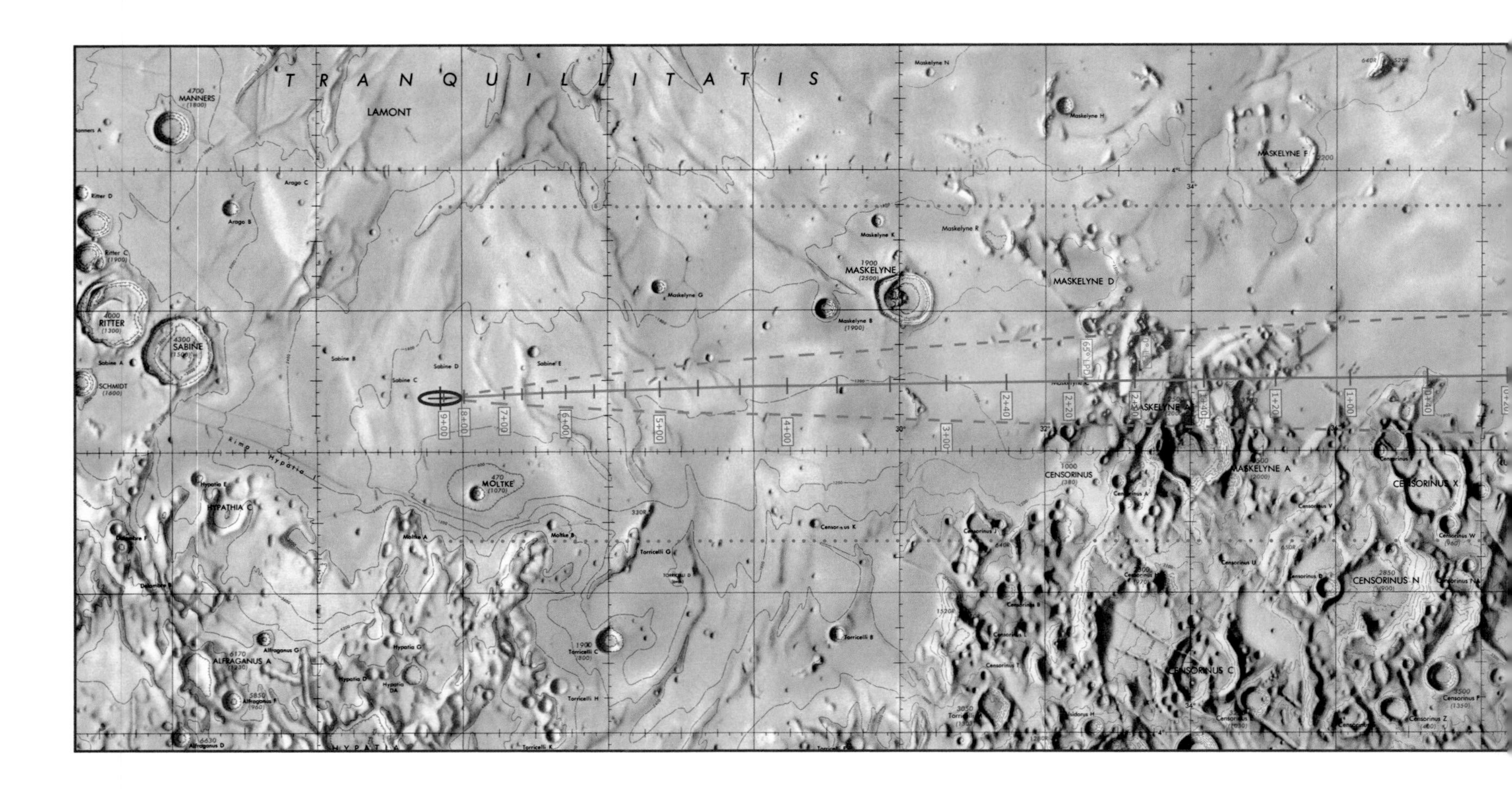

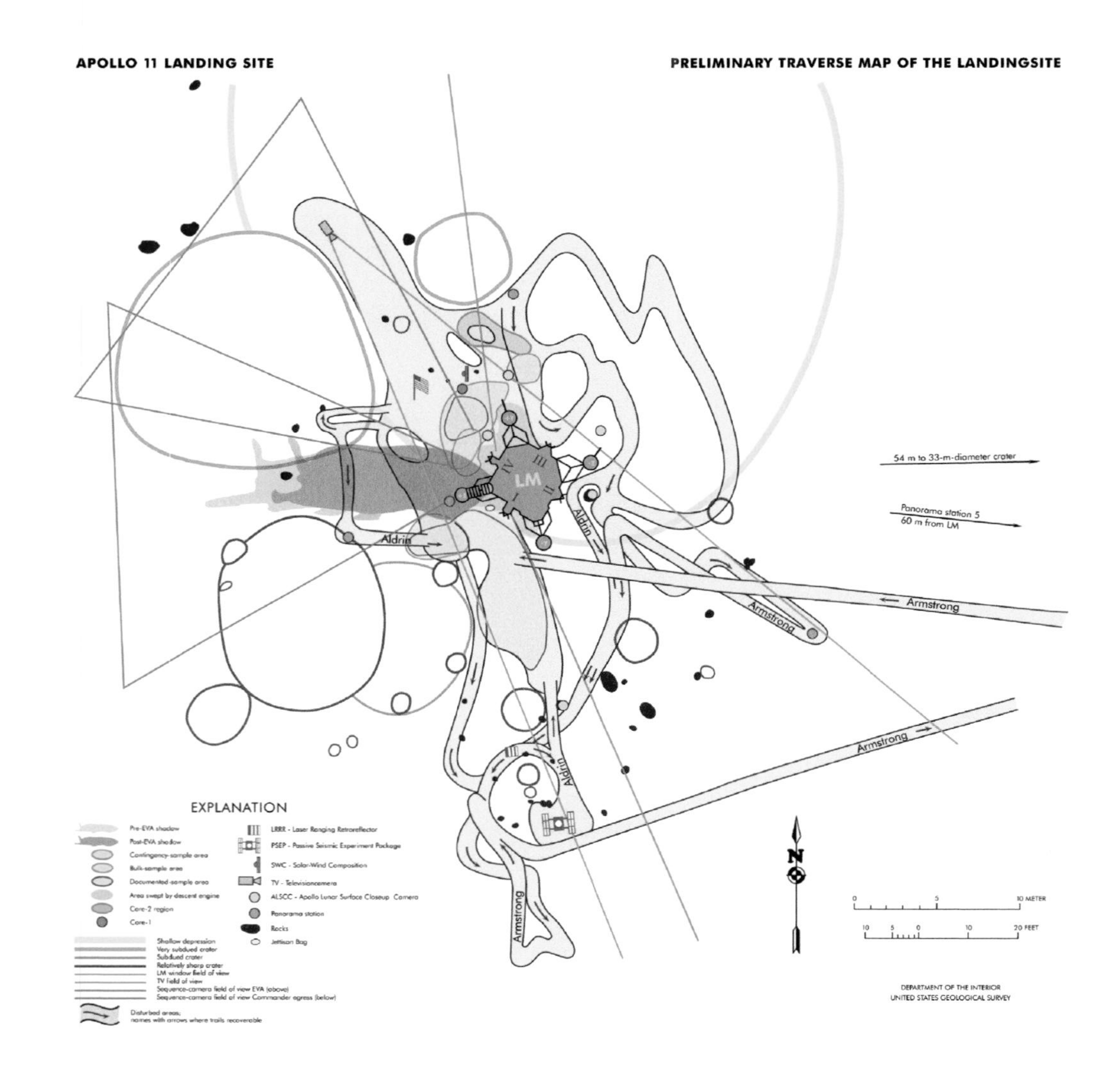

• 1969:

On July 20, 1969, human beings landed on the moon for the first time. The next day, U.S. astronaut Neil Armstrong climbed down the ladder of the Apollo 11 lunar module and set foot on the moon's surface. Apart from being a definitive moment in human history, the event ended the U.S.-Soviet space race that had dominated the 1960s, achieving a goal set by President John F. Kennedy in 1961. The three astronauts of Apollo 11 returned safely to Earth on July 24. **Left:** Contemporary reconstruction of Apollo 11 traverse map originally released in late October 1969 by NASA. The map shows the tracks the astronauts made, as reconstructed from still frames and TV footage.

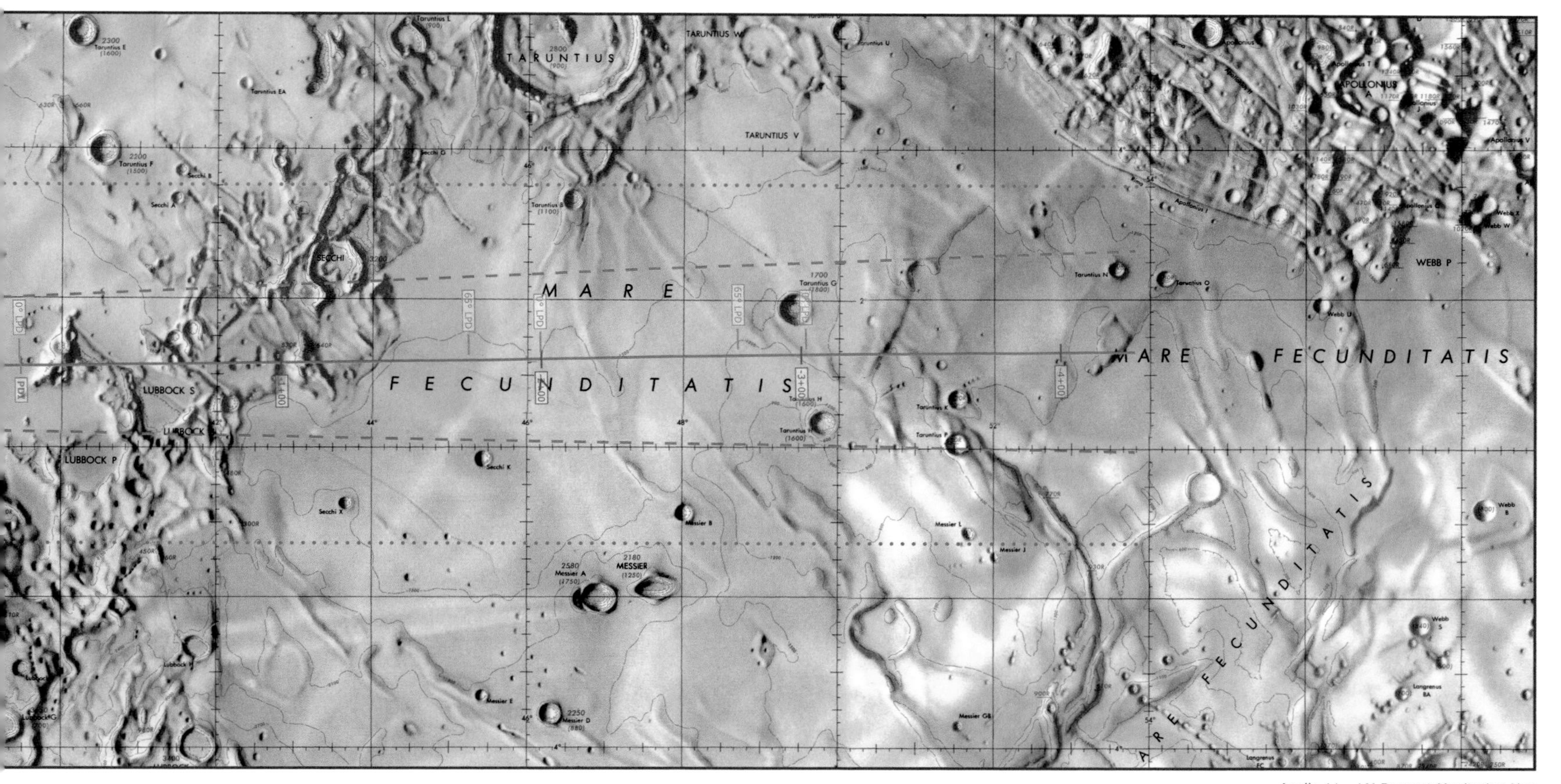

Apollo 11 – LM Descent Monitoring Map
1:1,000,000

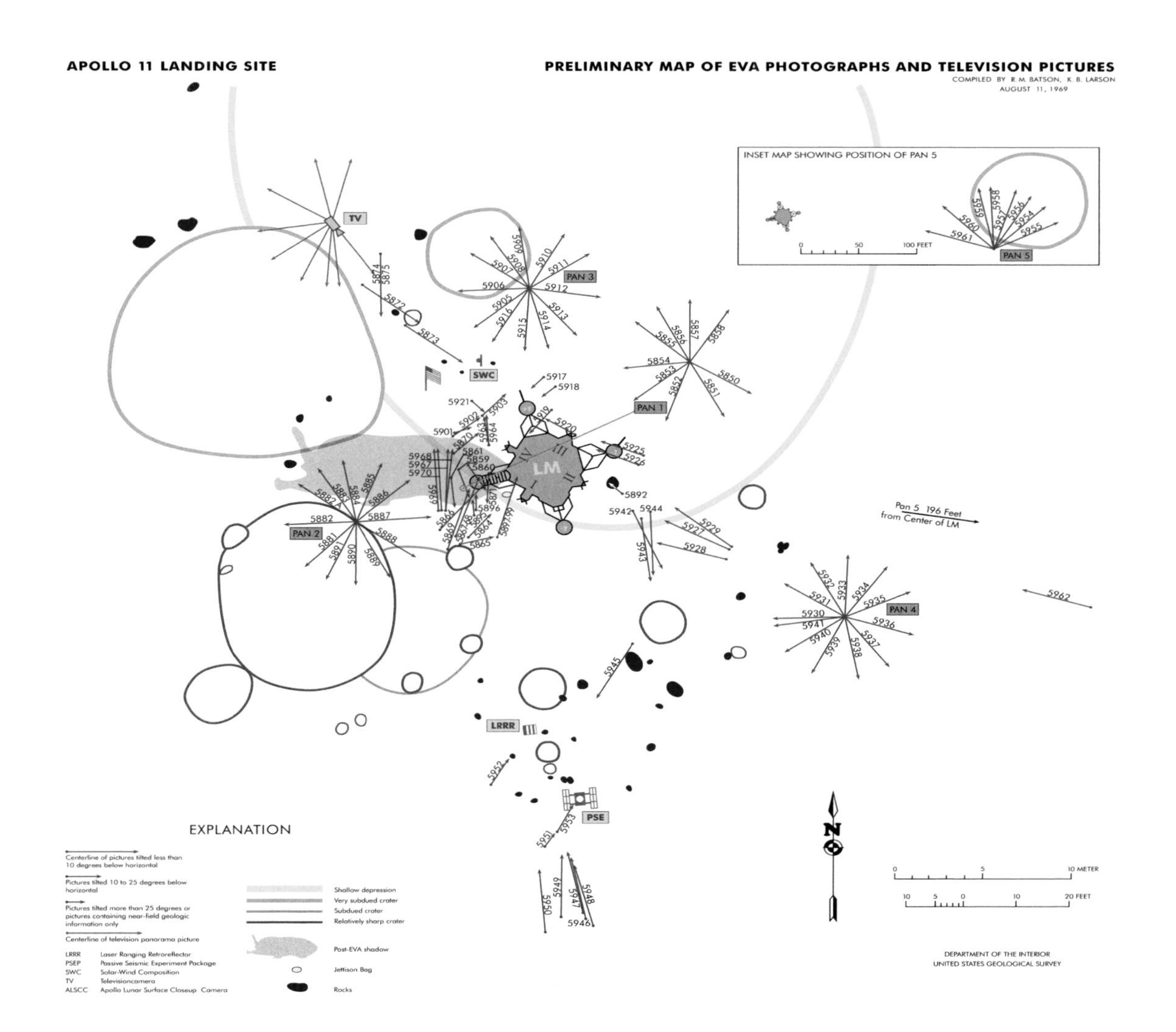

Above: Contemporary reconstruction of the Lunar Module Descent Monitoring Chart with landing ellipse, to a scale of 1:630,000. In the original flight chart used by the astronauts, the landing path is across a black-and-white photomosaic of the lunar surface. Here it is across a topographical map from the Lunar and Planetary Institute. **Left:** Contemporary reconstruction of a kind of media map of the landing site, with the location of all photographs and television pictures taken by the two astronauts. The original map was drawn by R. M. Batson and K. B. Larson and dated August 11, 1969. (All contemporary reconstructions on this and the facing page are by Thomas Schwagmeier and from the online Apollo Lunar Surface Journal.)

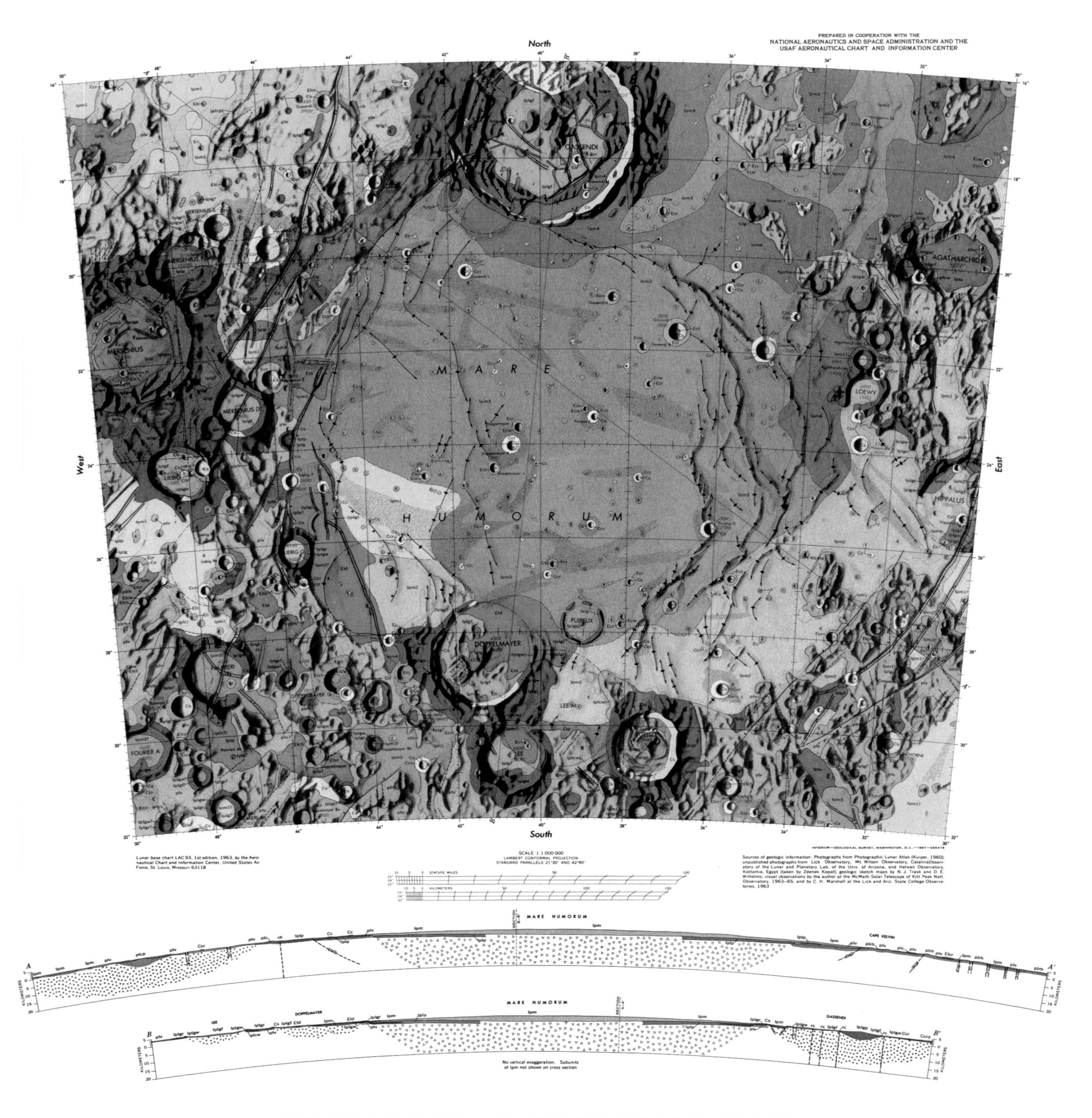

GEOLOGIC MAP OF THE MARE HUMORUM REGION OF THE MOON

By

S. R. Titley

1967

• 1967:

With the space age in full bloom, the U.S. Geological Survey began using the flood of data from robotic and manned missions to create highly elaborate geological charts of the moon. In this map of Mare Humorum, the two-hundred-mile-wide lunar "sea" in the southwestern quadrant of the Earth-facing side, color divisions delineate the characteristics and age of the various surface materials. The purple and gray-purple areas signify mare material; green and gradations of orange are for crater material of various assumed ages; yellow is for slopes and rays; and red is for plains-forming materials. The curved cross sections of the region at the bottom accurately reflect the surface curvature of the moon, with no exaggeration in relief. (See also pages 98–99 for astronomer-artist Trouvelot's depiction of Mare Humorum.)

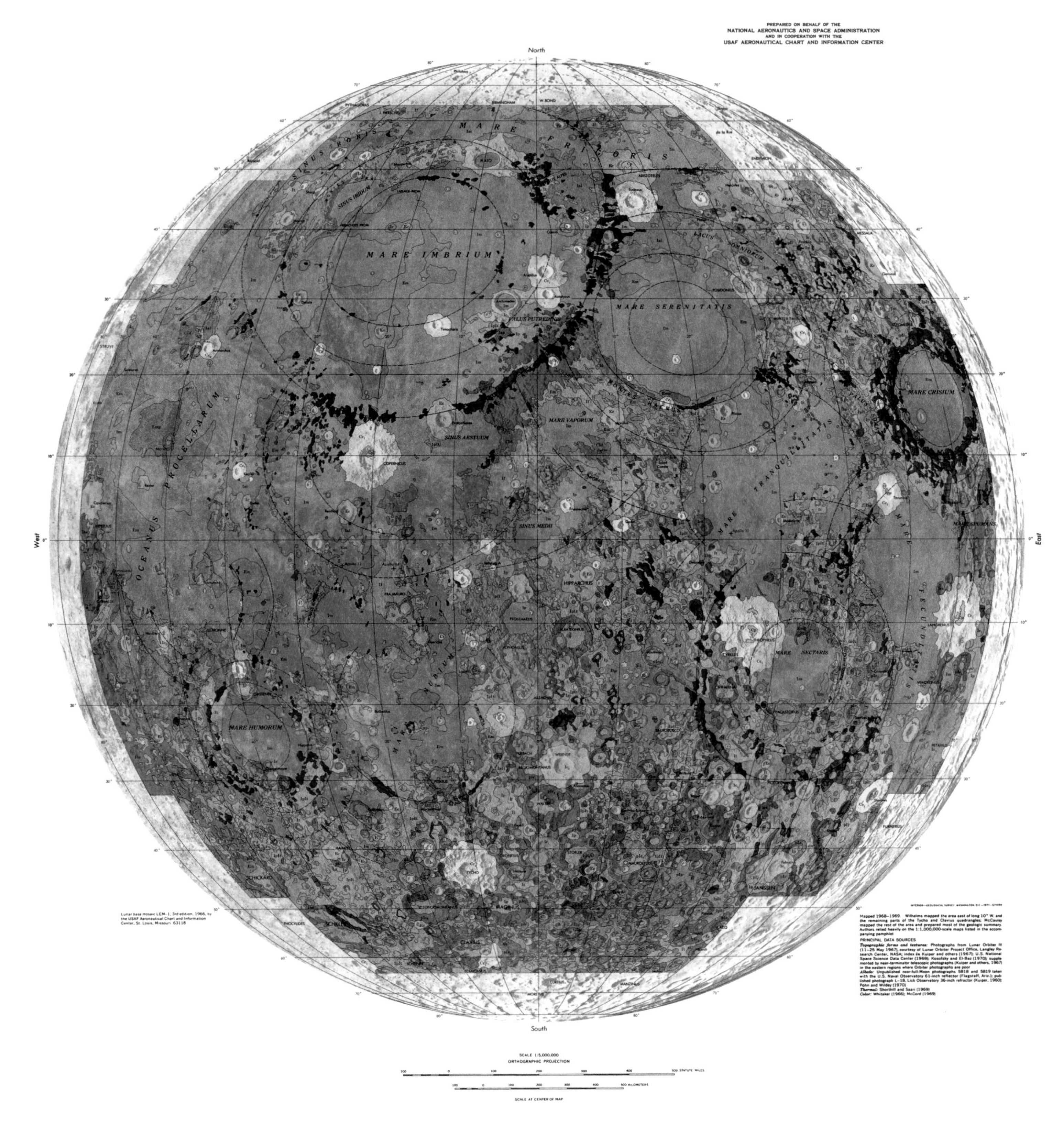

GEOLOGIC MAP OF THE NEAR SIDE OF THE MOON

By

Don E. Wilhelms and John F. McCauley

1971

• 1971:

Geological map of the moon's Earth-facing hemisphere. As the bull's-eye lines on this map indicate, the roughly circular "seas" of the near side are actually vast impact basins filled with ancient basaltic lava, giving a clear sense of the violent impacts the moon experienced during the first epoch after its formation about four billion years ago. In this highly nuanced map, green signifies the mares; blue the circum-basin materials; orange, red, and purple the plains plateaus and domes; and yellow and green the craters. Various gradations in the colors reflect the assumed ages of these materials. Extensive textual accompaniment to the map concludes that "Geological mapping reveals a Moon that is neither dominantly 'volcanic' nor dominantly 'impact' but rather one in which both processes have been operative."

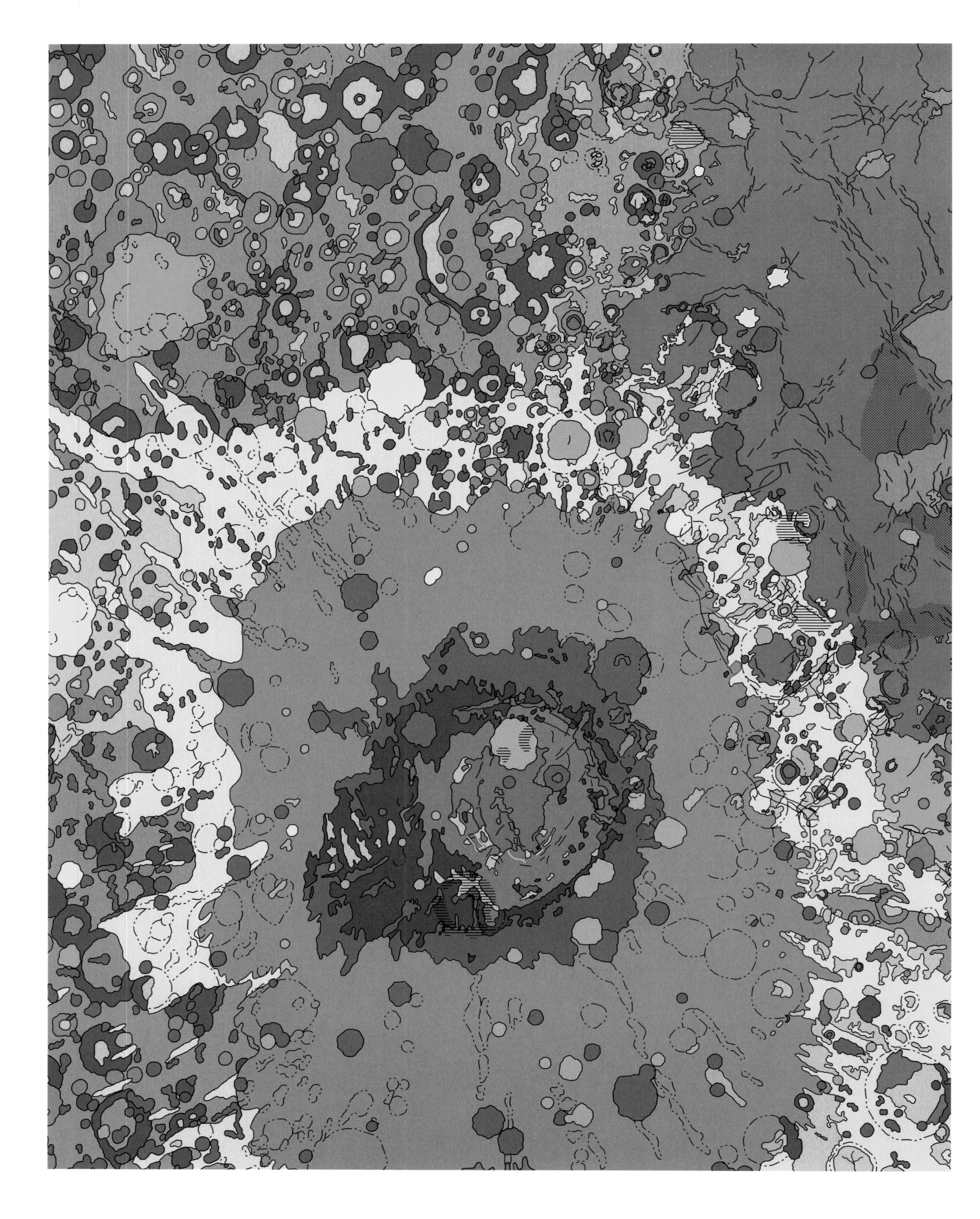

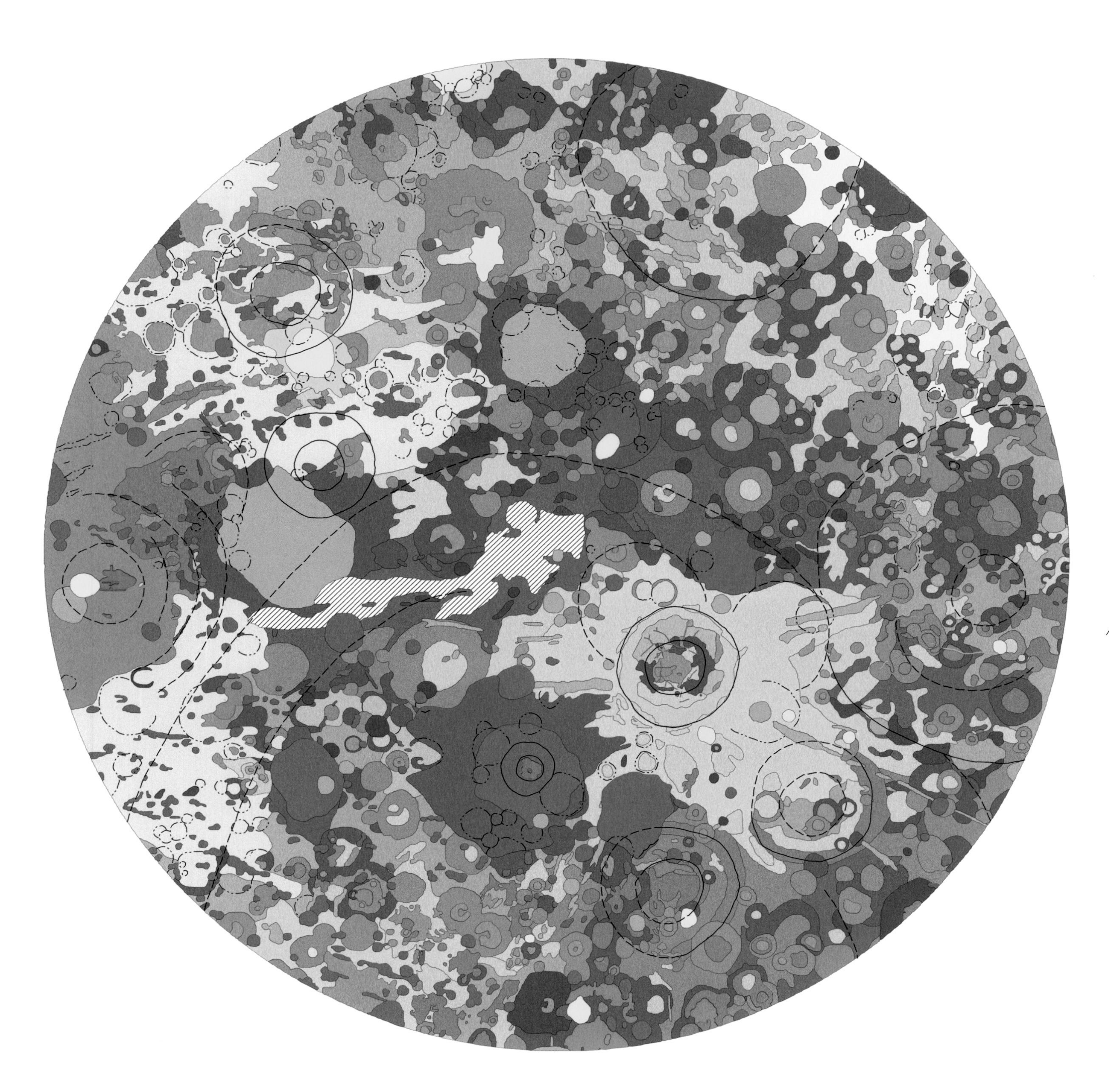

• 1977:

Facing page: The west side of the moon is dominated by the giant multi-ringed impact scar of the Orientale basin, which is more than six hundred miles across. Fringing Orientale in shades of scarlet on the upper right is the giant Oceanus Procellarum, the largest of the lunar maria (so large that it is not a mare, or sea, but rather an "ocean"—the only one on the moon, albeit bone-dry). In this geological map, blue signifies ridges and grooves radial to the Orientale impact, and red is flatter volcanic material.

• 1979:

This geological map of the south polar region of the moon follows the same color system as on the facing page. The Orientale basin continues to dominate the left side of the projection, with all the blue shadings there associated with it. The irregular oblong patch with denim diagonal lines near the center represents terrain still unsurveyed at the time this map was released. The large tan splotch just below and to the right of the South Pole is the two-hundred-mile-wide Schrodinger crater, one of the few areas on the moon that shows signs of relatively recent volcanic activity. The maroon patch within the baby-blue area at the center of the crater signifies what has been identified as pyroclastic deposits centering on a volcanic vent. The Shackleton crater, the small olive green patch located directly at the lunar south pole (and embedded in the area with diagonal lines), is thought to contain ice deposits in its perpetually shaded depths. Such deposits could become very important if we ever colonize the moon.

CHAPTER 4

The Sun

I know you love Manhattan, but
you ought to look up more often.
And
always embrace things, people earth
sky stars, as I do, freely and with
the appropriate sense of space.

—Frank O'Hara, *A True Account of Talking to the Sun at Fire Island*

Unlike the moon, which for all its mysterious allure rose only to the rank of secondary deity, the sun demanded full-scale worship as a major god throughout history. The Neolithic barge-borne sun became the Egyptian sun god Ra, who by the Fifth Dynasty ruled the rest of the gods, riding in his solar boat on a daily journey from east to west. In the Babylonian and Assyrian pantheons, sun god Shamash brought wrongdoers to justice, shedding a pitiless light on their crimes. In Indonesia, the solar system's central star was the mythological progenitor of the ruling family and the source of their authority. And in Aztec culture, the sun god Tonatiuh ruled all the other deities and required a steady diet of human hearts, which were excavated with spadelike chipped blades from the chests of his sacrificial subjects, before he would consent to budge in the sky. Tens of thousands of sacrifices a year may have been deemed necessary to placate the savage sunny Mesoamerican god.

The reason for all this solar power isn't particularly hard to understand. The sun drives the wind, determines the harvest, fuels most life, and lights our way. Its absence spells disaster, and too much of it also produces tragedy. The sun's moodiness is a leading manifestation of the capriciousness of nature and the vagaries of fate. To this day, most of the fuel our civilization requires derives from the sun: All the fossil fuel deposits under the Earth's crust are actually batteries filled with archival solar energy.

In many ways then, sun worship makes more sense than later religions. And in fact, there's good evidence that some of the major holidays associated with Judeo-Christian monotheism are linked to prior cults of sun worship. Contemporary birthday celebrations of the son of man, Jesus Christ, take place on the same day that the Romans celebrated the rebirth of the sun during the festival of *Dies natalis solis invicti* (Birthday of the invincible sun)—the winter solstice, or December 25, of the Julian calendar. A late-third-century mosaic in the Vatican Necropolis under Saint Peter's depicts a kind of fusion of the Roman god Sol Invictus (aka Apollo-Helios) and Christ. The rays behind his head, unmistakably a Christian halo, double as the sun. The vines framing the scene are equally interpretable as the source of Dionysus's drinking problem and John 15's True Vine; and the mosaic itself dates back to when the pagan gods were retreating up Olympus' slopes in the face of an ascendant Christianity. It's a transmigration of Sols.

Given these facts and informed suppositions, it's remarkable that the pre-Socratic Greeks settled on an Earth-centered universe in the first place. How could something that powerful and integral consent to orbit us? If all celestial phenomena can be explained by circular motion, as Plato suggested, does it necessarily follow that all that circularity centers on Earth—rather than this powerful definer of days?

Whatever the explanation, even fifteen centuries' worth of believers in a geocentric Ptolemaic system granted the sun real prerogatives. The ecliptic, or apparent path of the sun in the sky, functions as the equator of one of the celestial coordinate systems handed down from antiquity. The latitude and longitude system it's associated with is organized with respect to the ecliptic and the ecliptic poles, and its name derives from the fact that it's the only place where an eclipse is ever seen. The ecliptic also divides and defines the zodiac, a sixteen-degree-wide band in the sky within which the sun, moon, and planets can all be found at any given moment. The zodiac is in turn divided laterally into twelve constellations—a number deriving from the calendrical months. The etymology of the word *zodiac* is related to *zoo* because of the animal forms in its constellations. If it's a menagerie, its keeper is the sun.

Adherents of the Ptolemaic system—and in fairness there was no other game in town for a millennium and a half—conceded the sun's special role in numerous other ways, some of them subliminal. The blooming sun on page 115, for example, shines down on an Earth positioned at the center, causing a meandering inky shadow on the vellum substrate of the medieval encyclopedia *Liber floridus* (Book of Flowers). But while it's indubitably depicted as orbiting the Earth, its flowerlike petals make it seem much more prominent. Looking at this painting by Lambert, the Canon of Saint-Omer, one can easily imagine the red lines connecting the sun to the Earth, and presumably guiding it on its geocentric course, swinging the other way—a headstrong Ferris wheel now centered on that sunflower, and carrying the Earth from center to margin.

Or take the gold and gray discs a couple pages later, a late-fifteenth-century manuscript illumination by German miniaturist Joachinus de Gigantibus, from the treatise *Astronomia* by Christianus Prolianus. We're at least four decades before the publication of Copernicus's *De revolutionibus orbium coelestium* (On the Revolutions of the Celestial Spheres)—the 1543 book that ultimately drove a complacent Earth off its throne and gave the sun centrality. But although this early infographic is evidently meant to depict the magnitude, or brightness, of the celestial objects on display, it's hard to avoid understanding it as portraying their sizes as well.

How can one look at that page and not perceive that the large golden disc representing the sun holds much more than magnitudinal sway over the pale gray twenty-five-cent piece of Earth? And yet the very suggestion that the sun might lie at the center of the planetary system would have been met with incredulity by Prolianus. (There's another point of interest here as well, namely that the planets aren't portrayed as points, but rather discs. We're talking about a book from well over a century before the telescope. To de Gigantibus, the planets were only wandering stars.)

Soon after the telescope arrived, astronomers used it as a basis for creating the helioscope, a device allowing for the telescopic projection of the sun onto a sheet of paper. In the 1610s, Galileo Galilei and his rival, the appropriately named Jesuit Christoph Scheiner, both used helioscopes to study sunspots. Their helioscopic projection method allowed for very precise sketching, and so Scheiner's and Galileo's suns have an almost photographic accuracy, for the simple reason that they were made via a quasi-photographic process: all that was missing was the emulsion. (Sunspots occur when the sun's magnetic flux lines converge, producing slightly cooler areas on its photosphere, or visible surface layer.)

Apart from a fascination with sunspots, and the giant prominences that can be seen vaulting from its surface at the moment of totality during a solar eclipse, astronomers have always been preoccupied by the sun's terrestrial calendrical role. As previously mentioned, timekeeping was a major function of ancient astronomers, with the sun's seasonal oscillations measured since long before recorded history in order to calibrate the solar calendar.

In 1655 French-Italian astronomer Gian Domenico Cassini constructed a remarkable sundial on the floor of the Basilica of San Petronio in Bologna. The fourteenth-century church had already been used as a giant astronomical instrument, when a less ambitious sundial had been created by astronomer-priest Ignazio Danti in the previous century, but Cassini wanted to construct the most accurate such device on Earth. Using sophisticated engineering techniques to ensure an utterly level 219-foot-long meridian line—the longest in the world—Cassini's stated intention was to use his "heliometer" to monitor the accuracy of the Gregorian calendar. Today the most widely used calendar internationally, it had been put into effect by Pope Gregory XIII in 1582, replacing the old Julian calendar. The main motivation behind the Pope's reform was to ensure that the Church could in effect anchor the commemoration date of Easter, which had been drifting due to discrepancies between the solar and lunar years.

Cassini's sundial enabled him to measure the time between successive spring equinoxes with unprecedented accuracy, thus fulfilling its overt calendrical function. But it was also immediately employed in a sophisticated attempt to verify a deduction made by Johannes Kepler. The German astronomer discovered that Copernicus had assigned an eccentricity to the Earth's orbit twice as large as it should be, which would have made the subtle variations in the sun's apparent size too large as the Earth traveled around the sun in its off-centered orbit. Cassini measured the size of the sun's image on its noon crossings of his meridian line and confirmed Kepler's claim. By providing these measurements, Cassini's instrument in a Catholic church within the borders of the Papal States thus provided important empirical evidence for the veracity of Kepler's refined version of Copernicus's heliocentric theory. It did so already by late 1655—or twenty-two years after Galileo was found "vehemently suspect of heresy" in holding the opinion that the Earth moves around the sun.

The calendrical function of the Bologna heliometer subsequently inspired Pope Clement IX, who was crowned in November of 1700, to build a similar device in Rome within the immense Basilica of St. Mary—a church that had been designed by an elderly Michelangelo in 1561 to fit within the third-century Baths of Diocletian in Rome. Clement turned to a long-term member of the papal court, astronomer Francesco Bianchini, who had studied (and praised) Cassini's sundial. Bianchini ended up producing what many observers agreed was the most versatile and beautiful sundial of all.

The church had excellent southern exposure and very high walls, both necessary for the purpose, and it had been standing for so many centuries that its structure was unlikely to shift, which was also critical. Bianchini drilled a small hole high in the south-facing wall, producing a gnomon; and he had a thick bronze line constructed in the marble floor of the church, precisely along the meridian line of 12°30'.

Like Cassini's, Bianchini's sundial is still operating today. Because of the diameter of the gnomon hole and its distance from the bronze line, the sun shines through only at solar noon each day, hitting the line somewhere along its length. The device is precisely circumscribed by the maximum annual oscillation of the sun; at each solstice, it hits one or the other end of the line. At the two equinoxes, the sun's ray hits the same spot between the two ends.

There is, of course, a certain irony in Cassini and Bianchini turning Catholic churches into astronomical instruments dedicated to measuring, with exquisite precision, the location and apparent size of the sun that Copernicus had positioned at the center of the universe. Although Francesco Bianchini knew better than to express contrarian views on the heliocentric theory, there's an interesting illustration in a book on Venus that he published in 1728—long after his meridian line was finished. It's a plate depicting the planetary orbits, with nothing at its center.

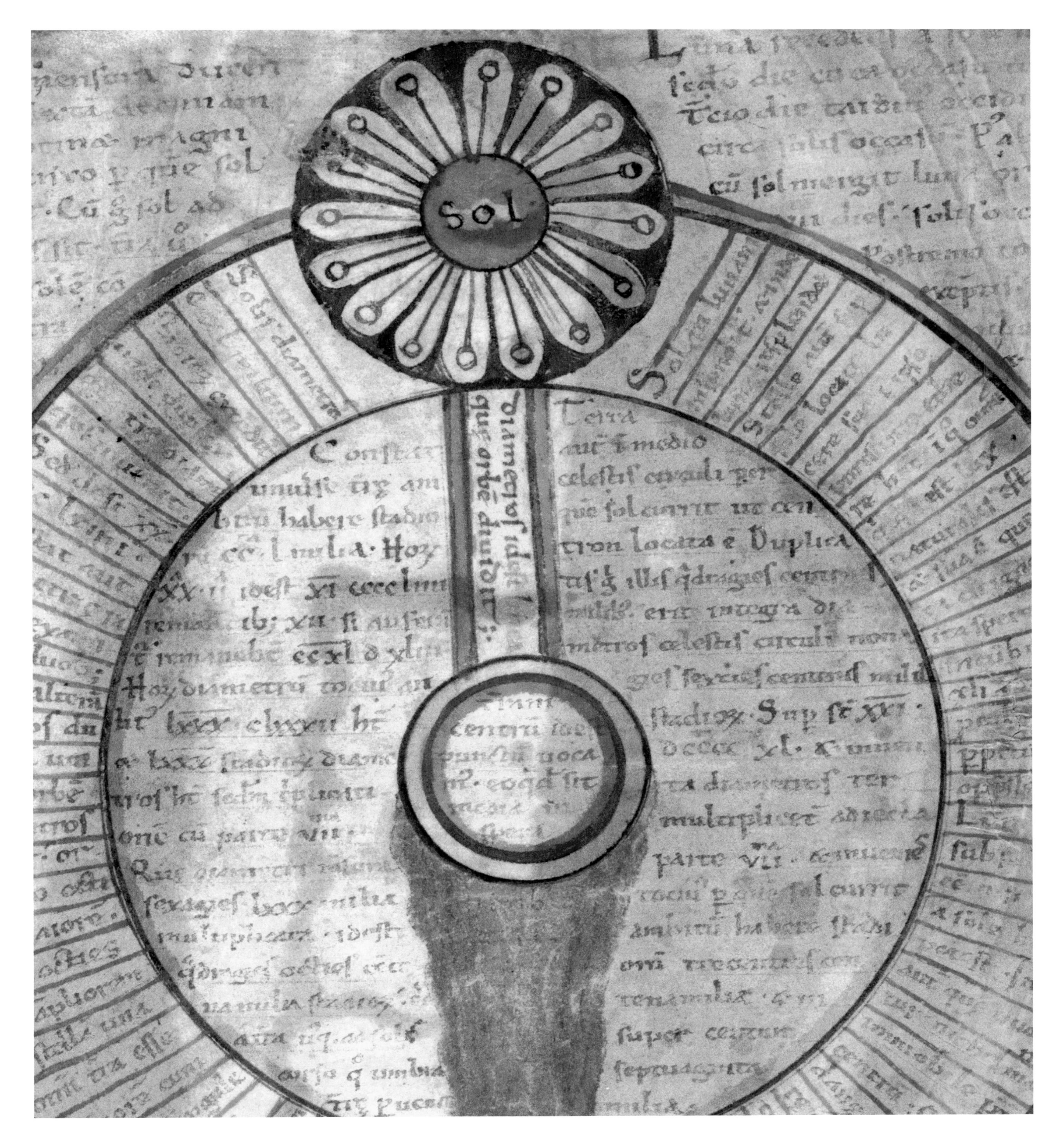

• 1121:

In this image of a blooming sun from the medieval encyclopedia *Liber floridus*, it shines down on the Earth at the center of the universe, causing a meandering shadow. The red ring extending out from either side and encircling the Earth is the plane of the ecliptic, or the apparent path, the sun makes in the sky. Because the moon also never strays more than five degrees from the ecliptic, occasionally the shadow of the Earth falls on it, producing an eclipse. Alternatively, the moon's shadow can fall on Earth, producing a solar eclipse. In fact, the term *ecliptic* was coined when ancient astronomers noticed that lunar and solar eclipses occur only when the moon crosses that imaginary line.

• 1440–50:

In these manuscript illuminations illustrating *Paradiso* in Dante's *The Divine Comedy,* Sienese artist Giovanni di Paolo depicts Dante and his guide, Beatrice, ascending to the sphere of the sun, where they are met by the souls of Thomas Aquinas and his mentor and collaborator, Albertus Magnus. Other great intellectual figures—referred to as "flamelets" by Aquinas—wait their turn to speak. These include Bede, who, in a work written about A.D. 723, provided information on how to calculate solar motion. For his part, Dante refers to the sun as "the greatest minister of Nature, who stamps the world with the power of Heaven, and measures time for us by his light." Di Paolo, who illuminated this parchment codex with gold leaf, manages to justify that term in reflected light. For other works by di Paolo, see pages 33, 75, 144, 178–79, and 256.

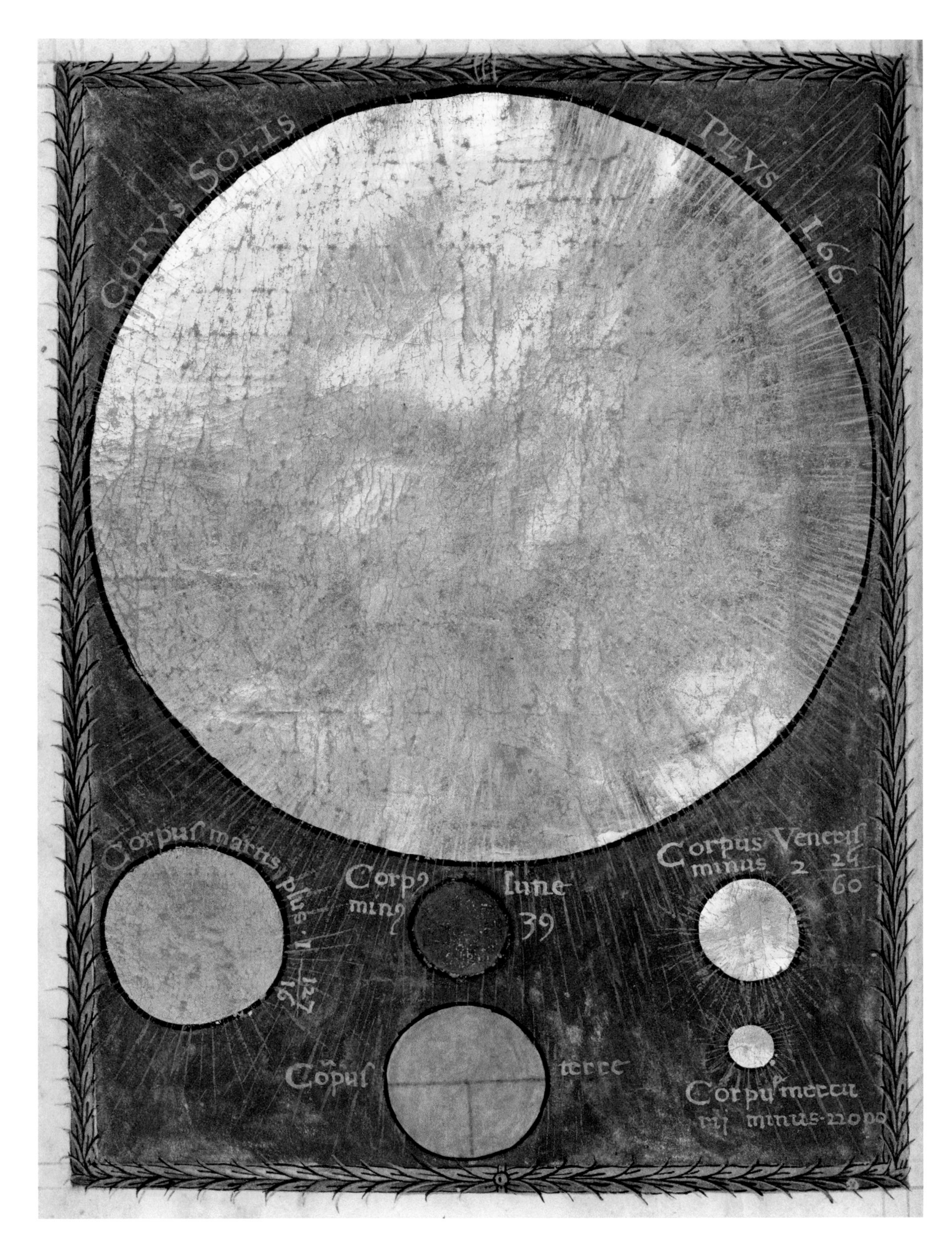

• 1478:

More gold leaf brightens this intriguing late-fifteenth-century infographic presenting the relative magnitude, or brightness, of the sun (the largest gold disc), the planets, and the moon by German miniaturist Joachinus de Gigantibus, from the scientific treatise *Astronomia* by Tuscan-Neopolitan humanist Christianus Prolianus. Golden Mars is below the sun and to the left; the moon, in gray, directly below it; and golden Venus to the right of the moon. At the bottom, Earth is also depicted in gray, and to its right, tiny Mercury shines with gold. Oddly, Mars is here depicted as being larger than Venus, which usually far outshines it, and the moon is given quite a somber color, though it is frequently brighter than anything else in the night sky. (The moon is in fact dark gray in color, intriguingly enough, but that would have been unknown to anybody in 1478.) While the relative diameters of the planets are quite incorrect, they are all presented as being far smaller than the sun, and subservient to it. It's true that the sun was understood to be nineteen times the diameter of the moon since antiquity (in fact, it's four hundred times the moon's diameter). And yet this is in an illustration created only five years after Copernicus was born, when the Earth was still at the center of the universe. Another fascinating aspect is that Mars, Venus, and Mercury are all portrayed as possessing discs in the first place, more than 130 years before telescopes first revealed them to be planetary bodies, rather than "wandering stars." For another illumination from this manuscript, see page 257.

• 1479:

Above: A good deal of mystery surrounds the famous Aztec Sun Stone, which was discovered under the main square of Mexico City in December of 1790, and is thought to date from around 1479. One thing that is clear is the sun played a central role in the Aztec culture that flourished in what's now Mexico from the fourteenth to the sixteenth centuries. Some sources indicate that the Aztecs believed that their sun god, Tonatiuh, required human sacrifices or the sun would stop moving through the sky. According to one theory, the stone was likely mounted faceup at the symbolic center of Aztec culture and used for human sacrifices, in which the heart of the sacrificial subject was then offered up, still beating, to the sun god. Whatever its function, as Mexican astronomer and archaeologist Antonio de Léon y Gama correctly pointed out in 1792, in the book this illustration came from, the stone could only have been made by a civilization possessing a good deal of knowledge of geometry and mechanics; the perfectly symmetrical stone is almost twelve feet wide and weighs more than twenty tons. The face at the stone's center is thought to represent Tonatiuh, and details of the surrounding stonework suggest calendric significance. Some sources claim that the stone contains both Aztec calendars, the secular and ritual (the first regulated agriculture, and the second was used by priests). It is possible that the stone played a role in the fifty-two-year cycle that was one foundation of Aztec civilization. The extent of our ignorance about its true function and the symbolic meanings encoded in the Sun Stone is a clear indication of the totality of the destruction visited on Aztec culture by marauding European conquistadores.

• 1540:

Facing page: Volvelle from Peter Apian's *Astronomicum Caesareum* (Caesar's Astronomy). This complex set of paper wheels could be used to monitor and also predict the exact position of the sun on the ecliptic. Unlike the planets, the sun does not retrogress in its apparent motion when viewed from the Earth—meaning it doesn't seem to reverse in its apparent motion in the sky for a period each year—but its motion along the ecliptic is a bit more rapid during the winter than the summer. As a result, the sun takes longer to move from the vernal equinox to the autumnal one than it does to move back. Apian's book doubled as a scientific instrument capable of predicting celestial motion. When turned, the changing angles in the lines visible on the pink and green volvelle seen here allow the user to compensate for these variations in apparent solar motion. For more from Apian, see pages 43, 77, 180, 227, and 262.

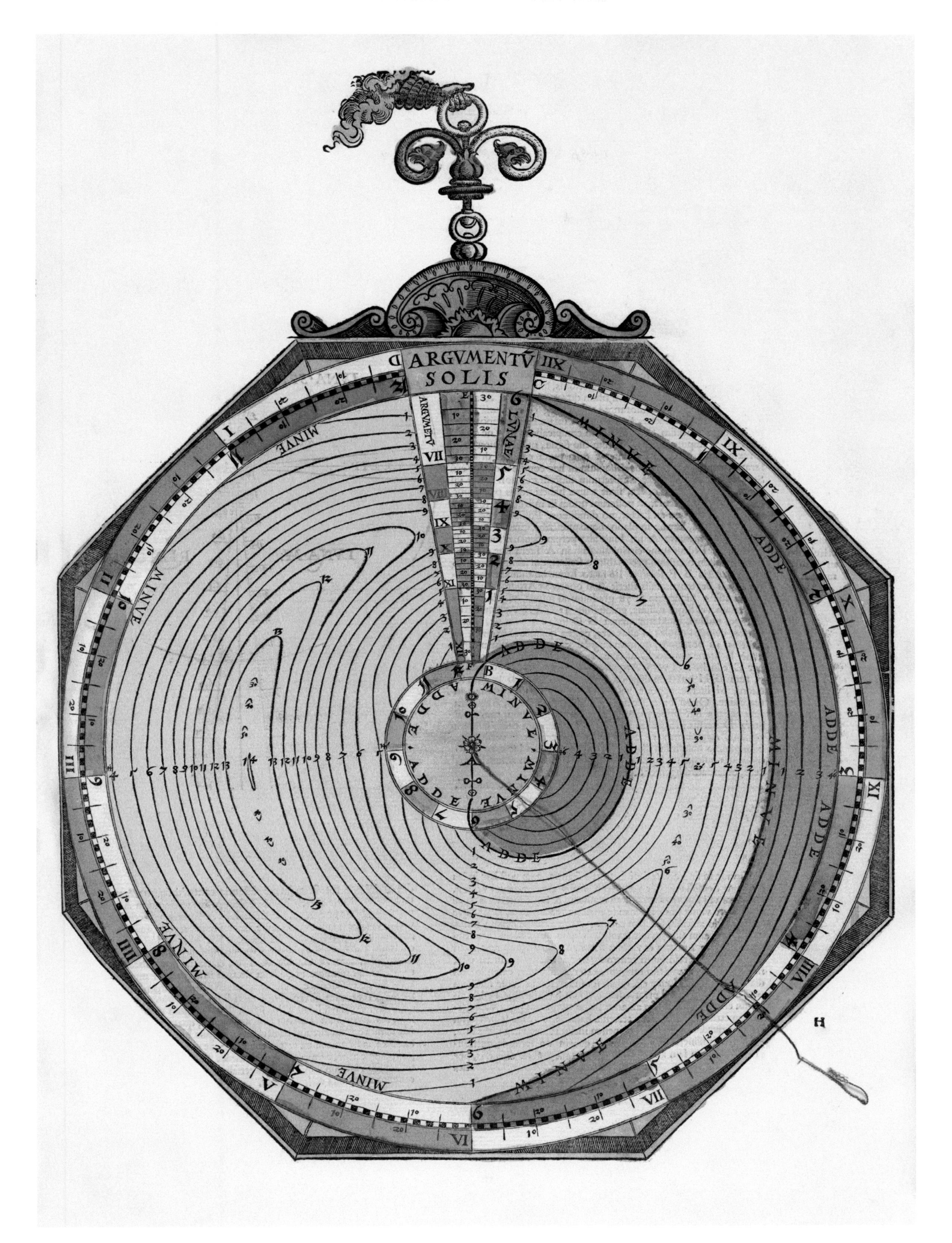
ARGVMENTV
SOLIS
ARGVMETV
LVNAE
MINVE
ADDE
H

• 1582:

Above left and right: As touched on in the previous pages, the sun contained religious and allegorical significance in many cultures. In European alchemical traditions, depictions of black, red, and white suns corresponded to stages in alchemical transformations intended ultimately to produce the philosopher's stone—a substance supposedly able to turn base metals into gold. These depictions of a black sun setting and an orange sun rising come from a well-known, heavily illustrated German alchemical treatise titled *Splendor solis* (Splendor of the Sun), attributed to Salomon Trismosin. In alchemy, the *sol niger*, or "black sun," refers to the first stage in this alchemical process, in which putrification and spiritual death leads to a cleansing. An orange or "rubedo" color denotes the fourth and final stage of the process leading to the philosopher's stone. These stages have been mined for psychological meanings, particularly in Jungian psychology, with its "dark night of the soul." This edition of the book, housed at the British Library and considered one of its most valuable possessions, has reportedly been studied by Yeats, Joyce, and Umberto Eco.

• 1613:

Facing page: The sun does in fact occasionally exhibit dark spots. These form in areas where magnetic field lines converge, producing cooler regions of "only" about 3,000–4,000 degrees Celsius (by contrast with their surroundings of 5,000 degrees Celsius or higher). Sunspots have been observed for more than two thousand years, but in the seventeenth century, astronomers devised new ways to view them, including a telescope-based projection device known as a helioscope. If this etching from Galileo's 1613 book, *Istoria e dimostrazioni intorno alle macchie solari* (History and Demonstrations Concerning Sunspots), is practically photographic in its precision, it's because it is in fact the result of directly tracing the projected image of the sun.

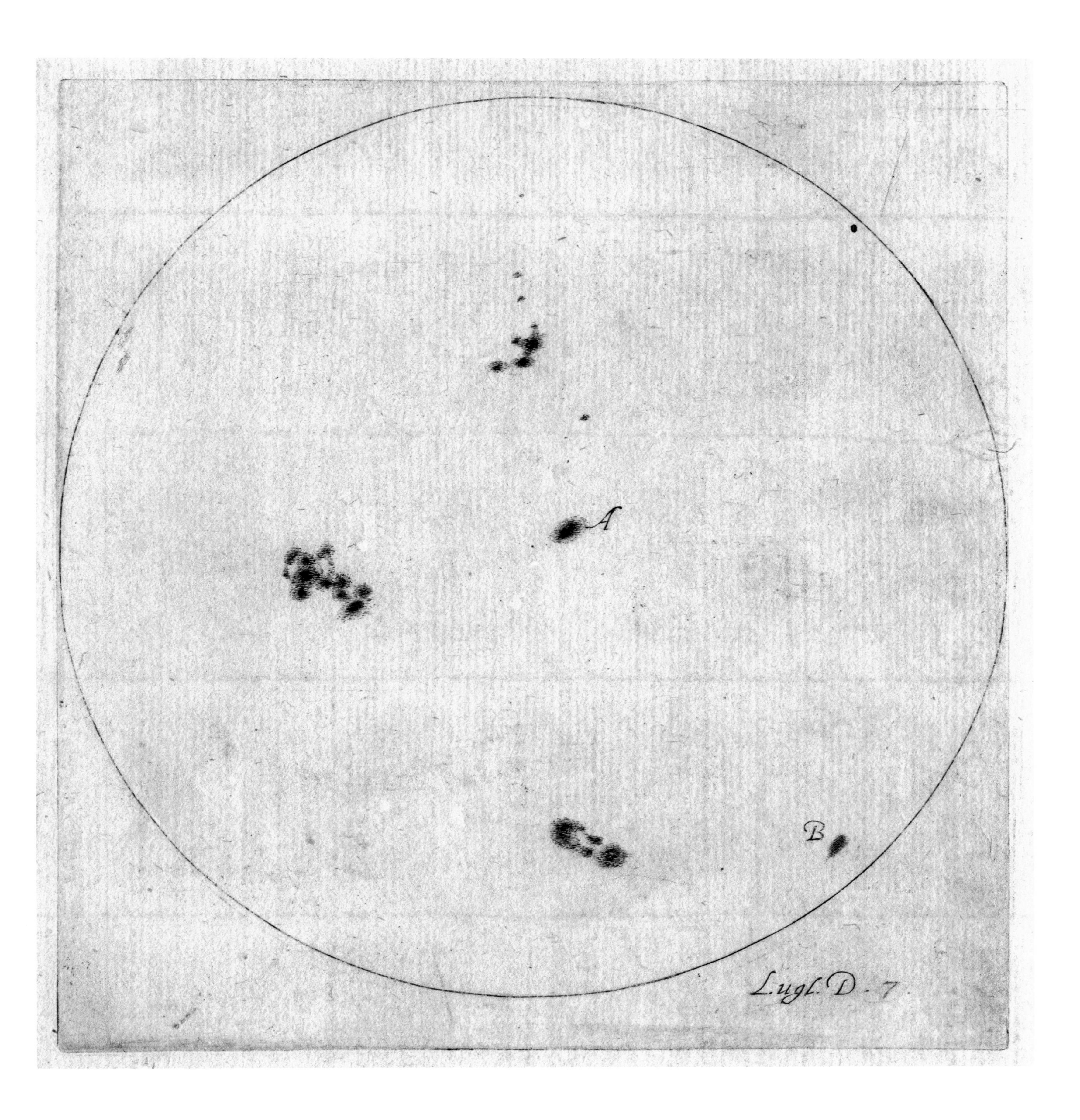
A
B
Lugl. D. 7

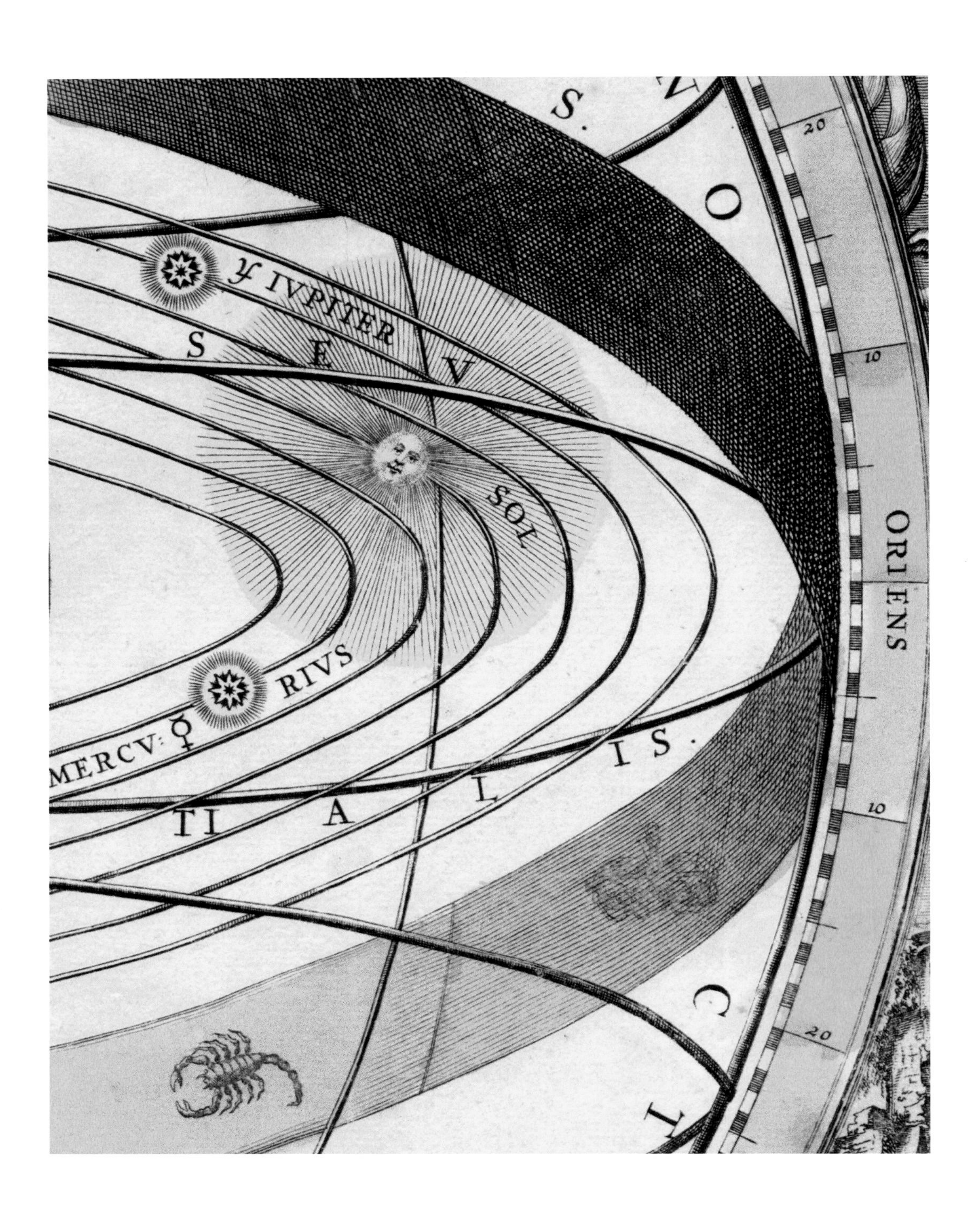

• 1660:

Above and facing page: These depictions of the sun in a Ptolemaic, geocentric cosmos, and in the alternative, helio-centric scheme proposed by Copernicus are from Andreas Cellarius's sumptuous *Harmonia macrocosmica*. Above, the sun occupies a slavish role, shuttling on a kind of wire along with the planets around a grand central Earth (for the full image, see pages 46–47). On the right, the sun has expanded radically in size, and its facial expression has acquired a solemnity in keeping with its enhanced stature. The four depictions of Earth denote the seasons. Note Cellarius's depictions of the moon, far smaller than Earth, and in an orbit seemingly as low as a communications satellite. All the other planets known in 1660 are also visible. For other maps from Cellarius, see pages 47, 147–48, 182–83, and 232–35.

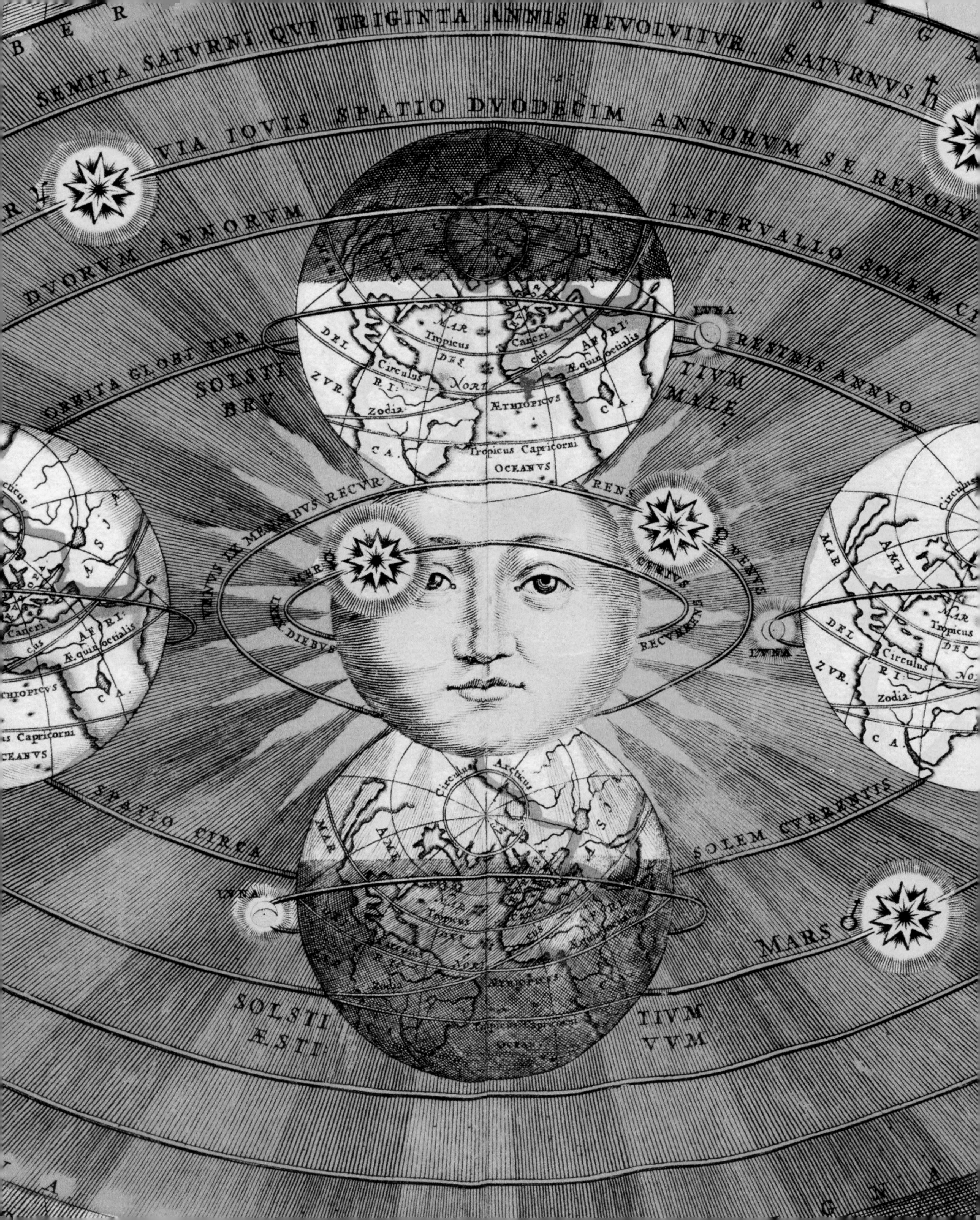
SEMITA SATVRNI QVI TRIGINTA ANNIS REVOLVITVR.
SATVRNVS
VIA IOVIS SPATIO DVODECIM ANNORVM SE
DVORVM ANNORVM
INTERVALLO SOLEM
RESTRVS ANNVO
SOLSTI
BRV
TIVM
MALE
LVNA
MAR
Tropicus
Cancri
DEL
AFRI
Circulus
Æquinoctialis
NORT
ZVR
Zodia
ÆTHIOPICVS
Tropicus Capricorni
OCEANVS
CA.
VENVS IX MENSIBVS RECVR
RENS
MERCVRIVS
DIEBVS
RECVRRENS
Circulus Arcticus
SPATIO CIRCA
SOLEM CVRREMVS
MARS
SOLSTI
ÆSTI
TIVM
VVM

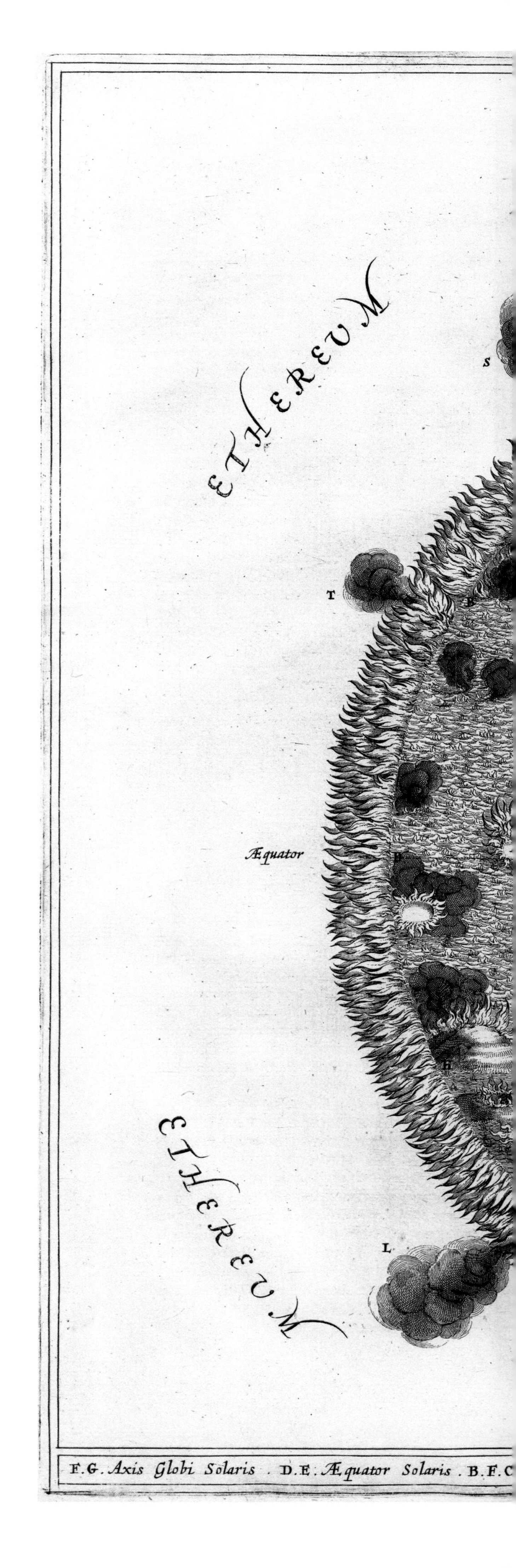

• 1664:

This incendiary depiction of the sun by German renaissance man Athanasius Kircher was widely reproduced, frequently without credit, for hundreds of years after. However, Kircher, memorably described in Encyclopaedia Britannica as "a one-man intellectual clearinghouse," frequently only glancingly cited sources fundamental to sections of his work, as well—though in fairness contemporary rules about such usage had not yet been established. In any case, as has been pointed out by some contemporary writers, there's a proto-postmodern, early steampunk quality to much of Kircher's work. This radiant image, from his book *Mundus subterraneus*, is no exception.

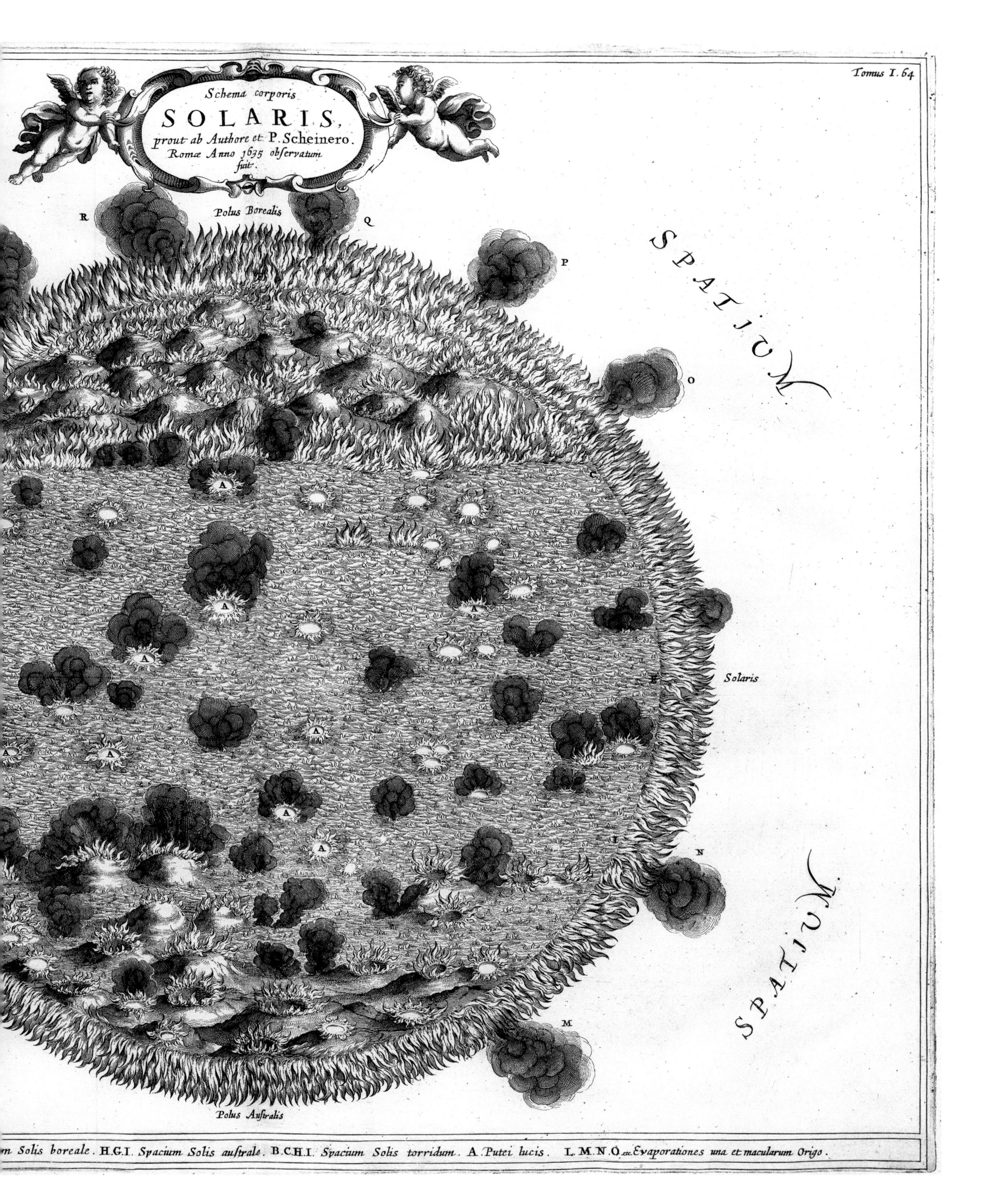
Tomus I. 64
Schema corporis
SOLARIS,
prout ab Authore et P. Scheinero.
Romæ Anno 1635 observatum
fuit.
Polus Borealis
R
Q
P
O
SPATIUM.
Solaris
N
SPATIUM.
M
Polus Australis
m Solis boreale. H.G.I. Spacium Solis australe. B.C.H.I. Spacium Solis torridum. A. Putei lucis. L.M.N.O. etc. Evaporationes una et macularum Origo.

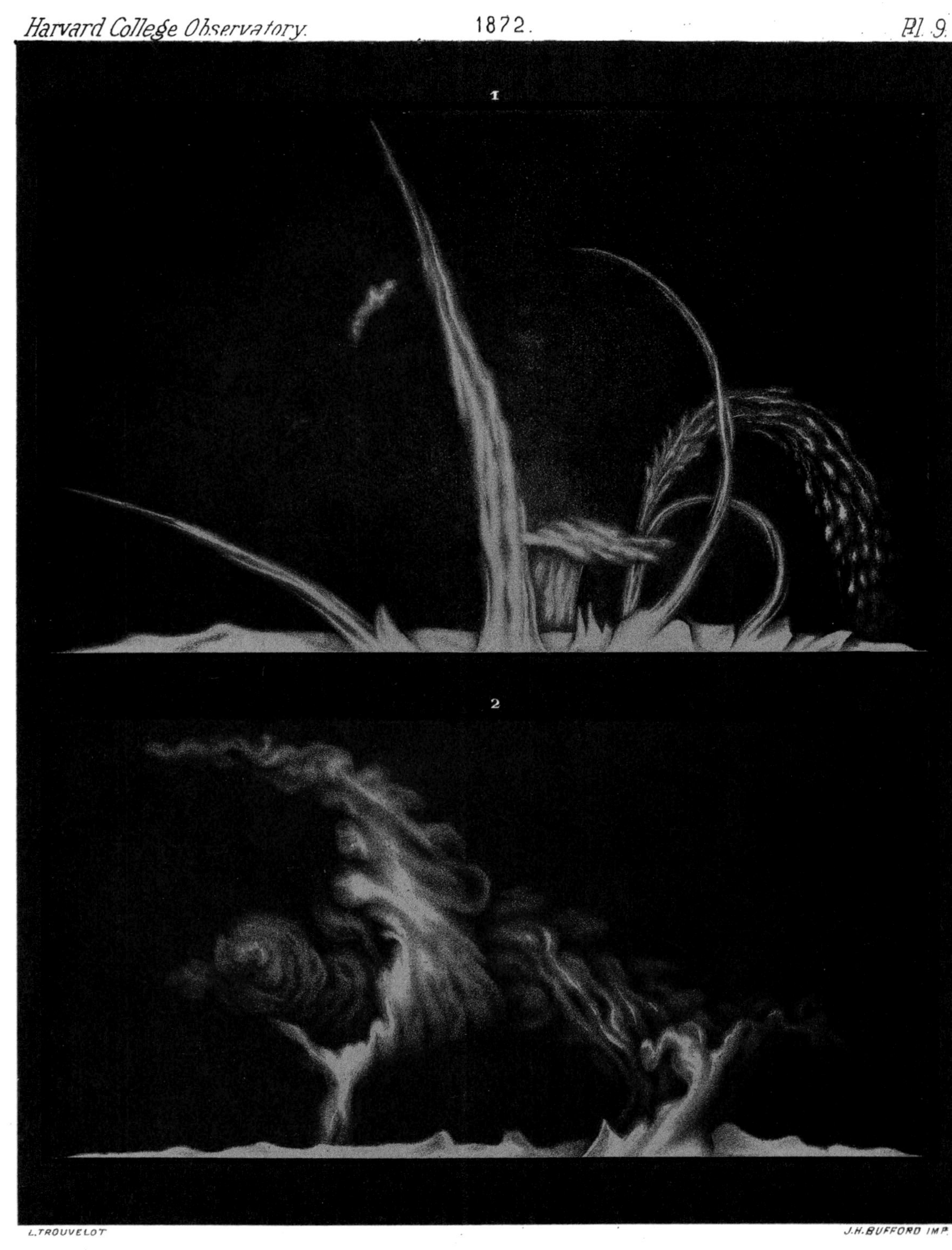

SOLAR PROMINENCES.

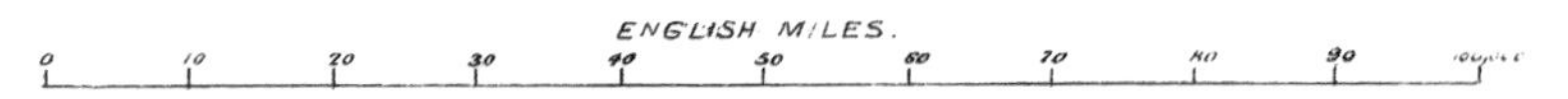

• 1752:

Facing page: The Greek sun god, Apollo, predates the Aztec version by two thousand years. In this oil sketch by Venetian master Giovanni Battista Tiepolo, Apollo—seen with the sun behind him—is about to embark on his course across the sky. Around him orbit allegorical figures meant to represent the planets. The four continents of Earth are depicted at the corners of the painting. By the mid-eighteenth century, Copernicus's heliocentrism was firmly established; no less than in Cellarius's overt representation of that cosmic scheme on page 123, it is the organizing principle of this painting. Titled *Allegory of the Planets and Continents*, Tiepolo's sketch served as the basis for the giant painting on the ceiling of the staircase in the palatial Residenz of Carl Philipp von Greiffenklau, the prince-bishop of Würzburg.

• 1872:

Above: During his first year on the staff of the Harvard College Observatory, artist-astronomer Étienne Trouvelot produced these engravings of solar phenomena for the observatory's *Annals*. Here Trouvelot depicted prominences on the sun's limb in nuanced detail. The legend at the bottom reveals that the distance between the two prominences in the lower part of the engraving is one hundred thousand miles—or more than twelve times the diameter of Earth. Although the *Annals of the Astronomical Observatory of Harvard College* didn't have a large circulation, illustrations such as these increasingly infiltrated more widely disseminated publications, and played a role in influencing public understanding of the visual qualities and staggering scale of phenomena being observed through telescopes. For more by Trouvelot, see pages 98–99, 128–29, 158, 188–91, 241, 270–71, 295, and 297–99.

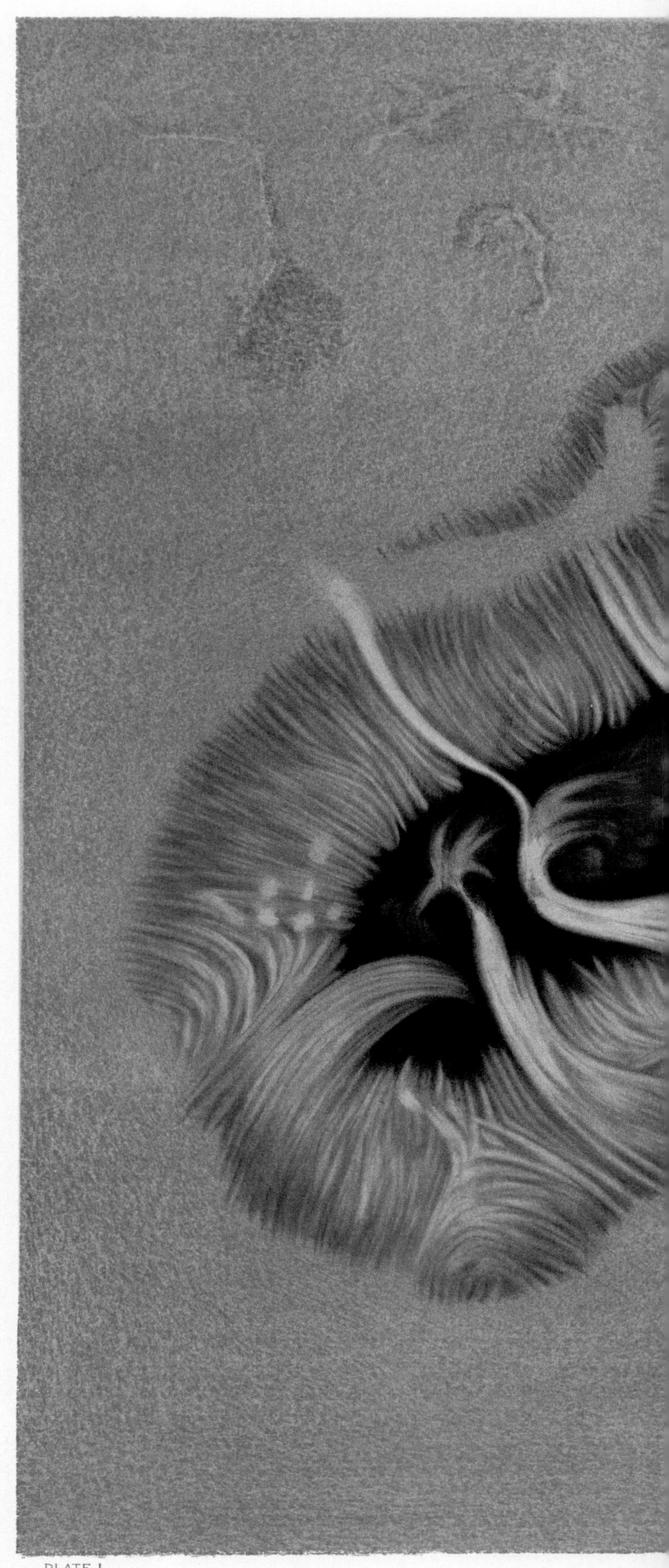

• 1881:

In this remarkable print, Étienne Trouvelot presented the most detailed look at sunspots made until that time, and for more than a century thereafter. Sunspots are slightly cooler areas of the solar photosphere, or visible outer layer of the sun, and form due to intense magnetic activity. They generally appear in pairs, as here, with each of the pair having an opposite magnetic polarity. From the 1881 Charles Scribner's Sons limited-edition collection of chromolithograph reproductions of Trouvelot pastels. For more by Trouvelot, see pages 98–99, 127, 158, 188–91, 241, 270–71, 295, and 297–99.

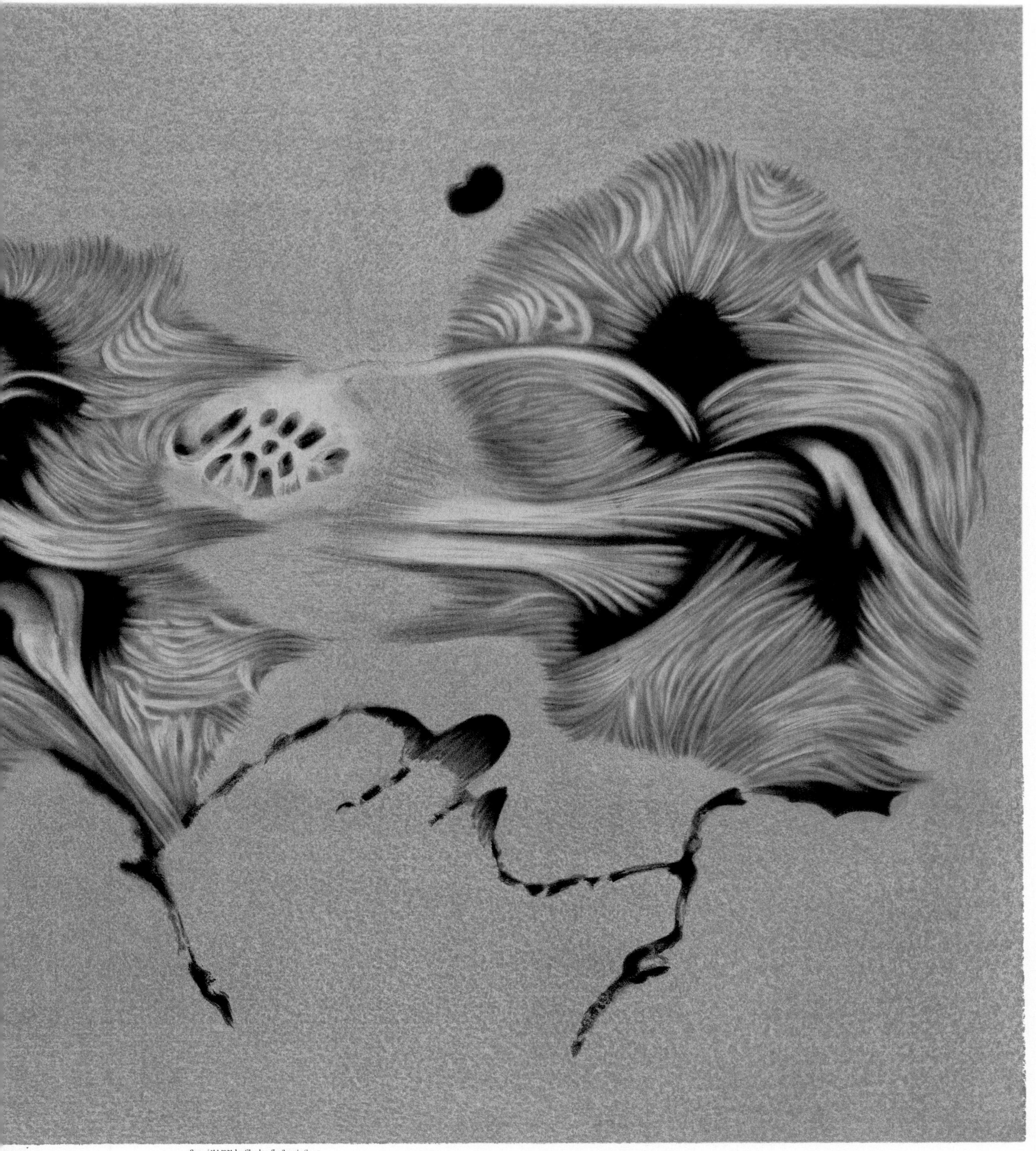

JP of SUN SPOTS and VEILED SPOTS.

Observed on June 17th 1875 at 7 h. 30 m. A.M.

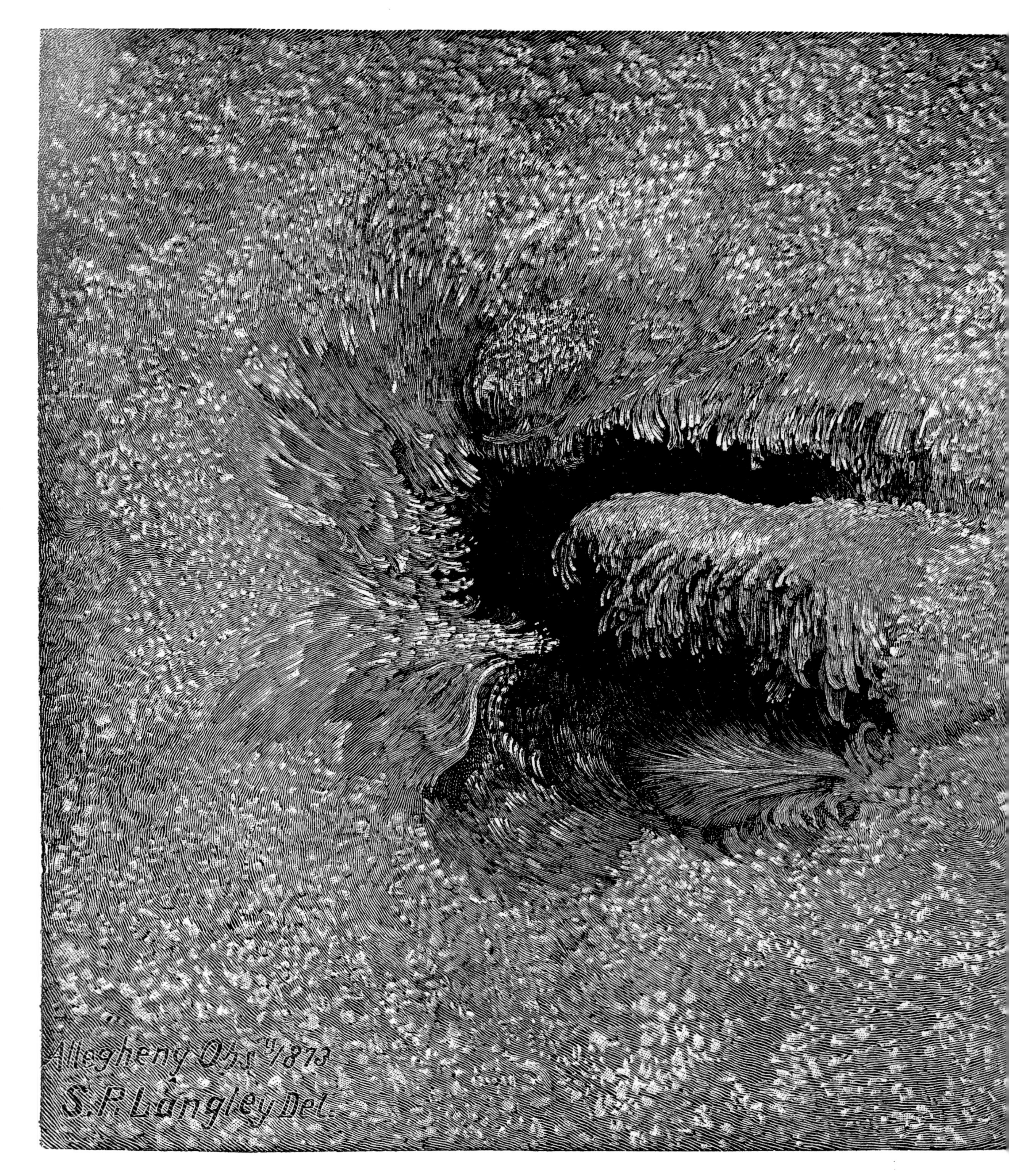
Allegheny Obs 1873
S.P. Langley Del.

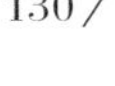
130/

THE EARTH AS IT WOULD APPEAR IN COMPARISON WITH THE FLAMES SHOOTING OUT FROM THE SUN.

• 1900:

Facing page: Detailed engraving of a sunspot from astronomer and aviation pioneer Samuel Pierpont Langley's *The New Astronomy*. Langley, who as a young man had apprenticed to architecture firms in St. Louis and Chicago, had excellent drafting skills. Years later, unsatisfied with the quality of photography of the sun, he relied on this training to document solar phenomena with precision. This image, based on observations he conducted at Allegheny Observatory, Pittsburgh, in 1873, comes close to equaling Étienne Trouvelot's sunspots on the previous pages.

• 1925:

Above: Astronomy is a science capable of providing a realistic sense of our diminutive scale within a vast universe. This sobering example of the popularization of astronomical observations comes from G. E. Mitton's *The Book of Stars for Young People*.

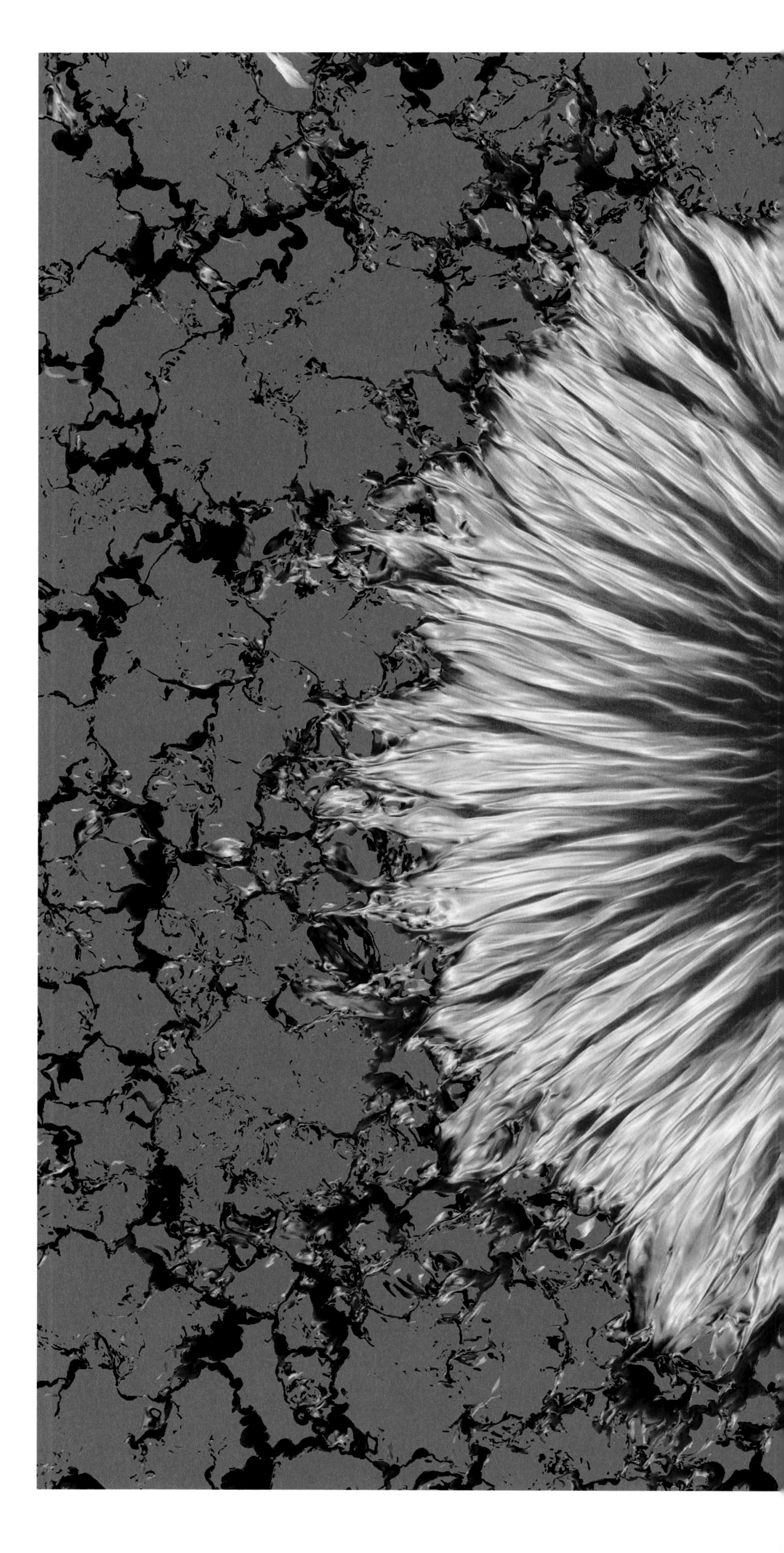

• 2009:

Supercomputer simulation of a sunspot. In this striking image created by researcher Matthias Rempel and collaborators, the highly complex filaments that flow between a sunspot's dark center and lighter outer region have been produced in exquisite detail by simulating the magnetic forces at play using a supercomputer at the National Center for Atmospheric Research (NCAR). Sunspots are frequently at the locus of solar prominences, such as have been illustrated in previous pages, with solar flares and coronal mass ejections all associated with the highly magnetically active regions where they occur. This image, effectively a still culled from the first comprehensive 3-D model of a sunspot, was created after NCAR received an IBM supercomputer capable of performing 76 trillion calculations per second.

• 2012:

Three supercomputer simulations of the complex flux and flow of magnetic field lines as the solar wind interacts with the Earth's protective, irregular magnetosphere. These images, created by a team led by Homa Karimabadi, reveal a highly complex and turbulent interplay between the solar wind and magnetosphere. The radiation unleashed by the sun is largely kept at bay by our planet's magnetic field, but it swoops in at the poles, triggering auroral lights. During major solar storms it can have an impact everywhere. In the top and bottom images, Earth is the gray sphere, and the multicolored spaghetti are the magnetic lines of solar radiation. The middle image depicts a solar wind current sheet, revealing cascades of turbulence at radically different scales. For depictions of the resulting polar auroras, see pages 309–15.

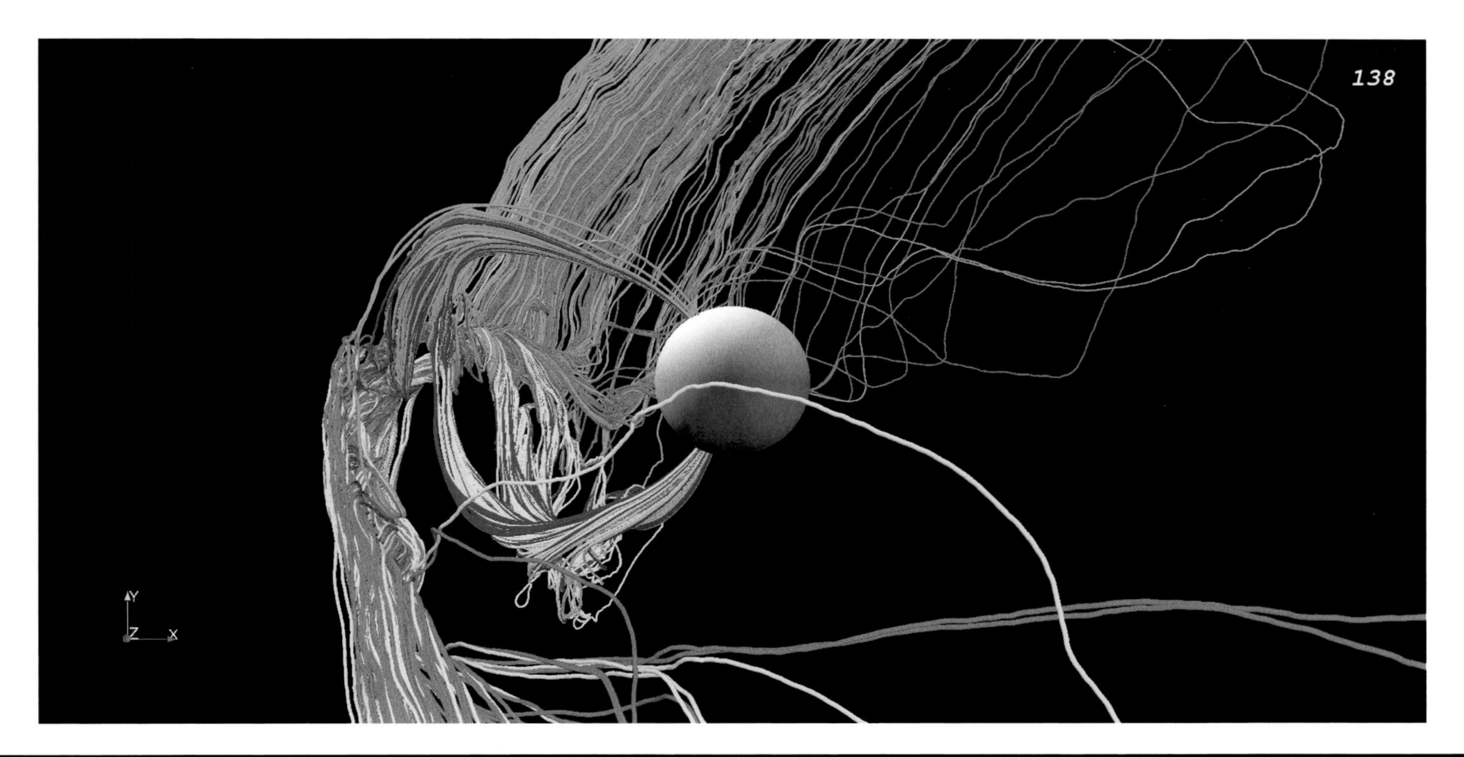
Y
Z
X

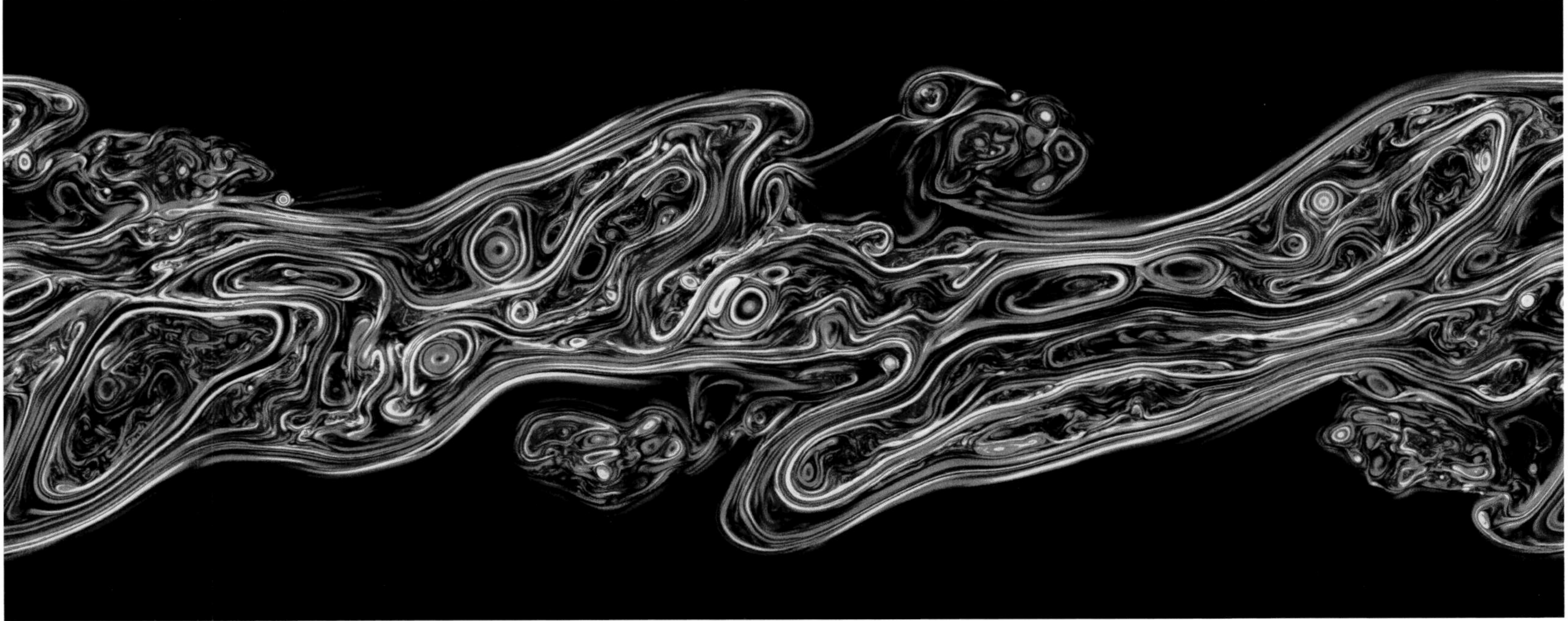

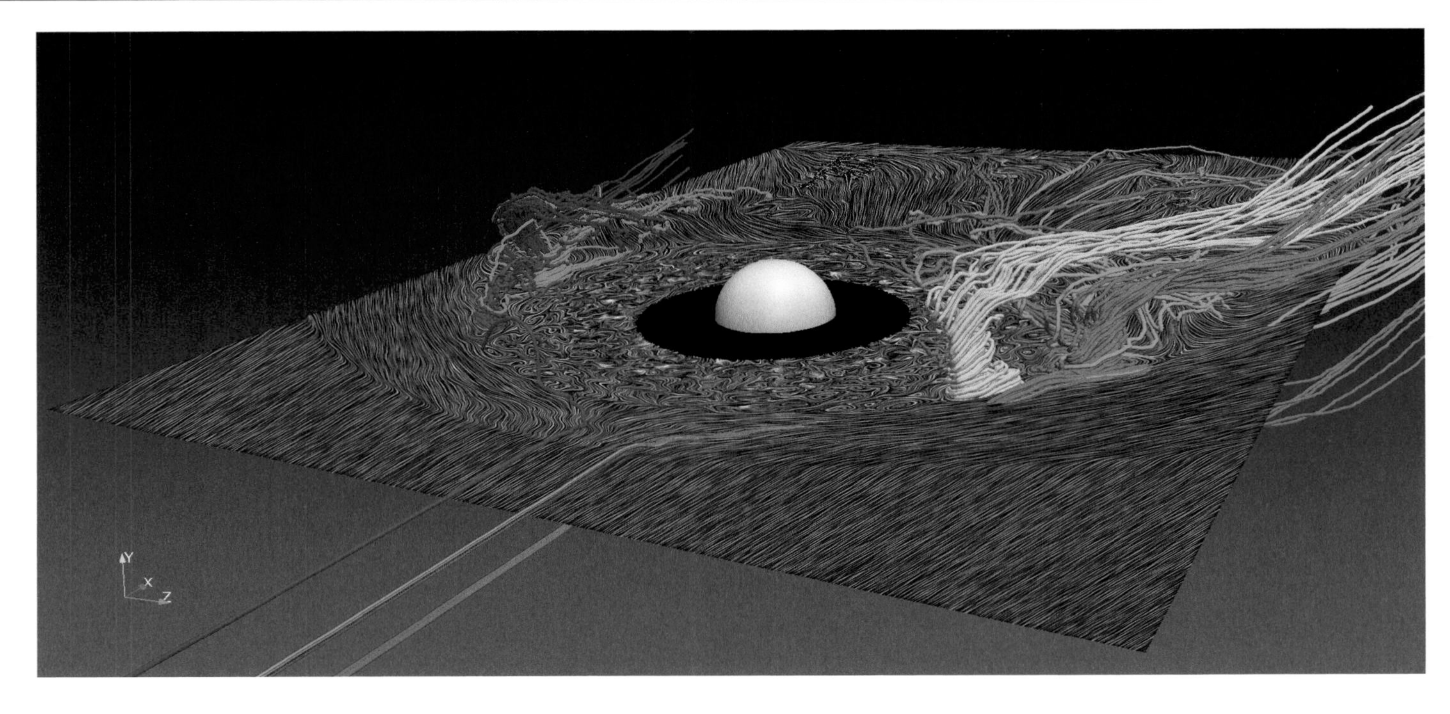
Y
X
Z

CHAPTER 5

The Structure of the Universe

They lived in a childhood
Prolonged from age to age. For them, the sun
was a farmer's ruddy face, the moon peeped through a cloud
and the Milky Way gladdened them like a birch-lined road.

—Czeslaw Milosz, *To Robinson Jeffers*

The Earth-centered design of the Aristotelian-Ptolemaic universe with its multiple celestial spheres remained in place as a kind of reigning cosmological ideology for more than 1,500 years—three times as long as the western Roman Empire. Ptolemy's bag of complex mathematical devices, his epicycles and deferents, had permitted astronomers to do quite a good job in predicting celestial movements. And the problems inherent in the Ptolemaic methodology—it had only partially tamed the complex variable motions of the planets, for example, and it had entirely failed to explain their counterintuitive, periodic westward motions against the regular eastward drift of the stars—were not a major concern during ten centuries of medievalism.

In contrast to many other cosmologies in the ancient world, the Aristotelian-Ptolemaic system was thoroughly grounded in centuries of actual observations of the motions of celestial objects, rather than simply associating them with gods or mythical characters. And though a sometimes capricious monotheistic god came to reign in Judeo-Christian heaven, his realm was seen as lying outside the outermost celestial sphere. So while we mortals were on the inside looking up, and he was on the outside looking in, the planets and stars arrayed in ever-shifting arrangements in between were seen as elements of a complex mechanism, one well worth studying.

The astronomical model worked out by the Ancient Greeks, and refined into something approaching mathematical exactitude by Ptolemy and his successors, predicted many celestial phenomena in satisfying ways. It both explained and scheduled lunar phases and eclipses. It allowed Eratosthenes to estimate the diameter of the Earth. It provided a method for modeling the motions of the planets, the sun, and the moon. And it tracked the deviations of the planets from the ecliptic—that imaginary line defined by the annual apparent motion of the sun against the celestial sphere. Although Earth-centered in design, it was a powerful and sophisticated tool.

All of these factors—the astronomical ephemerides data reaching back to about a millennium B.C.; the tradition of scrupulous observation of celestial phenomena; and the mathematical methodologies that had been systematized to handle the data—set the stage for what would later become known as the Copernican Revolution. The Polish astronomer based his astronomy on Ptolemaic principles, even if the 1543 publication of his book *De revolutionibus orbium coelestium* (On the Revolutions of the Celestial Spheres) ultimately resulted in the wholesale reordering of Ptolemy's celestial hierarchy. In essence, Copernicus tried to reposition a highly complex ancient methodology of technical astronomy so that it was centered on the sun, not the Earth—a change that allowed him to shed some of its more arcane and problematic features, while retaining many of its core principles. In the end, though, even what he retained was replaced by successor astronomers.

According to some historians of science, Copernicus's dissatisfaction with geocentrism originated in the ever-increasing complexity of Ptolemy's system of compounded circles. As astronomers across the centuries tried to account for the remaining irregularities between theory and observed celestial motion, they suggest, they added and mixed elements from Ptolemy's grab bag of nested, ever-spinning epicycles, equants, deferents, and eccentrics. Ultimately, Copernicus had an aesthetic objection to what had become an unwieldy Rube Goldberg contraption—one that didn't even have the virtue of producing absolute accuracy. If beauty is truth and truth, beauty, according to this view, Copernicus saw that something both ugly and dishonest was going on.

Another standpoint has it that rather than irritation over Ptolemy's mathematical devices, Copernicus may simply have been bothered by what remained unexplained in the ancient cosmological scheme. This included that mystifying occasional planetary retrograde motion. According to astronomer Owen Gingerich, Copernicus may also have had a problem with the essentially arbitrary arrangement of the planets in the Ptolemaic system. By contrast, Copernicus's heliocentric system imposed order:

> *What he realized was that the entire entourage of planets automatically arranged themselves so that the planet with the shortest period, Mercury, orbited closest to the sun, and lethargic Saturn, rounding the sun in 30 years, was farthest, and all the rest fell proportionately in between. There was something irresistibly beautiful about this layout. Furthermore, this arrangement explained something that was simply a mystery in the Ptolemaic astronomy: Mars, Jupiter and Saturn periodically stopped their eastward progress against the background stars, and moved westward for a few weeks, the so-called retrograde motion. Why did this always happen when the planet was directly opposite the sun in the sky?... Ptolemy couldn't explain it. But Copernicus could. With his sun-centered plan, retrograde motion occurred as the faster-moving Earth bypassed Mars (for example), and that happened when that planet was closest to the Earth and directly opposite the sun. The formerly mysterious coincidence now became a "reasoned fact." This arrangement provided a conceptual system. Copernicus had invented the solar system!*

Although he had assembled the observational backing to his theory over several decades, and word of its contents had spread across Europe, *De revolutionibus* only came out in the year of Copernicus's death. Most likely this is because he understood the controversy it might inspire, and possibly the threat it could pose to him personally from a Catholic Church committed to an Earth-centered cosmos. (When it finally was published, it was dedicated to Pope Paul III—a prophylactic gesture.)

Following its publication, its impact was only felt very gradually, because Copernicus's theory had to overcome a thoroughly entrenched worldview, one backed by Church doctrine, ingrained respect for Aristotle and Ptolemy, and millennia of human thought predicated on the notion that we were at the center of the universe. Well over a century passed before its general acceptance—although some astronomers understood its significance almost immediately, and became proselytizers.

If they lived in territories controlled by the Holy See, espousers of Copernicanism took their lives into their own hands. Italian philosopher and mathematician Giordano Bruno, an early backer of heliocentrism, went further than Copernicus. In 1584, he published two tracts putting forward the view not only that the Earth moved around the sun, but also that the sun was just another star, and that it, too, was in motion. In Bruno's cosmological design, neither the sun nor the Earth was assigned a particularly prominent role in a universe filled with innumerable stars and planets extending into infinity. Time, too, reached to infinity in both directions, with no beginning or end, and he also believed that intelligent life must exist throughout the cosmos. For Bruno, God permeated the material universe everywhere equally, rather than being a remote deity in another realm.

Bruno's vertiginously pantheistic perspective violated Church doctrine on many levels, and in 1592 he was picked up by the Inquisition. On January 20, 1600, after seven brutal years of imprisonment, interrogation, and a trial, he was convicted of heresy. Although the centrality of his cosmological views to his conviction have been disputed—he faced at least eight major charges—some scholars have cited Bruno's refusal to recant his belief in the plurality of worlds as one reason behind the verdict. According to a trial witness, he reacted to the conviction with a threatening gesture toward his judges, saying, "Perchance you who pronounce my sentence are in greater fear than I who receive it." On the morning of February 16, 1600, Giordano Bruno's mouth was immobilized by *una morsa di legno*—"a vice of wood"—and he was taken to the square of Campo de' Fiori in Rome and burned at the stake.

Although his cosmological views would prove to be largely correct, Bruno was a theorist, not an observational astronomer. His contemporary Galileo Galilei, on the other hand, was among the first to use the newly invented telescope to study the heavens, and he was the first to publish his results. In September of 1610, he observed that Venus exhibits phases similar to the moon's—the first real observational evidence in support of Copernicus's heliocentric theory.

Apart from confirming the premise that Venus orbits the sun, not the Earth, Galileo's Venus observations also shattered those crystalline terracentric Aristotelian-Ptolemaic spheres, because even if the sun continued to orbit the Earth with Venus in tow—something Danish astronomer Tycho Brahe had proposed years earlier—Venus would have to intersect the supposedly impenetrable spherical shell assumed to carry the sun on its course. (Brahe's late-sixteenth-century response to Copernicanism boiled down to admiration for many of its components, but disagreement with the idea that the Earth—"that ponderous, lazy body"—could ever move. Instead, he proposed a theologically expedient compromise system in which all the planets apart from Earth orbit the sun—which in turn orbits the Earth.)

Galileo's discovery of moons orbiting Jupiter further contributed to the discrediting of the Ptolemaic cosmos. His 1610 publication of *Sidereus nuncius* (Starry Messenger), the book that released these observations to the world, brought him to the attention of Church authorities. By 1616, he was commanded not to espouse or defend the view that the sun was at the center of the universe. But when he published a book titled *Dialogo sopra i due massimi sistemi del mondo* (Dialogue Concerning the Two Chief World Systems) almost two decades later, he attracted the fury of the Inquisition. Although the book gestured toward the 1616 injunction not to advocate for heliocentrism by presenting an ostensibly balanced debate between a Copernican, an Aristotelian, and an impartial scholar, in fact Galileo presented the Aristotelian as a dull-witted fool, and even named him "Simplicio."

By contrast, the advocate for Copernicanism was witty and persuasive as he made a detailed presentation of the theory, leaving no doubt of Galileo's allegiances. With the book a runaway best seller, the now elderly astronomer was ordered to Rome for trial. In 1633, he was found "vehemently suspect of heresy," and was required to recant his views, under threat of torture. By all contemporary accounts, he did so without defiance. That would have been too risky: He and all of Rome were well aware of Giordano Bruno's gruesome fate thirty years earlier. Galileo's cooperation led to his jail sentence being commuted to house arrest, and he remained confined to his villa near Florence for the rest of his life. His books were all banned, though they persisted in proliferating across Europe. And the Earth continued to move.

Another contemporary of Galileo who advocated vocally for Copernicanism was German astronomer Johannes Kepler, who did so in relative safety because he wasn't a Catholic and he lived outside the Holy See. Although he had been a salaried employee of Tycho Brahe, Kepler never adopted his mentor's unwieldy compromise between old-school Ptolemaic terracentrism and Copernican heliocentrism. In fact, Kepler's 1596 book *Mysterium Cosmographicum* (The Cosmographic Mystery), which predates Galileo's *Sidereus* by fourteen years (and also Kepler's employment by Brahe), has the distinction of being the first published defense of Copernicanism. In it, Kepler proposed what now seems a bizarre mechanistic underpinning to the Copernican cosmological scheme—even if the astronomer was asking a legitimate question: Why are the planets spaced as they are?

Copernicus had in effect disassembled the geocentric spheres, cleaned centuries worth of dust from their mechanisms, and reassembled them, only now with the sun at the center and the Earth just another planet. In this he had left the spheres model as the functional basis of the cosmos. Now Kepler proposed that the spacing of those spheres could be understood with reference to the Platonic solids—the five polyhedral shapes that had been studied since antiquity and admired for their symmetry and mathematical simplicity. In Kepler's heliocentric design, the planetary spheres could be circumscribed using these solids, resulting in a kind of matryoshka-doll solar system, in which the sphere of Saturn is circumscribed about the cube-shaped solid containing Jupiter's sphere, within which nests a tetrahedron containing Mars' sphere, and so on down the line. (See page 150.)

Kepler's real contribution, however, was the unmitigated brilliance of his three definitive laws of planetary motion, the first two of which were published in his 1609 treatise *Astronomia nova* (New Astronomy). He arrived at them after five years spent attempting to reconcile Tycho Brahe's exceedingly accurate measurements of the actual orbit of Mars with its positions as predicted by the Copernican mathematical model. In Copernican heliocentrism, planetary orbits were supposed to be uniformly circular, and the planets were presumed to travel along them at uniform speeds. Actual observational data failed to support these premises, however, which is why Copernicus had retained many of the Ptolemaic system's arcane compensatory techniques. Using Brahe's Mars data, however, Kepler deduced that the planets in fact don't orbit in perfect circles, but rather in ellipses. In a separate line of inquiry, he also determined that planetary motion is not uniform, but instead varies inversely based on the planet's distance from the sun. He had arrived at his first two laws.

Kepler's findings removed the need for all the Ptolemaic epicycles and equants, which were swept unceremoniously into the dustbin of history. Ten years later he published his third law, which explained the relationship between the length of a planetary year and its distance from the sun. As Thomas Kuhn wrote in his 1957 book *The Copernican Revolution*, "The Copernican astronomical system inherited by modern science is, therefore, a joint product of Kepler and Copernicus. Kepler's system of six ellipses made sun-centered astronomy work...."

Kepler's ellipses cleared the way for Isaac Newton's 1687 publication of *Principia*. The shattering of the Ptolemaic spheres required that something else keep planetary orbits on their elliptical tracks. In 1665 and '66, physicist Robert Hooke had argued for a principle of universal gravitation, and by 1670 he stated it applied to "all celestial bodies," but he failed to work out the proportion by which the mysterious force drops due to distance. Hooke's interest in the subject, and his cagey correspondence with his rival Newton, helped push Newton to focus on the problem.

In chapter 2, a tiny diagram from a section of the *Principia* that Newton chose not to publish in his lifetime illustrates a series of trajectories. (See page 53.) All fall to Earth except the last, which extends so far that it in effect falls *around* Earth—one of the earliest known depictions of the path to orbit. Using Kepler's laws as a basis, Newton deduced the rate of fall of the moon "toward" the Earth in order for it to remain in a stable orbit, and he did the same for the planets in their endlessly elliptical, terminally unresolved falls toward the sun. He then worked out that the gravitational attraction exerted by the sun on its planets decreases inversely as the square of the distance separating them from the sun.

Newton went on to extrapolate that if the source of gravitation is assumed to be the Earth's center, then his inverse-square law could be used to measure the difference in fall rates between such disparate objects as the moon and a stone. Eventually, he also applied his inverse-square principle to Kepler's first two laws, demonstrating that it covered elliptical planetary orbits and variable planetary speeds as well. With a kind of majestic finality and exactitude, he had explained the mechanism behind Kepler's laws, and in fact all the known behaviors exhibited by both terrestrial and celestial motion. Newton's universal laws of gravitation make him the greatest physicist of all time, and certainly one of the greatest scientists of history. He was knighted in 1705, and died in 1727.

Although Newton's laws were revelatory, by themselves they didn't reveal the structure of the universe, even if they did explain the physics it ran by. But to underline the magnitude of his accomplishments, observational astronomy also wouldn't be the same without him. Newton's work in optics in the late 1660s led him to the conclusion that the refractor telescopes of the day had reached the limits of their effectiveness. This was due to a distorting phenomenon produced by their lenses known as chromatic aberration. He decided that curved mirrors would solve the problem, rather than polished glass, and designed the first functional reflecting telescope. Most optical telescopes in use today are Newtonian reflectors, including the Hubble Space Telescope.

It took quite some time, however, for telescopes to reach a size capable of discerning much detail in the Milky Way, let alone in the vague gray blobs of opalescence—the "nebulae"—that could sometimes be discerned both in and outside of its hazy band. Since Galileo's observations in the early seventeenth century, our galaxy was known to consist of vast numbers of tightly packed stars. But the modern concept of the galaxy was unknown, let alone the idea that the Milky Way was one of many. The sheer number of stars in the sky—and the dawning awareness that, instead of ornamenting a crystalline sphere, they were probably suns like our own and dispersed throughout the universe—contributed to the notion that a plurality of worlds was possible. But although Giordano Bruno and Nicholas of Cusa had argued for such ideas in the sixteenth and fifteenth centuries respectively, neither had ever arrived at a structure.

The man who did, English astronomer and mathematician Thomas Wright, is discussed at length in the introduction. Wright's 1750 proposal that the Milky Way may be shaped like a flattened disc was the first description of our galaxy's actual form. (See pages 153 and 237.) He also argued that some of the ghostly blobs visible to eighteenth-century astronomers—shapes that had been taken for nebulae within our own Milky Way—were galaxies like our own.

It would take observational astronomy almost two centuries to confirm Wright's theory, in part through the labors of observational astronomers like William Herschel, his sister Caroline, and eventually his son John, who together in the eighteenth and nineteenth centuries amassed a catalog of nebulae, many of which ultimately turned out to be galaxies outside our Milky Way. During the second half of the nineteenth century, increasingly large telescopes started to reveal more of such enigmatic objects in the deep sky. In 1845 William Parsons, the 3rd Earl of Rosse, installed a six-ton reflector telescope in his residence at Birr Castle, in County Offaly, Ireland. With a six-foot aperture, it was nicknamed "Leviathan," and was the largest telescope in the world until 1918. Parsons was intrigued to discover that a number of "nebulae" exhibited spiral shapes.

Despite these developments, the prevailing view as late as 1920 held that the Milky Way *was* the universe. Confirmation of Thomas Wright's "celestial mansions" finally came with the construction of the one-hundred-inch Hooker Telescope at Mount Wilson Observatory in Los Angeles, and the arrival there of an astronomer named Edwin Hubble. The extraordinary resolving power of the new telescope, combined with the increasing sophistication of photographic astronomy, produced high-resolution plates of what was then called the Andromeda Nebula. Hubble began tracking the brightness levels of a Cepheid variable star within Andromeda—a so-called standard candle, or star whose intrinsic brightness could be determined according to the speed of its fluctuations. He discovered that the star was more than a million light-years away—putting Andromeda without a doubt very far outside the Milky Way. This was no nebula, but a vast galaxy comparable to our own.

By the mid-twentieth century, it was well accepted that the universe is filled with galaxies to a dazzling degree, with the Milky Way one of many billions. The chain of demotions initiated by Copernicus's 1543 reassignment of centrality to the sun had seemingly run its course, with no universal center anywhere visible. But this didn't mean that no structure could be discerned. In the late 1950s, French astronomer Gérard de Vaucouleurs presented a theory, based on thousands of observations, that nearby galaxies are organized into vast superclusters, including a Local Supercluster containing the Milky Way. Although the Inquisition had largely ceased its activities by the nineteenth century, making a trial out of the question, de Vaucouleurs was immediately labeled as delusional by some of his fellow astronomers. But his work was so fastidious, and backed by so much hard evidence, that it ultimately won acceptance.

In 1987, astronomers R. Brent Tully and J. Richard Fisher expanded on de Vaucouleurs's work with their pioneering *Nearby Galaxies Atlas*, the first attempt to chart the structure of the local universe. (Published with a bright red cover and spiral binding, it's similar visually to a standard American road atlas, and seems only to require an intergalactic spaceship.) In an echo of Thomas Wright describing the Milky Way's shape, Tully and Fisher found clear evidence that nearby galaxies "tend to align themselves parallel to the same plane, and . . . that this plane is incredibly extensive." Only months after the atlas came out, Tully announced that he had discovered a supercluster complex encompassing millions of galaxies, including our own, and extending across about 10 percent of the observable universe. He called the structure the Pisces-Cetus Supercluster Complex. At about one billion light-years long, it remains among the largest structures yet seen. (For Tully's most recent discovery, see the introduction.)

By the first decade of the twenty-first century, tens of thousands of galaxies had been catalogued, and estimates as to how many exist in the observable universe were ranging well over 150 billion. Attempting to depict such immensity on a piece of paper would seem utterly impossible. But in the early 2000s, Princeton cosmologist J. Richard Gott III and researcher Mario Juric, judging that previous attempts to compress all of known space-time within a single map projection had so far been both rare and unsatisfactory, set about conceptualizing one adequate to the task.

The result was the conformal map of the universe reproduced on pages 164–65. Based on a logarithmic scale, in which units of measurement increase exponentially, the map squeezes a span of space and time extending from the Big Bang to the day of publication, from the Earth's warming surface to the farthest microwave echo of observable reality, all within one very tall and narrow projection. In plotting hundreds of thousands of galaxies from the Sloan Digital Sky Survey using their logarithmic technique, Gott, Juric, and their collaborators discovered a giant wall of galaxies about one billion light-years away. They estimated it to be 1.38 billion light-years long—about one-sixtieth the diameter of the visible universe—and they called it the Sloan Great Wall.

It was the largest single structure yet seen anywhere in space and time, and they had discovered it while making a map.

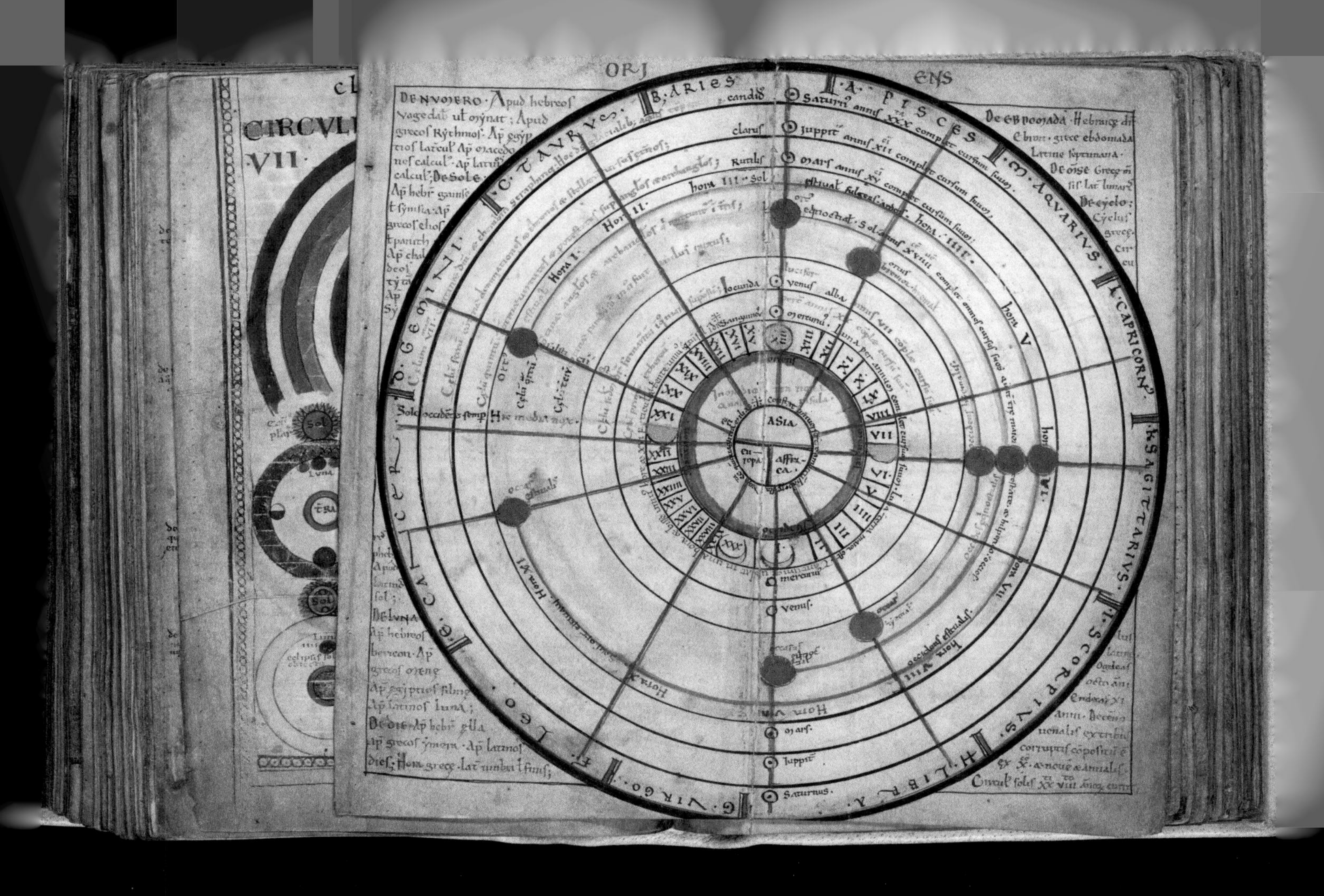

ORI
ENS
ARIES
PISCES
AQVARIVS
CAPRICORN
SAGITTARIVS
SCORPIVS
LIBRA
VIRGO
LEO
TAVRVS
ASIA

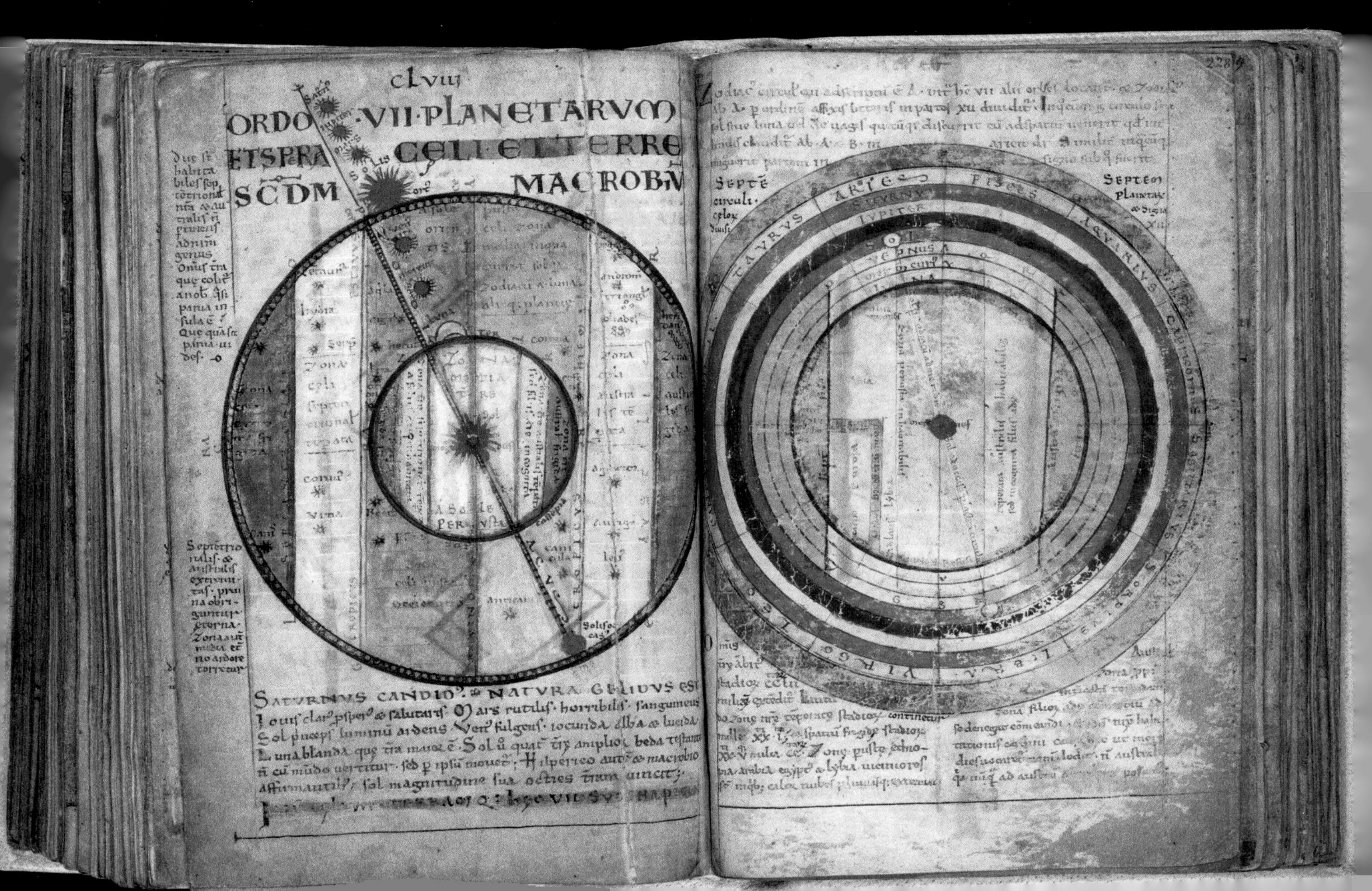

ORDO VII PLANETARVM
ET SPERA CELLI ET TERRE
SCDM MACROBIV
TAVRVS
ARIES
PISCES
AQVARIVS

• 1121:

Previous page, top: Graphic depictions of the design of the universe as understood in the twelfth century. In these pages from the medieval encyclopedia *Liber floridus*, Lambert, the canon of St. Omer in northern France, depicts the sun nine times, probably to reflect seasonal variations in its position on the ecliptic. The five planets are visible like beads on a string above and below the sun at the top and bottom. Because the sun appears to move more slowly against backdrop stars from the vernal to the autumnal equinox than it does from autumnal to vernal, predicting its position has never been an easy matter. At the center, the Earth is divided between Asia at the top, Europe on the lower left, and Africa on the lower right—the so-called T-O medieval world map scheme attributed to eighth-century Spanish monk Beatus of Liébana, and based on a description of the world by seventh-century scholar Isidore of Seville. A ring encircling Earth depicts lunar phases, and the five planets known at the time are also depicted in this classic early medieval depiction of a Ptolemaic universe.

Previous page, bottom left:
A depiction of the Ptolemaic solar system along the plane of the ecliptic. In keeping with the conventions of the time, Earth is positioned with the poles on the left and right. The equator and the tropics of Cancer and Capricorn extend out into the solar system in a kind of cross section of an armillary sphere, here extending only as far as the farthest extensions of the sun's orbit around the Earth. The sun is depicted three times: at each of its farthest extensions and at exactly halfway between them, putting it directly in front of the Earth. As a result, Canon Lambert presents us with a novel view, on animal-skin parchment, of the sun directly in front of the Earth as it cuts diagonally across our planet's equator. Meanwhile, the crescent moon is on the other side of our planet, partly bisected as it transits behind the Earth's limb. The diagram is titled "The Order of the Seven Planets" (meaning the five visible planets, the sun, and the moon).

Previous page, bottom right: The facing page provides a polar view of the same, with the planetary orbits now seen face-on rather than edge-on, and the zodiac encircling and divided into twelve equal zones.

• 1375:

These outstandingly beautiful depictions of the medieval cosmos come from the *Catalan Atlas*. Attributed to Jewish Majorcan cartographer and astronomer Abraham Cresques, it is the most significant atlas from Catalonia of the Middle Ages. Although star maps of the so-called golden age of celestial cartography from the seventeenth to the eighteenth centuries have deservedly been given a great deal of attention in recent years, these three extraordinary illuminated manuscript leaves prove that, at their peak, medieval attempts to depict the cosmos equaled or even exceeded later efforts, both aesthetically and in the efficiency with which they convey information.

Far right: In previous medieval depictions of the terracentric Earth, depictions of bearded wise men within the terrestrial sphere were usually meant to represent God the Creator. Here, in a likely sign of Cresques's worldview, God has been replaced by a sage wielding an astrolabe—an instrument used to determine celestial positions. (Unfortunately, he has been partly obscured due to the age of the manuscript.) In short: an astronomer! Earth is surrounded as usual by rings signifying the four elements; the seven spheres carrying the planets, moon, and sun; the sphere of fixed stars; and the zodiac. An outer blue ring represents the phases of the moon, and at the four corners, allegorical figures are meant to represent the seasons. This is a rather secular chart: Massed ranks of angels have been dismissed in favor of highly granular calendric information; this is a perpetual calendar.

Right: This leaf contains astronomical and astrological information in Catalan; a twenty-four-hour tide table oriented with the north down; a calendar for determining the dates of the movable feasts (Easter, the Pentacost, and the weeks of Carnival); and a Zodiac Man inscribed with astrological signs. For much of history, the signs of the zodiac were thought to govern regions of the body, and this human form doubles as a bloodletting figure, accompanied by instructions on when to stage surgical interventions or take medicine, as guided by the table to the lower left. That table graphs the lunar year, and shows which signs of the zodiac control each day. Taken together, then, the information in the lower third of the page deals with medical astrology, in which the structure of the universe is directly linked to the structure of the human body. The text to the left of the figure mentions Ptolemy in his role as astrologer.

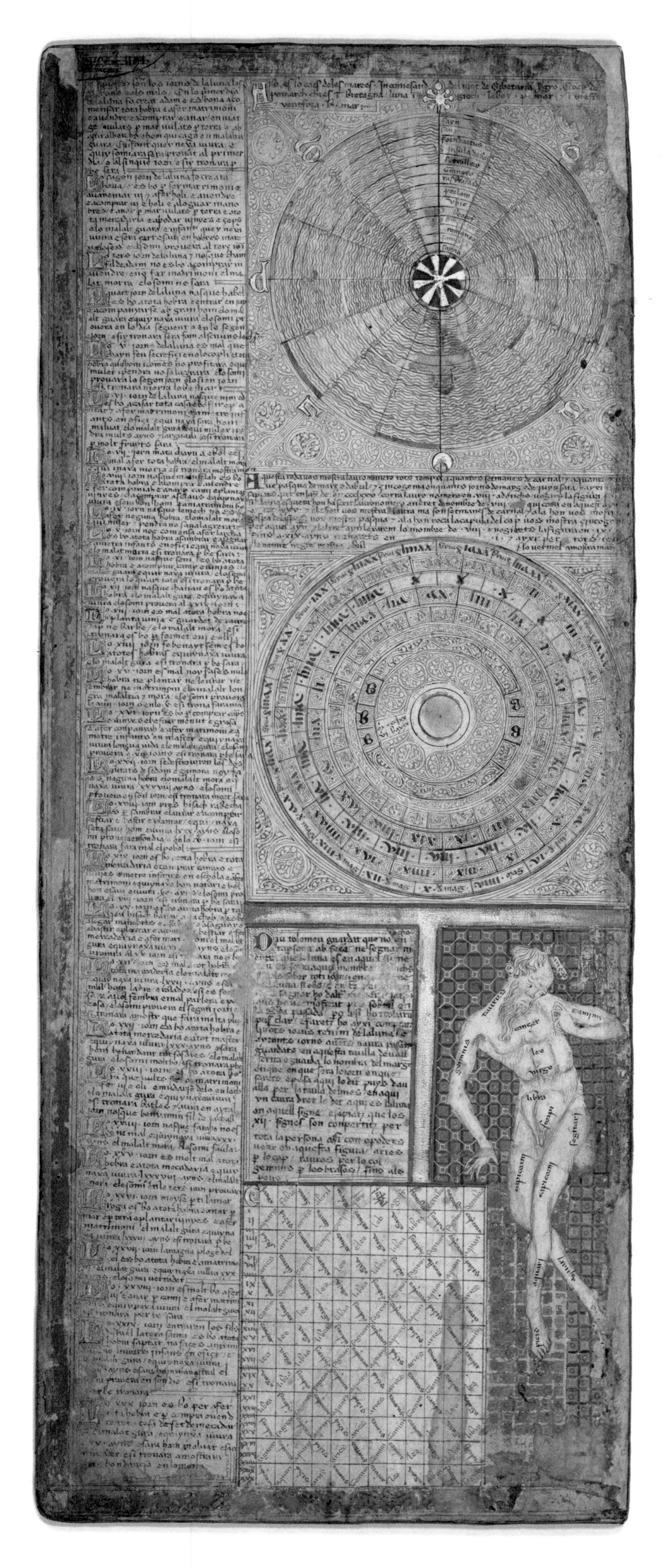

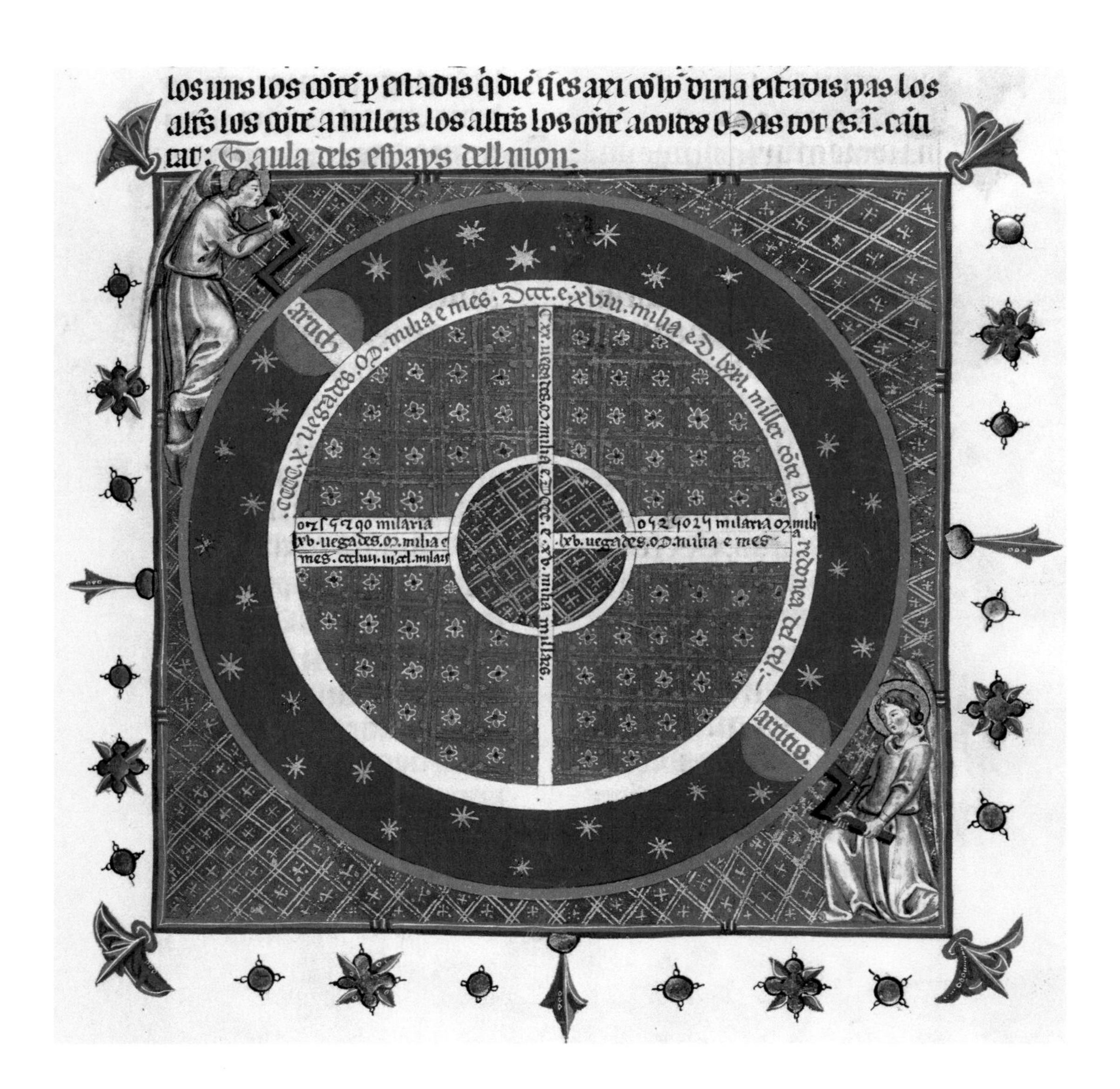

• 1375–1400:

Facing page, top: In the medieval Ptolemaic cosmos, which inherited the concept from Aristotle and Plato, everything contained within the sphere of the moon was changeable and corruptible, whereas everything beyond its orbit was changeless and perfect. The elements recognized from antiquity—fire, water, earth, and air—were all contained within this sublunary sphere. In this manuscript illumination from Matfré Ermengau of Béziers's *Breviari d'Amor* (Abstract of Love), angels use cranks to turn the sublunary sphere on its endless round—changeless supernatural beings winding the clockwork of temporality.

• 1440–1450:

Facing page, bottom: Another example of an illumination from *The Divine Comedy* by Giovanni di Paolo. At the opening of *Paradiso*, already grappling with a dawning awareness that he, too, is ascending toward the celestial spheres, Dante describes witnessing a rising sun joining "four circles in three crosses." This description has been understood to refer to the moment at sunrise on the day of the equinox when three celestial circles (the ecliptic; the celestial equator; and the equinoctial colure, or the circle of the celestial sphere that passes through the celestial poles) meet at the horizon, forming three crosses. (See the left-hand spherical shape in this illumination.) As Dante and Beatrice loft miraculously toward heaven, to the music of the spheres, Beatrice explains the universal hierarchy. Intriguingly, di Paolo inverts the classical visual ordering of the pre-Copernican universe. Because we're looking *toward* the heavens that our protagonists are rising toward from the vantage point of the Earth, rather than at an Earth positioned at the center of the universe, in this revolutionary depiction the *outer* rings contain water, air, and fire, and the inner rings contain the familiar celestial spheres. The winged "putti" figure at the center of this set of spheres confirms that the timeless and changeless empyrean, rather than the temporal and corruptible Earth, centers the spheres in di Paolo's sleight-of-hand visual reversal. (For a regular outside-in Ptolemaic ordering by di Paolo, see his *Expulsion* on page 33.) This complete inversion of the classical ordering due to the point of view of the viewer (and of our levitating protagonists) is original and probably unprecedented. For other works by di Paolo, see pages 33, 75, 116, 178–79, and 256.

• 1550–1600:

Above: A single angle holds the entire assembly of celestial spheres in this manuscript illumination probably painted in western Iran in the second half of the sixteenth century. Arab cosmographer and geographer Zakarīyā' ibn Muhammad al-Qazwini's highly popular and influential book *Wonders of Creation and the Oddities of Existence*, like his European contemporary Johannes de Sacrobosco's *Tractatus de sphaera*, went through innumerable editions for hundreds of years after it was written in 1270. But unlike Sacrobosco, whose relatively short text transmitted Ptolemaic astronomical principles to the early medieval world of Europe, al-Qazwini's illustrated compendium of universal knowledge covers geography and natural history as well as astrology and Ptolemaic astronomy. Arab astronomy inherited the spheres of the Greeks, as seen here, while frequently infusing it with a mysticism well represented in this image.

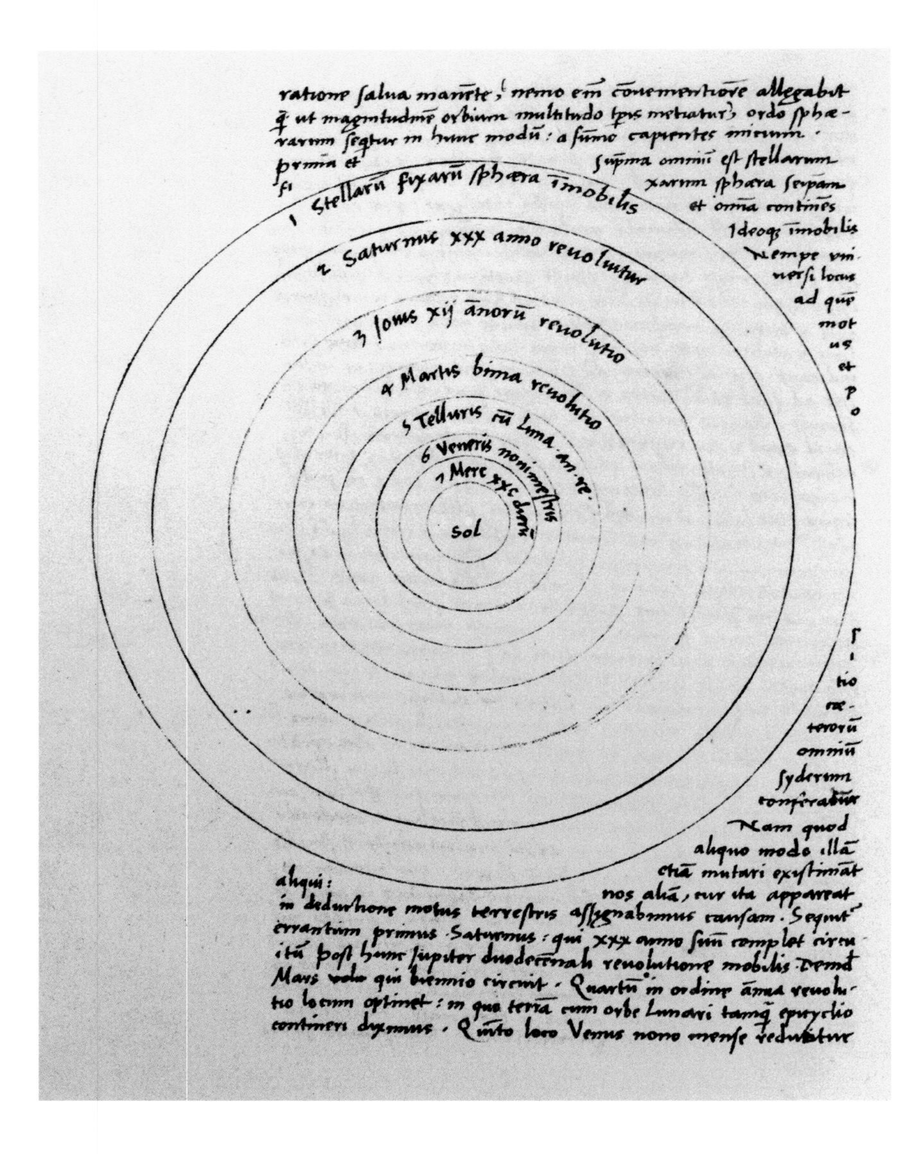

• 1520–41:

Above: Although Nicolaus Copernicus's *De revolutionibus orbium coelestium* (On the Revolutions of the Heavenly Spheres) was only published in 1543, he had assembled the observational backing to his heliocentric theory over several decades. At first glance, this simple diagram surrounded by neat handwriting might seem just another depiction of the Ptolemaic cosmos—and a rather simple one at that—until one notices the word *sol* at its center, for "sun." Not so simple. Marveling at the "skilled draftsmanship, the precise hand, and, above all, the way in which he has elegantly written his text around the famous diagram of the heliocentric system," astronomer-historian Owen Gingerich has called Copernicus's original manuscript "perhaps the most priceless artifact of the entire scientific renaissance." Effectively, Copernicus had disassembled the celestial spheres first conceptualized by the ancient Greeks, and later refined by Ptolemy in the A.D. 100s, and found them wanting. And so he reassembled them again—this time with the sun at their center.

• 1660:

Right: A radiant transformation of Copernicus's spare diagram. In this depiction of the Copernican universe from Andreas Cellarius's *Harmonia macrocosmica*, the sun dominates the solar system. Note that the moon is shown patrolling the sublunary sphere; Copernicus didn't throw out Ptolemaic concepts, he reordered them. The legend in Latin outside the Earth's sphere reads, in part, "Sublunary sphere with the four elements themselves about the Sun." To the lower right sits Copernicus, looking rather pleased with himself. Across from him sits an astronomer of the ancient world, either a supplanted Ptolemy or just possibly, the Greek astronomer Aristarchus of Samos, who first suggested that the sun was at the center of the universe prior to 200 B.C. For other maps from Cellarius, see pages 47, 84, 122, 148, 182–83, and 232–35.

COPERNICANVM
Systema
TIVS CREATI
THESI
CANA IN
EXHIBITVM.
SATURNI
IUPI TER
MARS
VENUS
ORBIS MERCURII
MERCU RIUS
GLOBI SUBLUNARIS CUM LUNA ET QUATUOR ELEMENTIS SE CIRCA SOLEM VERTENTEM
LUNA CUM MUNDO SUBLUNARI
TRIA SIGNA ÆSTIVA
TRIA SIGNA VERNA
TAURUS
GEMINI
CANCER
LEO
VIRGO
DODECA MORION ZODIACI

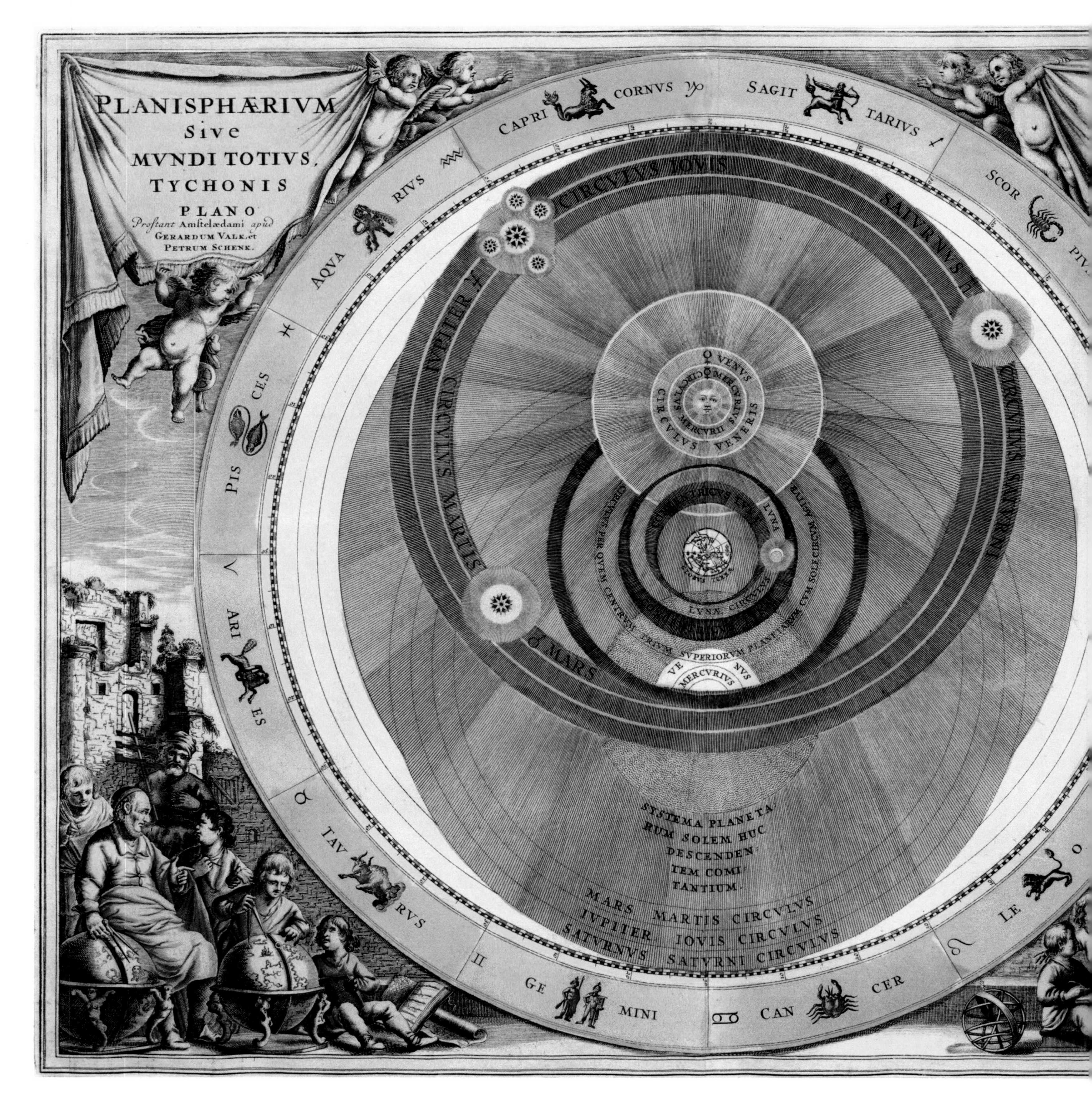

• 1716:

In the second half of the sixteenth century, Danish astronomer Tycho Brahe did his part to dismantle Aristotelian ideas about an immutable universe beyond the orbit of the moon—in part by proving that comets passed directly through the supposedly crystalline spheres as easily as Dante and Beatrice did in *The Divine Comedy*. (For more on this, see chapter 9.) He had problems with the Copernican system, though, and in particular refused to accept anything other than the centrality and immovability of the Earth, which he termed "lazy." Instead, he proposed an alternative that historian of science Thomas Kuhn called "precisely equivalent mathematically to Copernicus's system." In it, all the planets except for the Earth orbit the sun. The sun, however, orbits the Earth, with all its planets in tow. This depiction of Brahe's mixed geocentric-heliocentric system is from Andreas Cellarius's *Harmonia macrocosmica* (Cosmic Harmony). Tycho's system had the political virtue of preserving the theological propriety of astronomers mindful of Catholic doctrine concerning the immovable Earth, but it had the side effect of shattering the Aristotelian-Ptolemaic crystalline spheres, because, as can be seen in the inset above, in the Tychonic system the orbit of Mars intersects that of the sun, and the sun's sphere passes through that of Mercury and Venus, none of which would be possible with hard-shell spheres. Tycho's "geo-heliocentric" system had numerous problems, and although it was accepted by a large number of seventeenth-century astronomers uncomfortable with Copernicus's revolutionary reordering of the universe, it faded away under mounting evidence confirming a more total heliocentrism. For other maps from Cellarius, see pages 47, 84, 122, 146–47, 182–83, and 232–35.

• 1651:

Italian astronomer and Jesuit priest Giovanni Battista Riccioli proposed his own adaptation of Brahe's cosmology, as visible in this frontispiece from his formidable 1,500-page book *Almagestum novum* (New Almagest). Here Urania, the starry muse of astronomy, weighs the Copernican system on the left and Riccioli's adaptation of Brahe on the right: This being Riccioli's book, the latter system has more weight. (Riccioli places Mercury, Venus, and Mars, in orbit of the sun, which in turn orbits Earth; Jupiter and Saturn remain in their Ptolemaic geocentric orbits). Many-eyed Argus, on the left, holds a telescope, which has revealed the wondrous new celestial objects being held by a squadron of *putti* on the upper left and right. Ptolemy, reclining near the ground below, has been superseded; his cosmological system, on the lower right, isn't even being weighed.

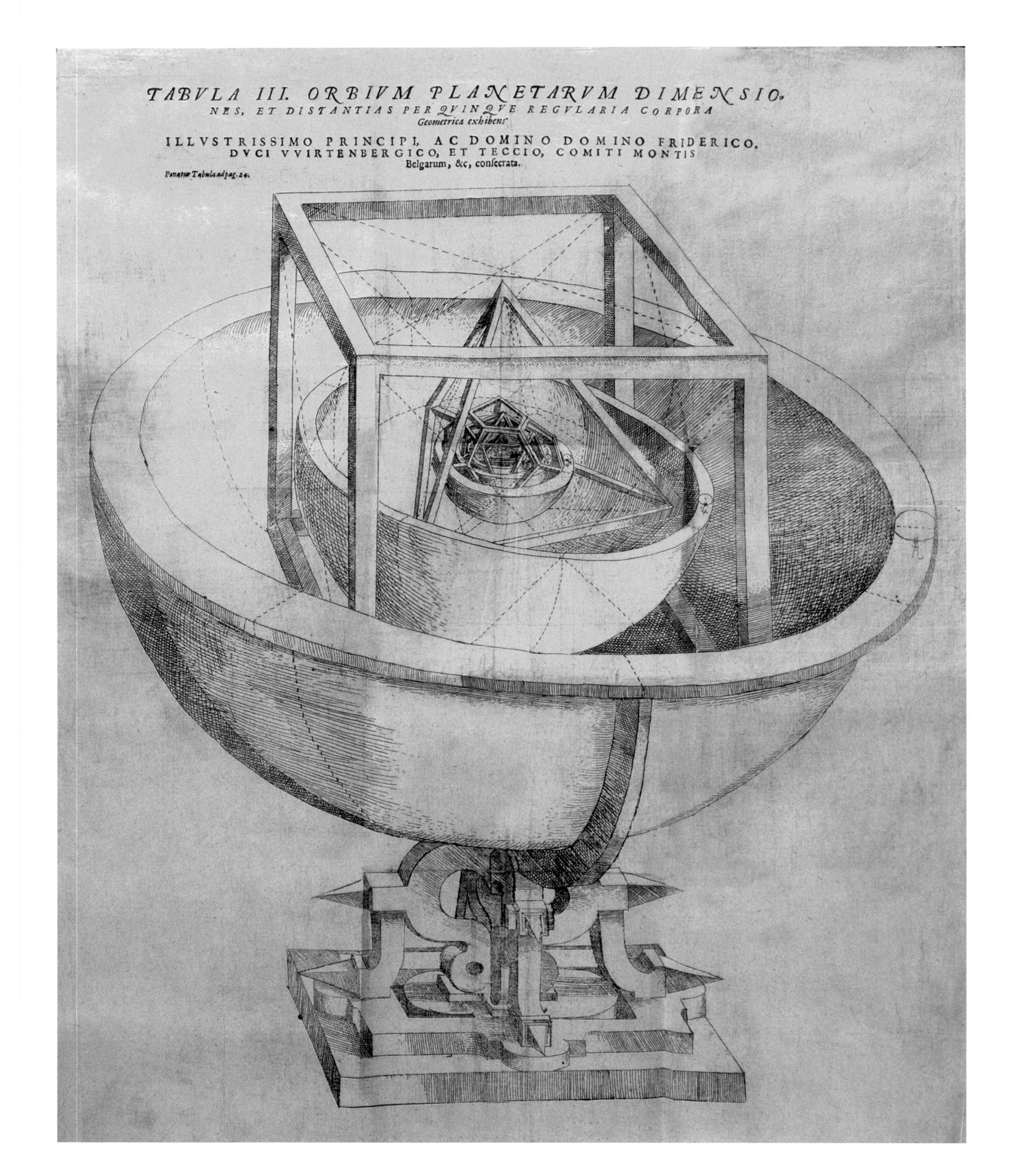

TABVLA III. ORBIVM PLANETARVM DIMENSIO-
NES, ET DISTANTIAS PER QVINQVE REGVLARIA CORPORA
Geometrica exhibens
ILLVSTRISSIMO PRINCIPI, AC DOMINO DOMINO FRIDERICO,
DVCI VVIRTENBERGICO, ET TECCIO, COMITI MONTIS
Belgarum, &c, consecrata.
Ponatur Tabula ad pag. 24.

• 1595:

Tycho Brahe's colleague and collaborator Johannes Kepler was a fervent Copernican, and never embraced Brahe's mixed model of the solar system. In this etching from Kepler's 1595 book *Mysterium cosmographicum* (The Cosmographic Mystery), which has the distinction of being the first published defense of Copernicanism, Kepler did however present a modification of the celestial spheres model handed down from antiquity. He relied on another finding from the Greeks, one familiar from Euclidian geometry: the five platonic solids. These geometrical shapes are regular convex polyhedrons with identical faces, in which the same number of faces meet at each vertex. Advancing Plato's theory that the classical elements were made of these solids, Kepler postulated that the planetary spheres could be circumscribed using the solids. The result was Kepler's matryoshka solar system, as seen here. The sphere of Saturn, then thought to be the outermost planet, is circumscribed about the cube-shaped solid containing Jupiter's sphere, within which nests a tetrahedron containing Mars' sphere, and so on down the line to a central sun. While from a contemporary perspective this complicated quasi-mechanical construction can seem like a geometrization conceived by a purist mathematician, as historian of science Thomas Kuhn points out, the same drive to explain the universe through pure deduction led Kepler to the brilliance of his three definitive laws of planetary motion.

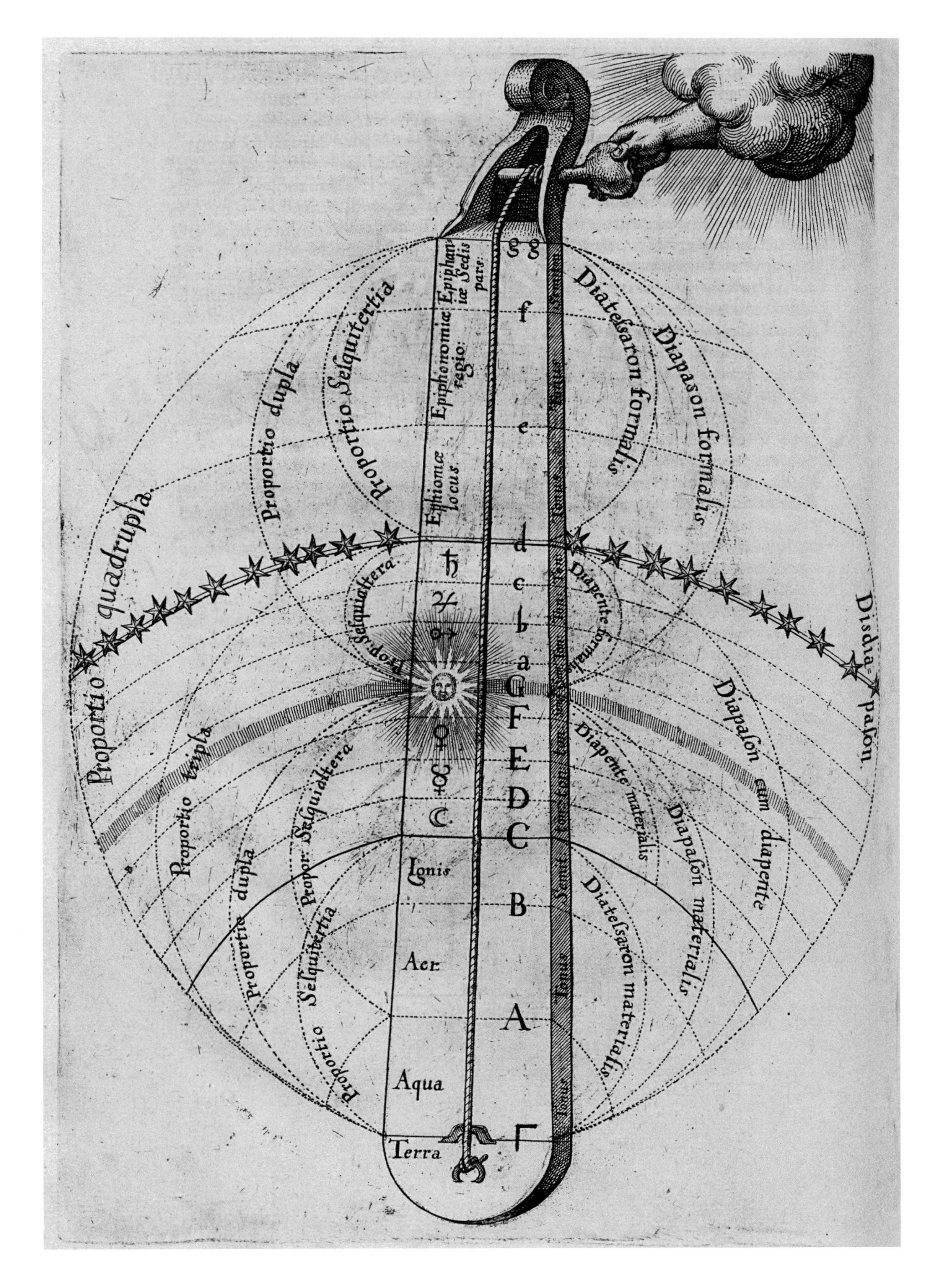

• 1617:

The Keplerian tendency to explain the natural world in musical terms—one of his later works was titled *Harmonice mundi*, or "Harmony of the World"—may have found its purest expression in this illustration from English physician and cosmologist Robert Fludd's *Utriusque cosmi, maioris scilicet et minoris, metaphysica, physica, atque technica historia* (The Metaphysical, Physical, and Technical History of Two Worlds, the Macrocosm and the Microcosm). Described by Fludd as "the monochord of the universe," it depicts a musical instrument with a single string extending between the temporal and the empyrean worlds. Above, the hand of God tunes the string. At the opposite end we see the word *Terra*, for "Earth," above which the familiar ancient elements climb toward the sun and the ecliptic. A starry band marks the outer limit of the spheres. Above that are the angelic orders, divided into three sections. The harmonic proportions of the universe are written as ratios on the left and as numerals on the right. The single string stretches between fifteen musical notes, which together produce a kind of universal harmony. It's the music of the spheres with a celestial choir on top.

• 1644:

Toward the end of the first half of the seventeenth century, French philosopher and mathematician René Descartes, the inventor of analytical geometry, was increasingly drawn to questions of cosmology. Because Aristotle decreed the principle of *horror vacui*, or "nature abhors a vacuum," for centuries the space between the spheres was said to be filled with a mysterious substance called ether. Descartes proposed instead that everything in the universe was made of tiny particles he called "corpuscles," and that they also swirled in the space between celestial objects. With Aristotle's spheres increasingly discredited as a cosmological model, and in an effort to understand the mysterious forces that seem to hold celestial bodies together and govern their movement, Descartes developed a theory in which the innumerable corpuscles ostensibly filling the vast spaces between celestial objects must be filled with vortices. In this illustration from his book *Principia philosophiae* (Principles of Philosophy), Descartes depicts a comet buffeted by vortices as it traces a sinuous path through the solar system. He developed four laws of motion based on his corpuscular cosmology.

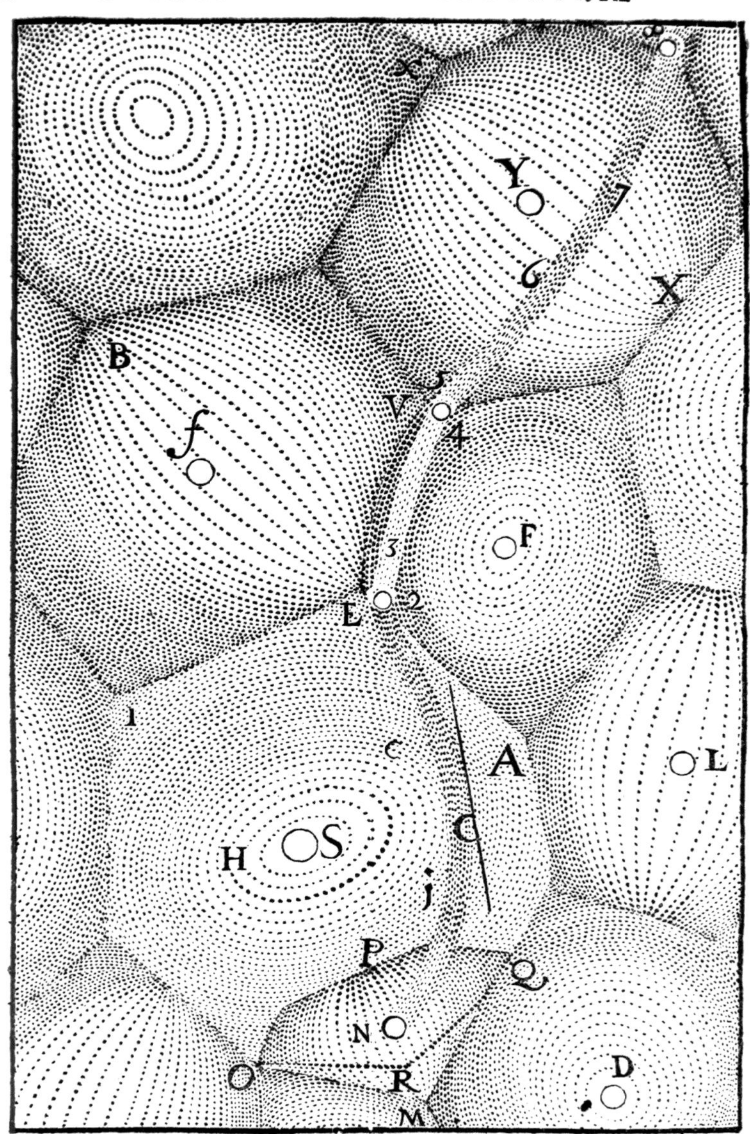

• 1673:

The radical modification of long-held views about the structure of the cosmos gradually caused by Copernicus's heliocentric theory, followed by the revelations being published by Galileo and other early telescopic astronomers, revived a speculation with ancient roots: Could the universe contain more—possibly many more—worlds than were known in the solar system? Might the stars be suns like our own, with planets of their own? And all that being true, could extraterrestrial life exist, either among the sun's planets or beyond? This detail of a print by French master engraver Bernard Picart illustrates the theory that a plurality of worlds could exist throughout the universe. Something close to the modern conception of the universe was being born.

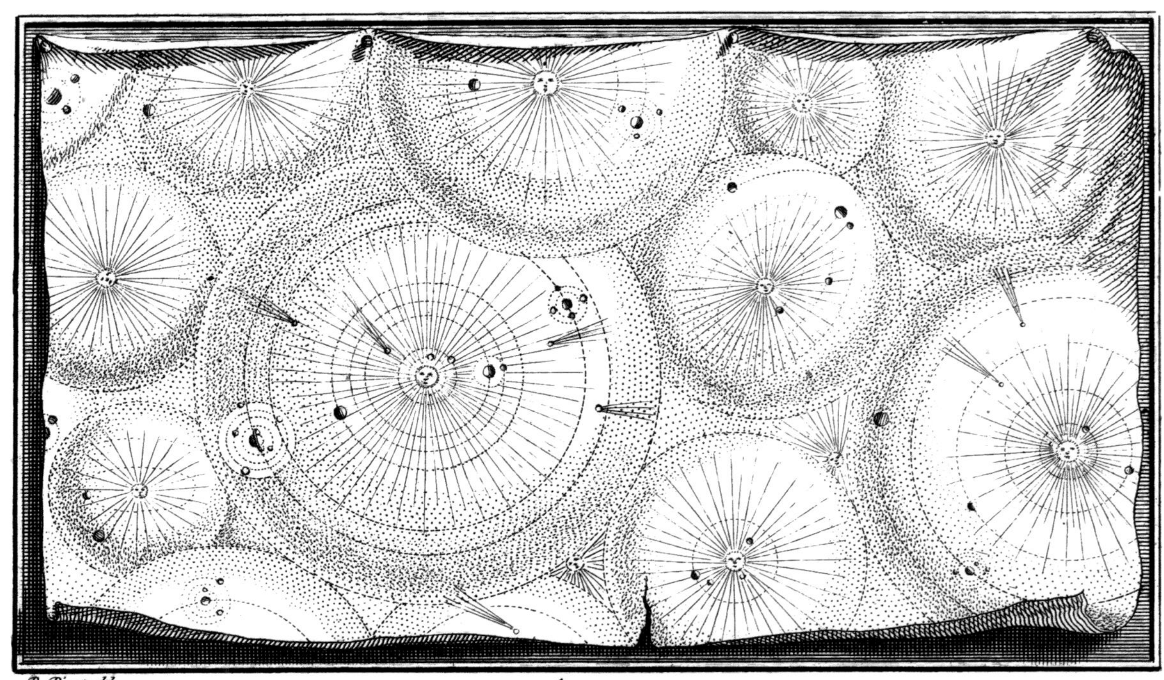

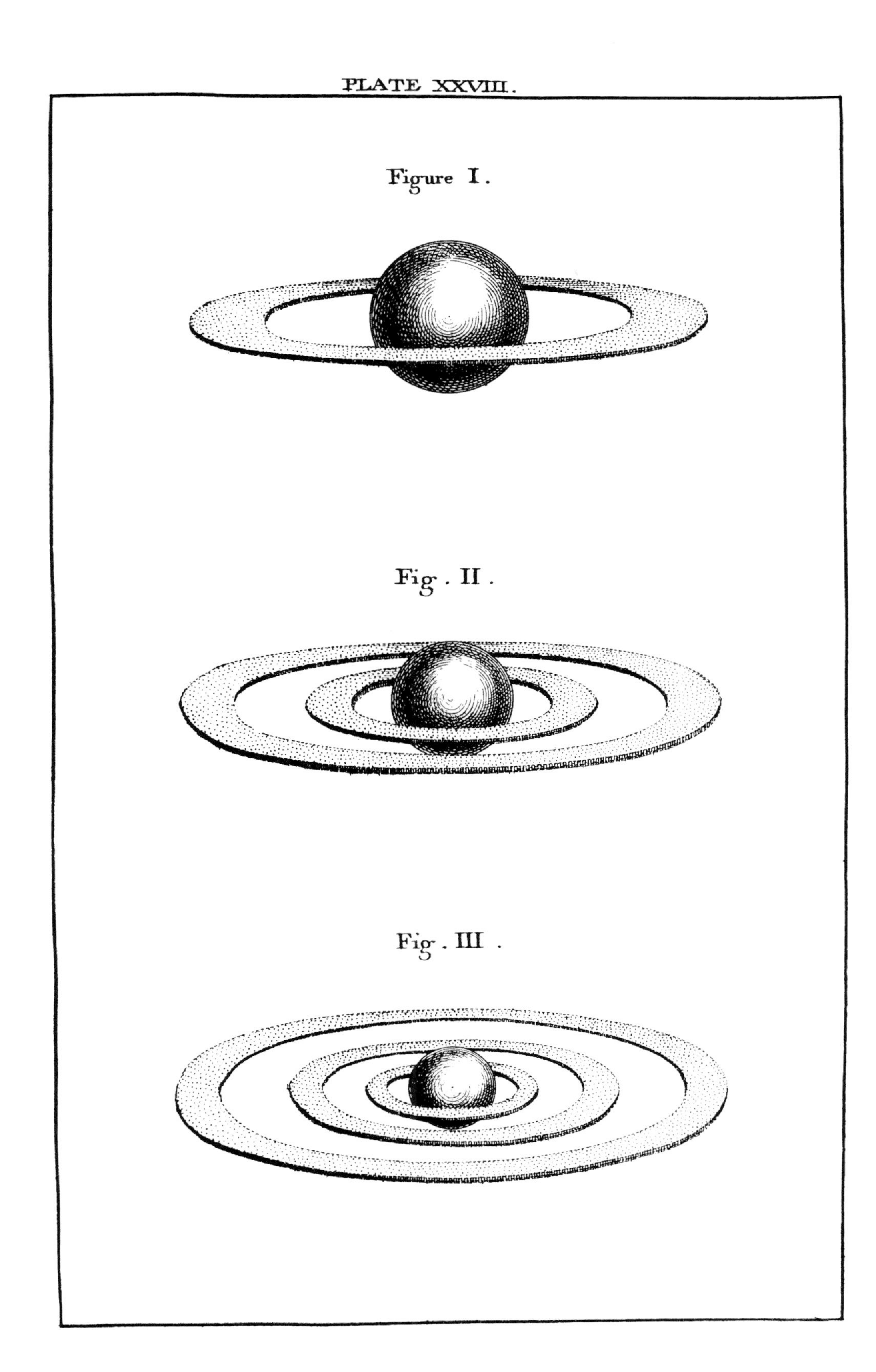

• 1750:

The first depiction of the Milky Way's flattened-disc form in history. By the mid-eighteenth century, the stage was set for a significant epiphany about the building blocks of the cosmos. In 1750, English astronomer Thomas Wright published a book titled *An Original Theory or New Hypothesis of the Universe*. Influenced both by the shape of Saturn and of the solar system, Wright proposed that the Milky Way galaxy may be structured on one plane like a disc with a large central nucleus, as in the above mezzotint plate. He also came up with an alternative hypothesis: It could be in the form of a giant sphere. In this he not only arrived at the first general description in history of the form of spiral galaxies like our own (albeit without hitting on their spiral arms: instead Wright imagined rings of stars), he also conceived of the external appearance of elliptical galaxies—another common galaxy shape, many of which appear almost perfectly spherical. He then proceeded to argue that the ghostly blobs seen dotted across the night sky—shapes that had been taken for nebulae within our own Milky Way—were "myriads of . . . celestial mansions," or galaxies like our own. In one book, published prior to the American Revolution, Wright blew prior cosmological schemes wide open, arriving at an almost contemporary picture of the universe.

Plate from Thomas Wright's *Original Theory*. The idea that stars are suns like our own, and orbited by planets, dates back at least as far as the late sixteenth century and Italian philosopher Giordano Bruno, and a more general concept that a plurality of worlds exists in the universe can even be found in surviving accounts of concepts devised by the pre-Socratic philosophers. In this stunning mezzotint plate, however, Wright arrives at an entirely new conception: of a universe filled with multiple galaxies. (Their spherical shapes represent one of two galactic forms he proposed; see the previous page for the other.) For more by Wright, see pages 153 and 236–37.

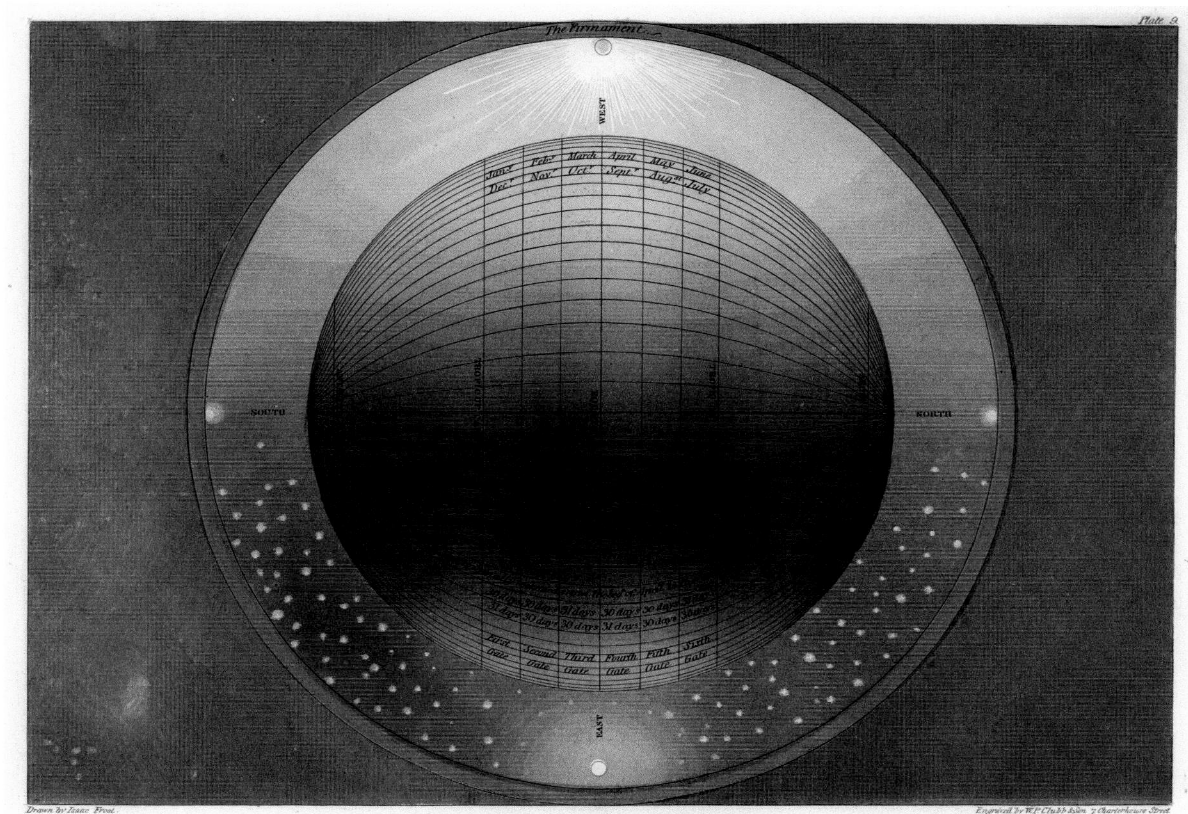

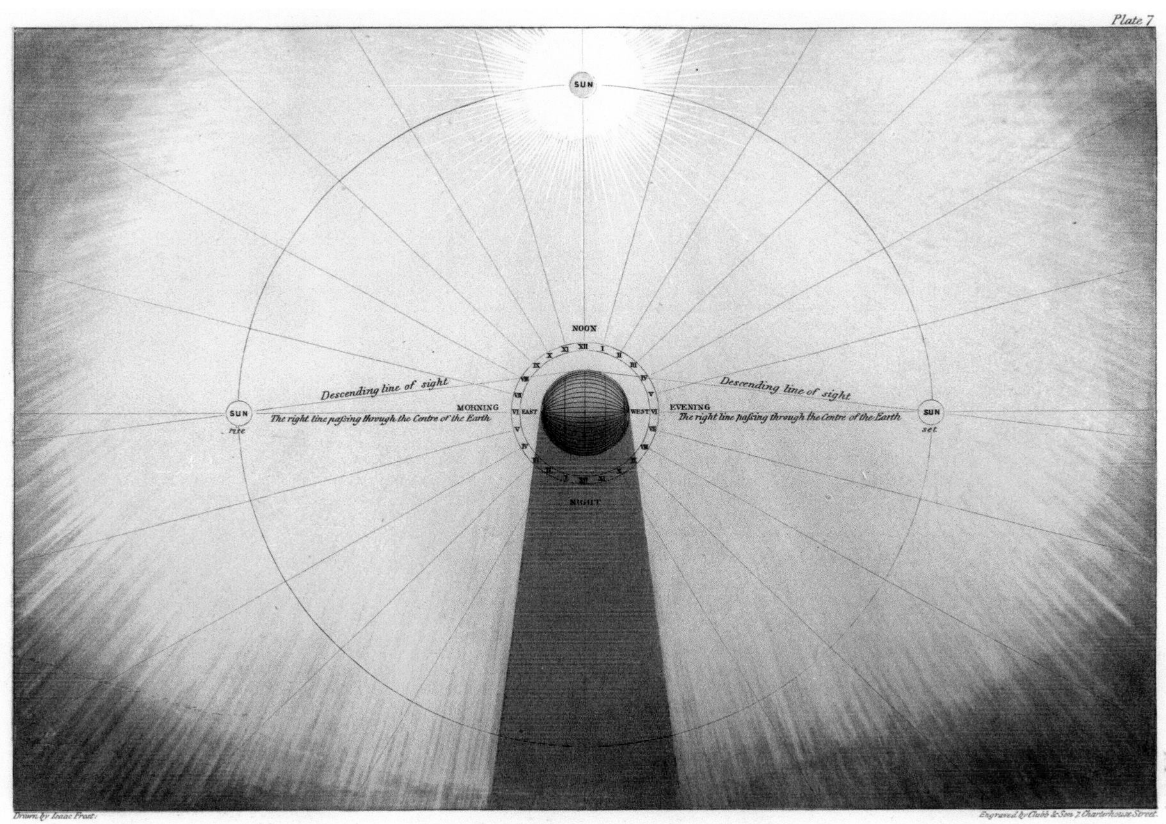

• 1846:

From Thomas Wright's daring vision to a more reactionary view. Named after their founder, Lodowicke Muggleton, the Muggletonians were an obscure English Protestant sect that got its start in the mid-seventeenth century. They insisted on a literalist understanding of the Bible and therefore condemned Copernican ideas about the centrality of the sun, let alone a plurality of worlds. In fact, their views of science were clear: One Muggletonian principle was "There is no Devil but the unclean Reason of men." These prints, based on drawings by nineteenth-century Muggletonian Isaac Frost and taken from his 1846 book *Two Systems of Astronomy*, take us back to a pre-Copernican universe, but in a graphically innovative way.

THE GREAT SPIRAL NEBULA

• 1845:

By the nineteenth century, more than two hundred years after the first use of the telescope in astronomy, the instrument had expanded radically in size. In 1845, Anglo-Irish astronomer William Parsons, the 3rd Earl of Rosse, installed a giant six-ton telescope in his residence at Birr Castle, in County Offaly. With a six-foot aperture, it was soon nicknamed the "Leviathan," and was the largest telescope in the world until 1918. Although Irish skies provided at best sixty good viewing nights annually, Parsons managed to glimpse intriguing spiral structures in what were still generally understood as nebulae within the Milky Way. This print, based on Lord Rosse's drawing of one such "nebula," M51, caused a sensation when it was reproduced in the UK and across Europe. M51, now popularly known as the Whirlpool Galaxy, is about 23 million light-years away. A major spiral galaxy, it is about the size of the Milky Way.

• 1889:

By 1879, a reproduction of the print of M51 based on Lord Rosse's drawing had made its way into a best-selling French book on astronomy, Camille Flammarion's *L'Astronomie populaire* (Popular Astronomy), a copy of which may have been acquired by the Saint-Paul de Mausole Asylum in Saint-Rémy de Provence, southern France. The above ink-on-paper study by Vincent van Gogh, made after his oil painting *The Starry Night*, is thus very likely an indirect result of the need by astronomers to produce graphic documentary evidence of their observations. It's widely believed that Van Gogh, then a patient at the asylum, was intrigued by Flammarion's reprint of Lord Rosse's drawing, seeing it either in the asylum or while in Paris. The full-color *Starry Night*, on view at the Museum of Modern Art in New York, is probably the single best-known artistic representation of the night sky.

HARVARD COLLEGE OBSERVATORY 1874. PL. 33

L. Trouvelot, del. J.H. BUFFORD'S SONS LITH BOSTON

THE ANDROMEDA NEBULA.

• 1874:

In the late nineteenth century, the possibility that the Milky Way may be one of many galaxies scattered throughout the universe had not yet been widely accepted, and many enigmatic objects visible in the increasingly powerful telescopes of the day were still considered nebulae within our own galaxy. Some were in fact giant galaxies, as in this depiction of the Andromeda "nebula." Andromeda is the largest galaxy in our Local Group, and the nearest major spiral to the Milky Way. Although its galactic shape is well known today due to contemporary astronomical photography, the details of its structure are the result of long-duration time exposures. When seen "live" through the eyepiece of even very large telescopes, most galaxies appear as indistinct silvery-gray clouds, as here. Still, the human eye is especially sensitive to changes of intensity, and hence the black streaks, now known to be dust lanes distinguishing the spiral arms, that stand out in sharp relief in this drawing. The central nucleus, far more brilliant than the rest of the galaxy, is now known to harbor a super-massive black hole. This print, by artist-astronomer Étienne Trouvelot, was published in the *Annals of the Astronomical Observatory of Harvard College*.

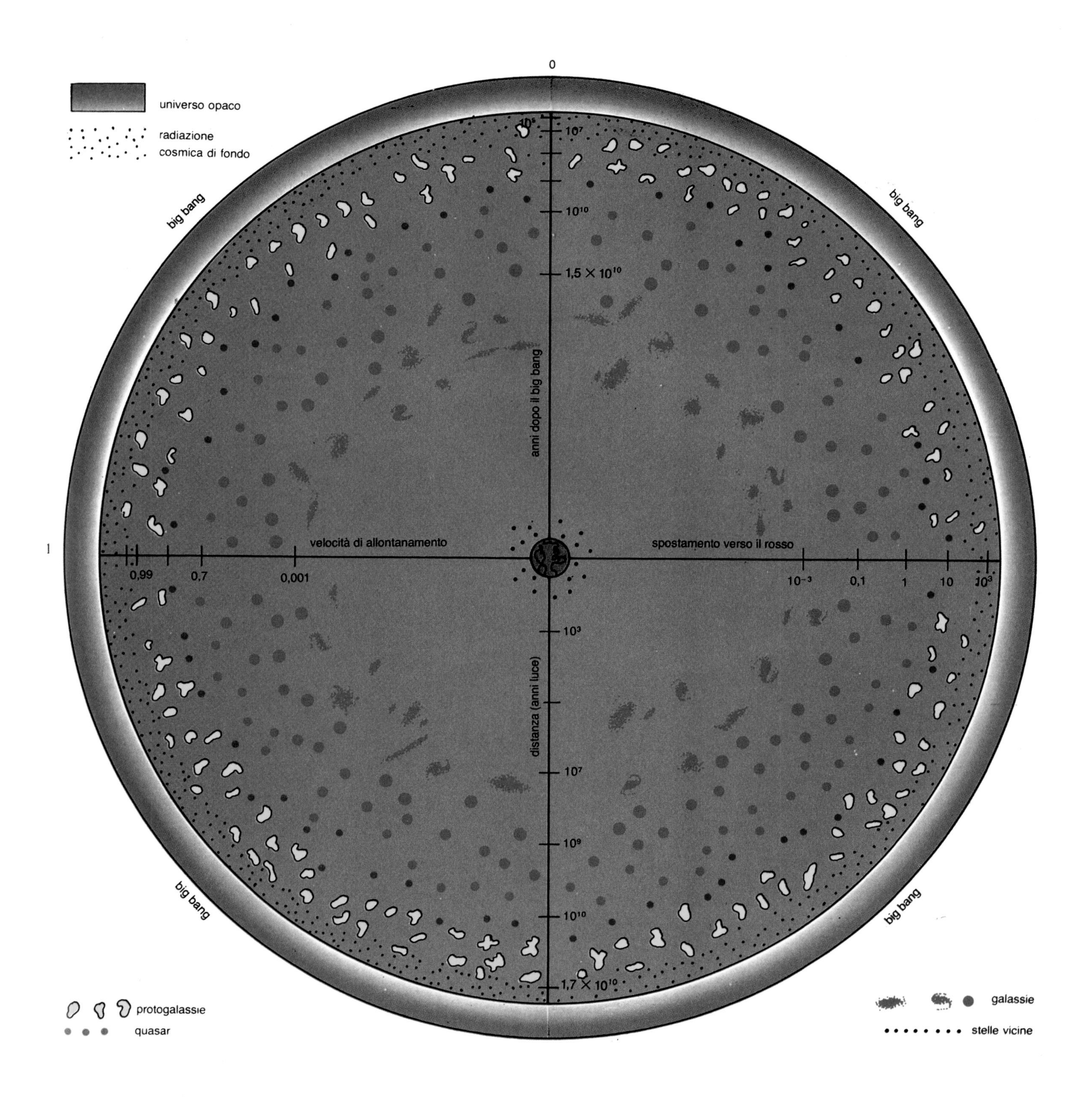

• 1982:

If this depiction of the universe by Italian astronomer Francesco Bertola appears to return to a geocentric scheme, that's because it does—albeit as a cartographical strategy informed by a contemporary cosmology which decrees that no center exists. If the universe is homogeneous and isotropic, as the cosmological principle states—meaning it will look more or less the same, at least on large scales, to any observer anywhere in it, looking in any direction—then any center will do, because all places function as a local center of space-time. First published in *Scienza e tecnica, annuario* (Science and Technology Yearbook), this graphic representation of what we might call a one-sphere cosmos contains the entire timeline of the universe, from the Big Bang (the outer black line of the circle) through a period of foggy opacity known as the recombination epoch (represented by the entire outer ring of the schematic), then on to protogalaxy formation (the yellow kidneys), then to the emergence of the first quasars (the red dots), and finally, a progression of increasingly mature galaxy morphologies (the blue shapes). The vertical axis contains a logarithmic scale of the distance in light-years to the edge of the observable universe, which doubles as the estimated number of years since the Big Bang. The horizontal one uses a similar logarithmic scale to tabulate the velocity of the expansion of the universe based on the redshift of the objects seen at different distances. (*Redshift* is a phenomenon in which light from a receding object, be it a jet plane or a galaxy, shifts to longer wavelengths, or the red side of the electromagnetic spectrum.)

• 1987:

By the mid-twentieth century, it was well accepted that the universe is filled with galaxies to a dazzling and almost incomprehensible degree, with our own Milky Way only one among many billions. So it seemed logical to attempt to understand the distribution of these galaxies by starting with relatively nearby intergalactic space. In the late 1950s, French astronomer Gérard de Vaucouleurs presented the theory, based on thousands of observations, that nearby galaxies are organized into vast superclusters. One of them, he said, is a Local Supercluster containing the Milky Way. Although his ideas were widely dismissed as speculative, they proved to be entirely correct, and by the 1970s a new generation of astronomers were building on de Vaucouleurs's work, including R. Brent Tully and J. Richard Fisher, who used radio telescopes to conduct an all-sky survey intended to locate nearby galaxies. In 1987, they published their pioneering *Nearby Galaxies Atlas*, the first attempt to chart the structure of the local universe. (Published with a bright red cover and a spiral binding, it's entertainingly similar visually to a standard American road atlas.) They found clear evidence that nearby galaxies "tend to align themselves parallel to the same plane, and . . . that this plane is incredibly extensive." The Tully-Fisher atlas uses ten plates to chart 2,367 galaxies, with the first shown to the right. Only months after it came out, Tully announced that he had discovered a supercluster complex encompassing millions of galaxies, including our own, and extending across approximately 10 percent of the observable universe. He said he determined its approximate size by trying to locate the edge of the Local Supercluster seen here. That edge turned out to be a lot farther away than he'd suspected.

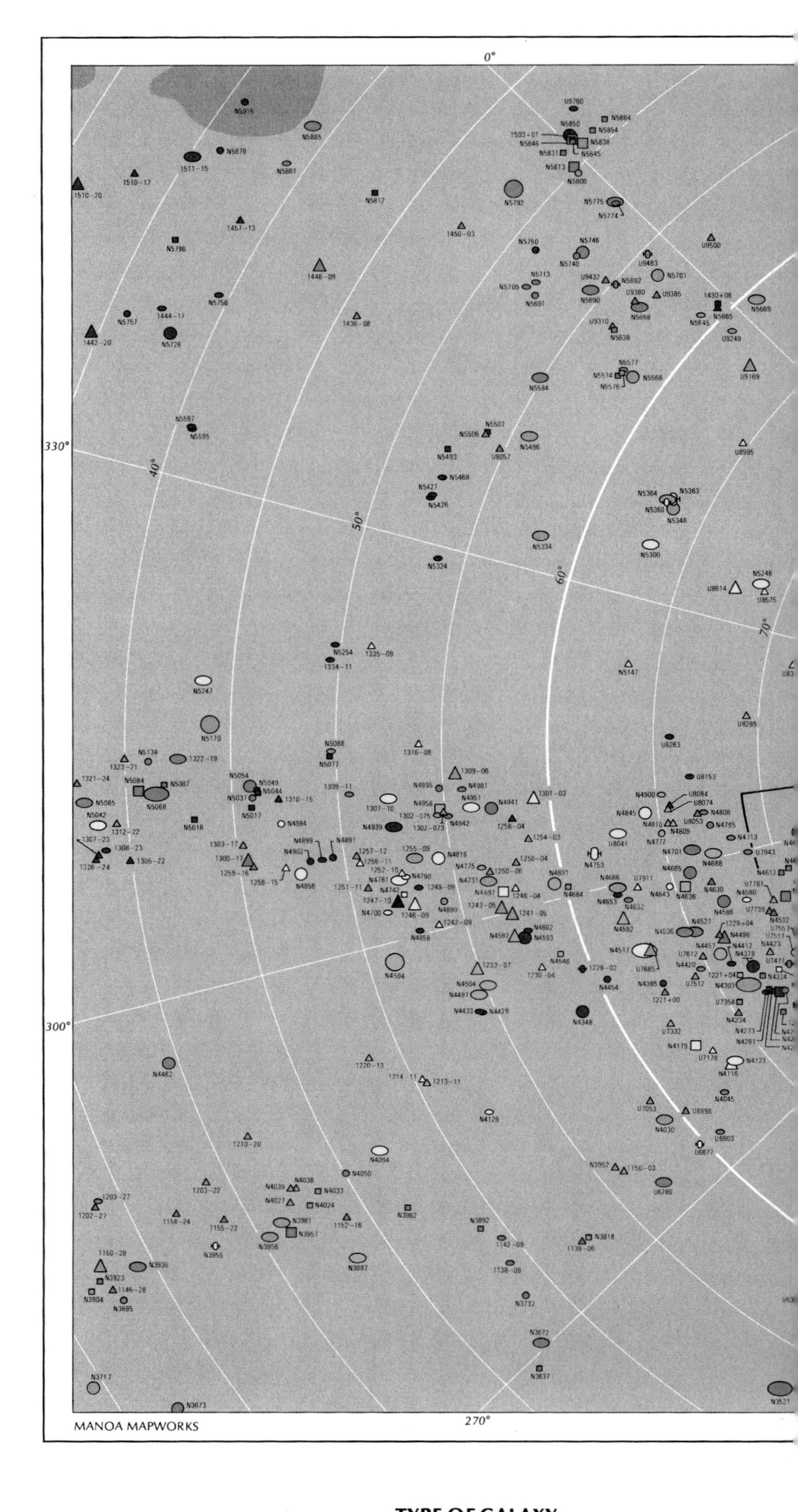

APPARENT SIZE OF GALAXY:	TYPE OF GALAXY: E,S0	S0/a–Sb	Sbc–Sd	Sdm–Im	Peculiar or Unidentified
$D_{25} \leq 3.0$	▪	●	⬬	▲	◆
$3.1 \leq D_{25} \leq 6.7$	■	●	⬬	▲	◆
$6.8 \leq D_{25} \leq 15.0$	■	●	⬬	▲	◆
$15.1 \leq D_{25} \leq 50.0$	■	●	⬬	▲	

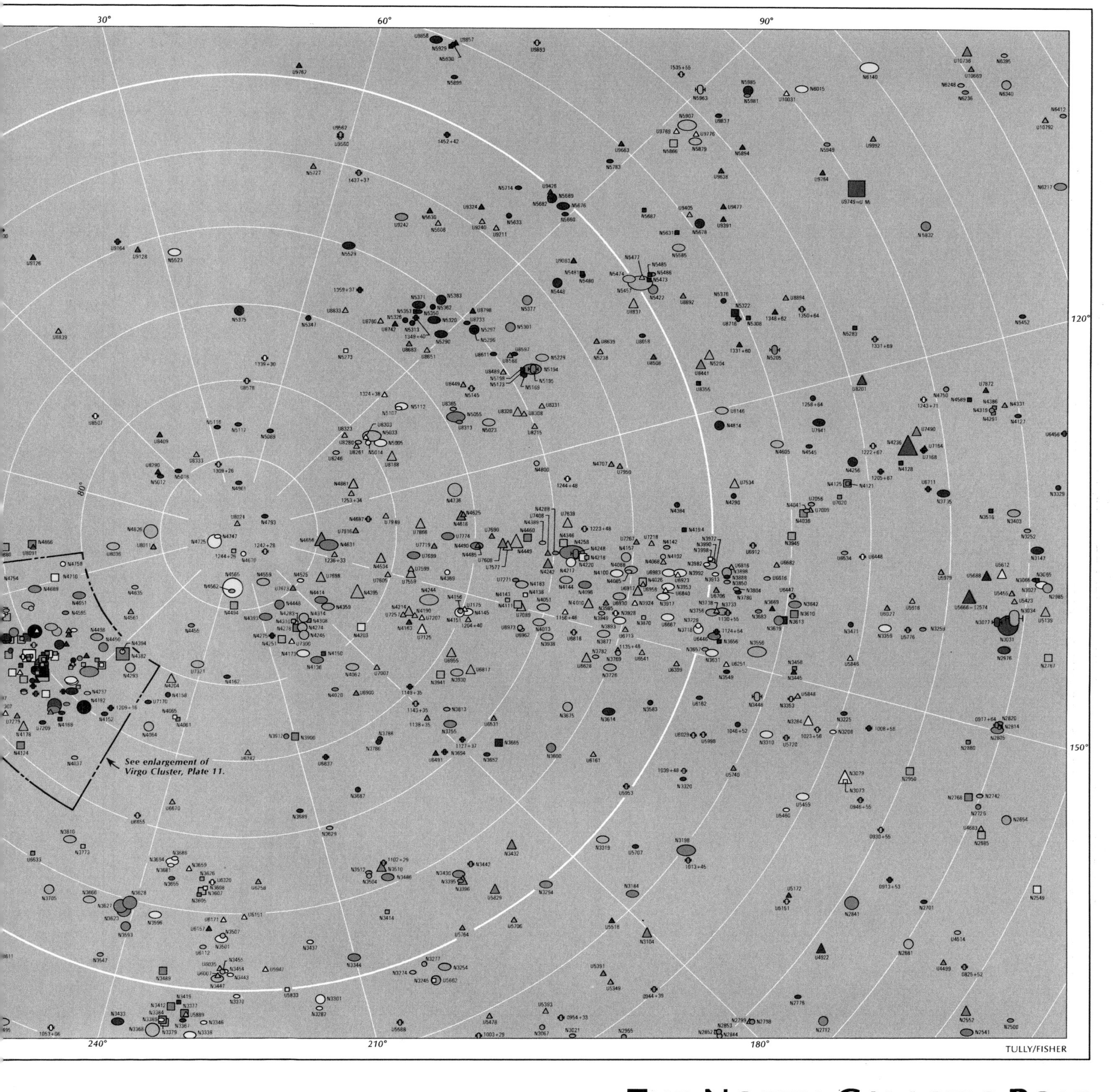

THE NORTH GALACTIC POLE

– the heart of the Local Supercluster; featuring the Virgo and Ursa Major clusters.

Y WAY

RECESSIONAL VELOCITY / DISTANCE:

(in kilometers per second)

- $V_0 < 0$
- $0 \le V_0 < 250$
- $250 \le V_0 < 500$
- $500 \le V_0 < 750$
- $750 \le V_0 < 1000$
- $1000 \le V_0 < 1250$
- $1250 \le V_0 < 1500$
- $1500 \le V_0 < 2000$
- $2000 \le V_0 < 3000$

PLATE
1

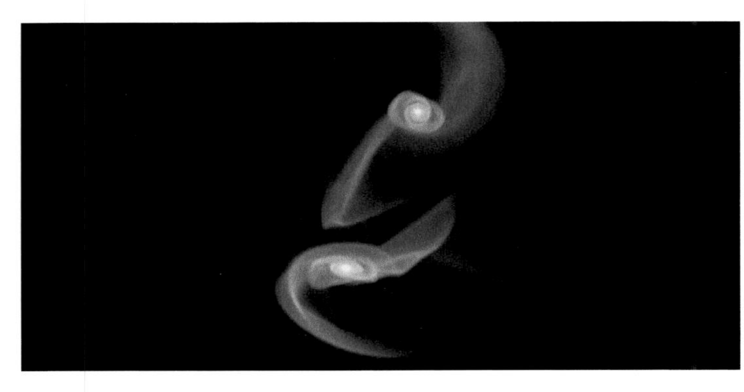

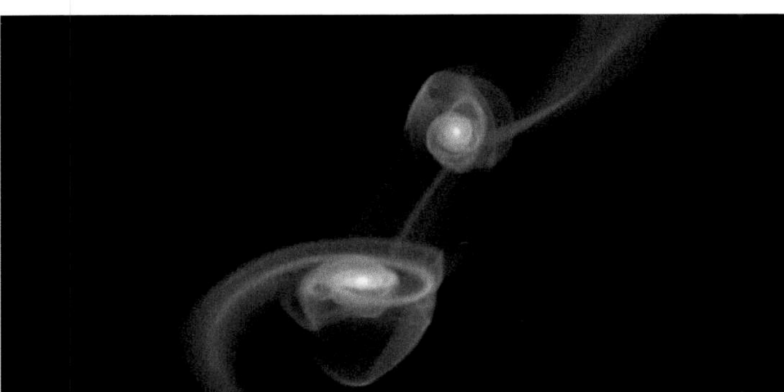

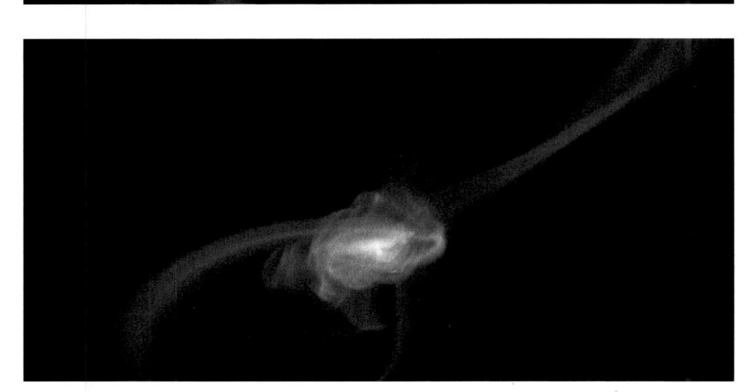

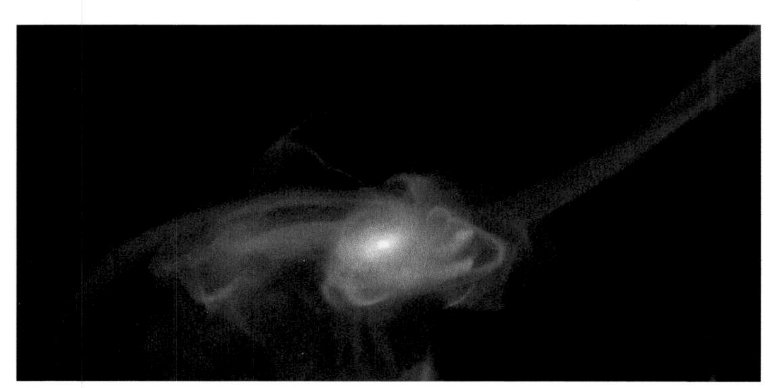

• 2006:

Right: Although supercomputers are increasingly capable of handling the trillions of calculations per second necessary to simulate collisions of individual galaxies, a computer-generated replication of galaxy clusters—some of the largest scale structures in the universe—requires use of a technique in numerical analysis known as "adaptive mesh refinement," as visible in this visualization produced by Daniel Pomarède. In 2006, he and fellow astronomer Romain Teyssier used the supercomputers of the COAST "Computational Astrophysics" Project in Saclay, near Paris, to simulate the large-scale clustering of galaxies. The work served as a basis for computing the density of both the baryonic gas and the dark matter halo in and among these clusters. The existence of dark matter can only be inferred by the gravitational effects on its environment. The image to the right is a single frame from a movie of galaxy cluster formation and evolution designed to study the effects on galaxy clusters of sometimes turbulent baryonic gas flows. It shows the grid of refined cells where the simulation code has solved the equations describing the physics of this giant system. Simulations of such vast structures continue to take current computing technologies to the limits of their capabilities.

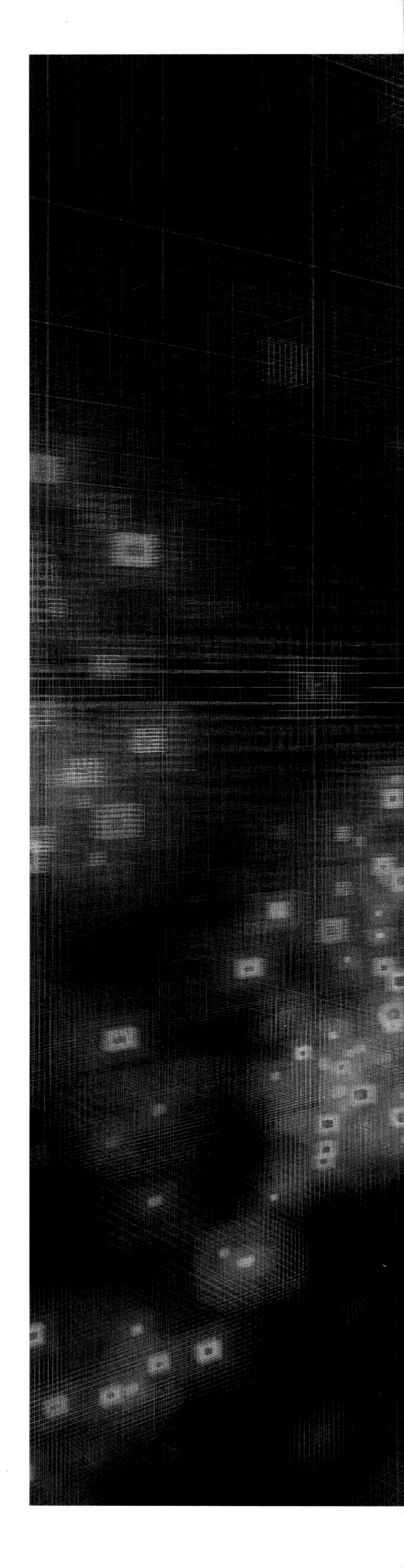

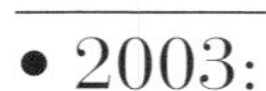

• 2003:

Above: By the early twenty-first century, supercomputers were getting powerful enough to produce simulations of the complex dynamics that unfold when galaxies collide. These four frames are from a film by Canadian astronomer John Dubinski, depicting the collision and merger of the Milky Way and the Andromeda Galaxy—a seemingly inevitable event, albeit one scheduled for about 3 billion years from now. In this "n-body simulation," in which the actions of dynamic systems of particles are mimicked while they are under the influence of physical forces such as gravity, more than 300 million particles were subject to Newton's laws of gravity in an attempt to see how the future collision may evolve. The resulting images, which end in the confused turbulence of a single merged galaxy—call it the Andromeda Way—look uncannily like photographs of galaxy collisions taken by the Hubble Space Telescope.

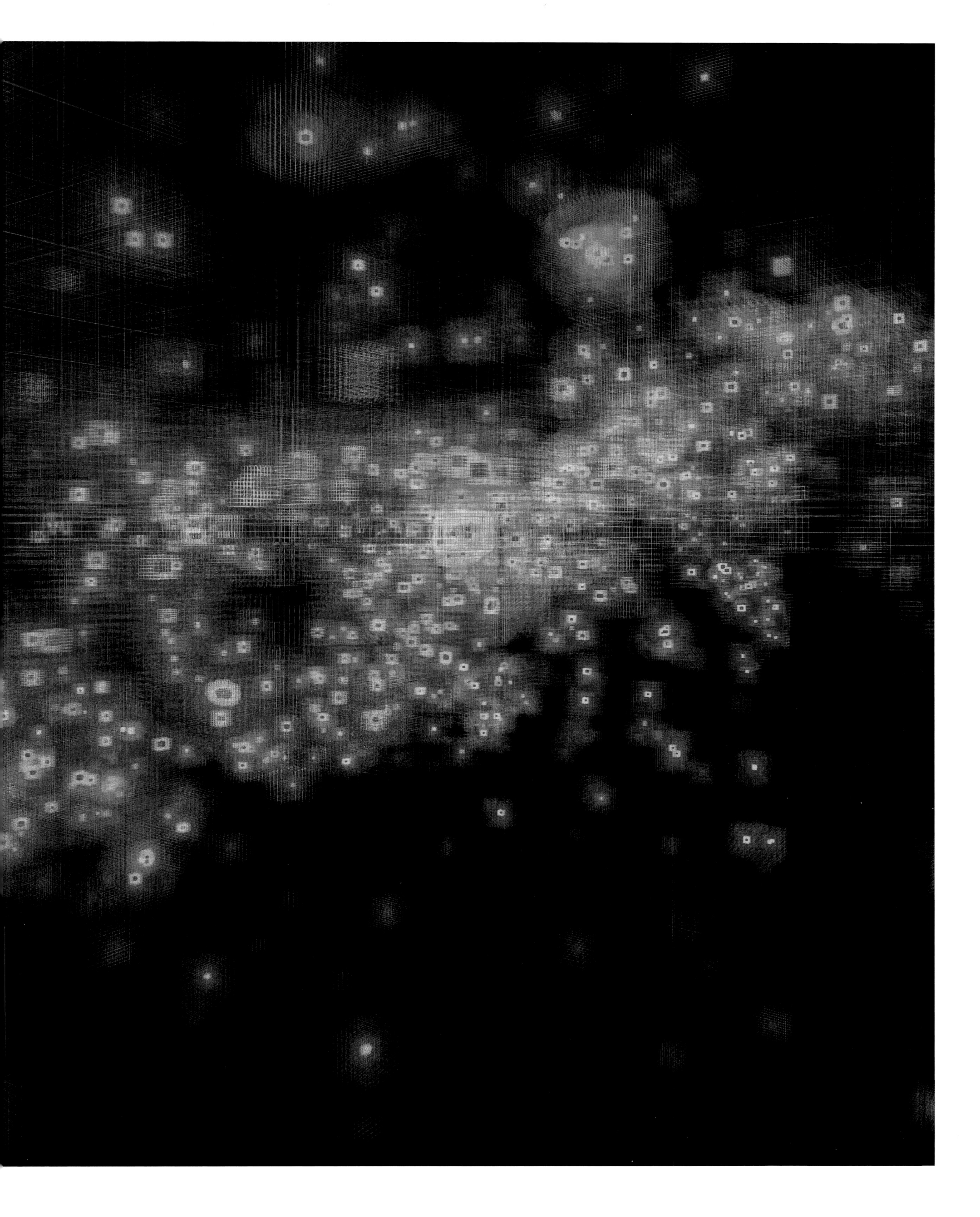

Distance from Earth (Megaparsecs)
rsecs)
10
100
1000
10^4
z=0.01
z=0.1
z=0.5
z=1
z=2
z=4
z=6
z=10
z=20
SDSS galaxies
SDSS quasars
First stars
Comoving future visibility limit
Big Bang
Cosmic microwave background
SDF z=6.578 galaxy
SDSS z=6.42 QSO
QSO 0957+561
QSO 0957 lens
3C 273
Coma Cluster
GRB 990123
Unreachable z=1.69
Decelerating
Accelerating z=0.76
Zone of avoidance
Sloan Great Wall
CfA2 Great Wall
Great Attractor
Antennae
M87
Sombrero galaxy
Whirlpool galaxy
M101
M81
M77
M74
Distance from Earth center (R_{Earth})
10^{17}
10^{18}
10^{19}
10^{20}

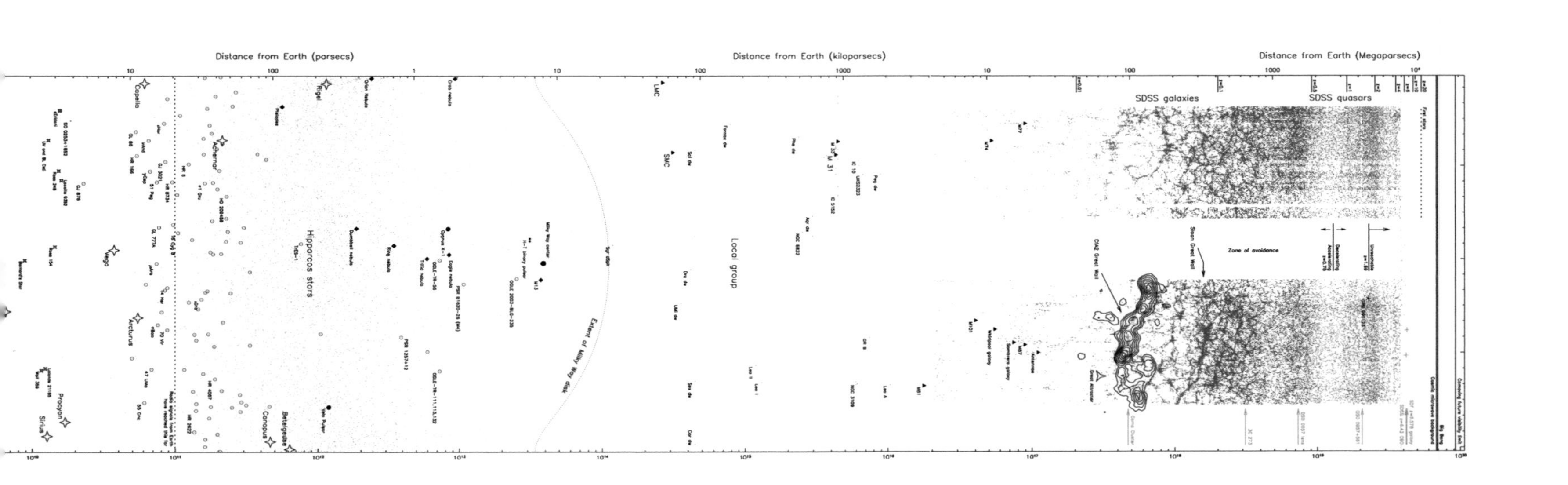

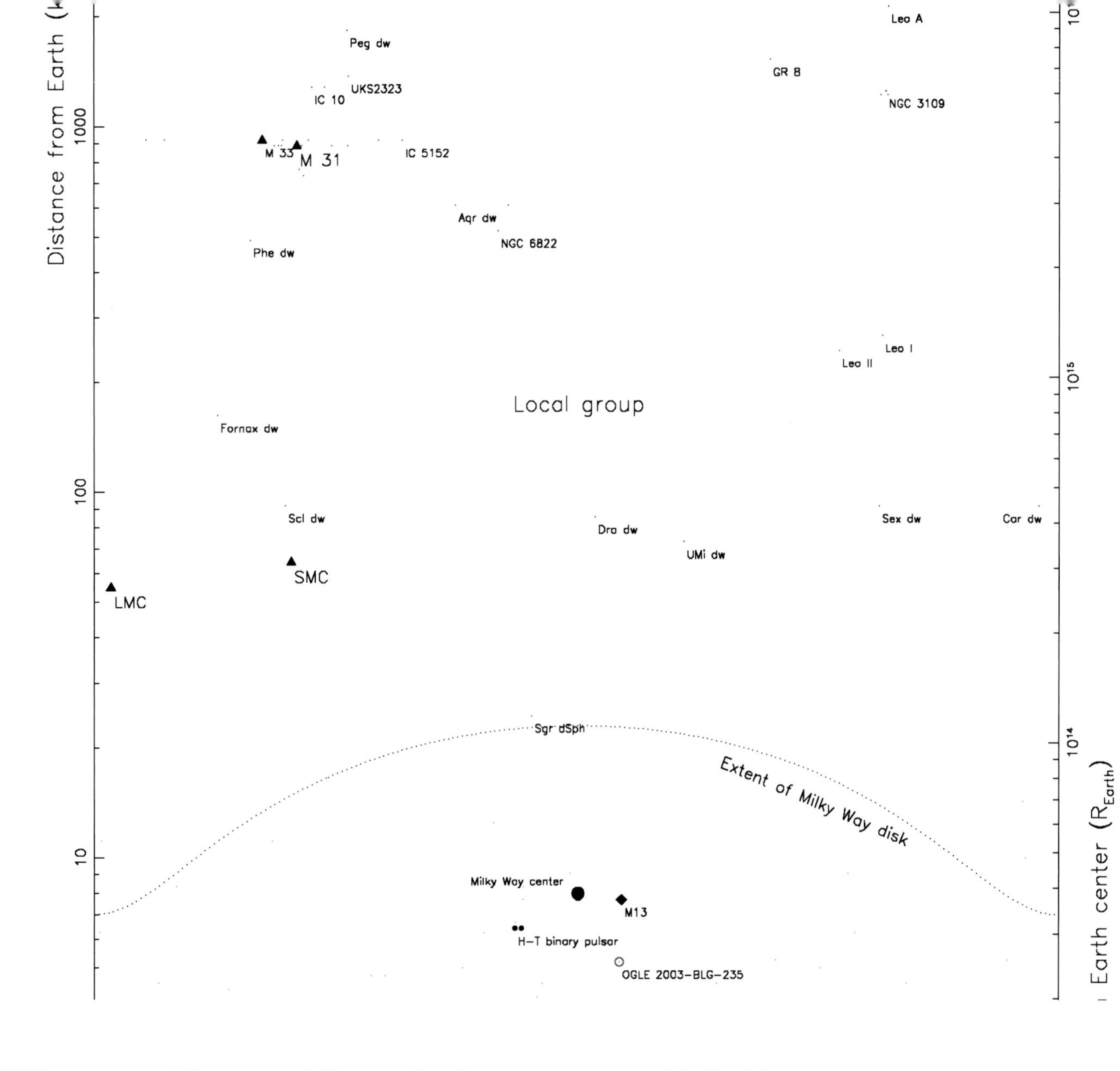

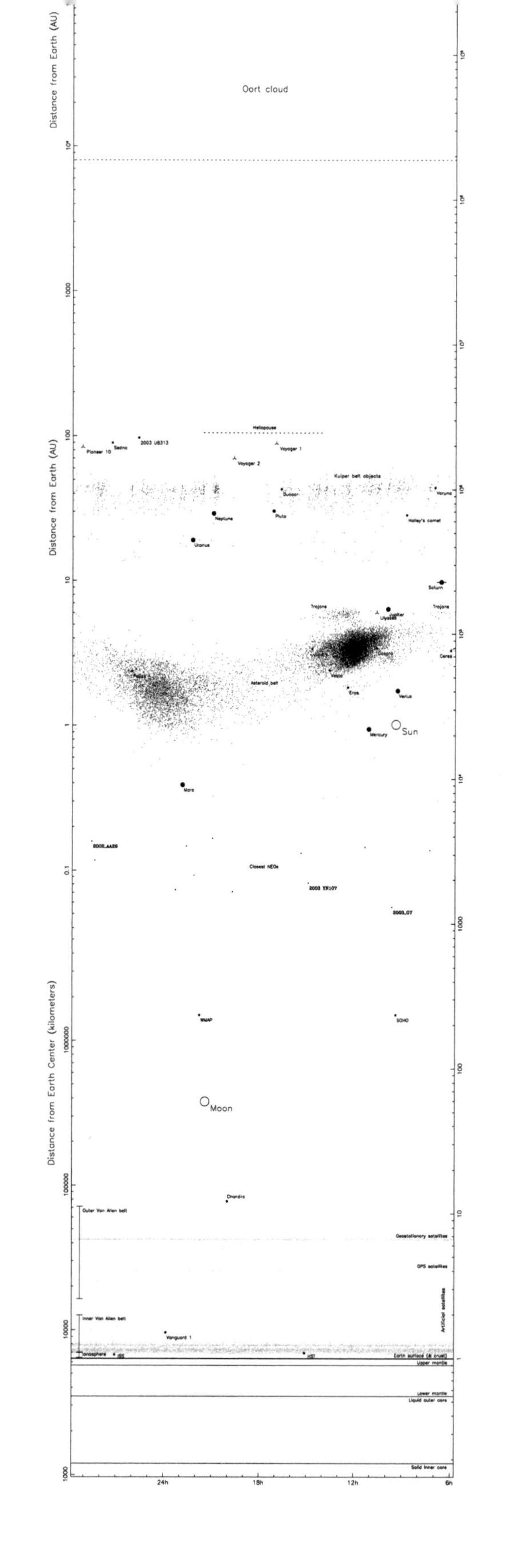

• 2003:

Right: Conformal map of the universe on logarithmic scales. **Above:** Detail view containing the top of that map, terminating in the first stars, cosmic microwave background radiation, and the Big Bang. Because depictions of the universe in a single projection had been rare and unsatisfactory, in 2005, Princeton cosmologist J. Richard Gott III and researcher Mario Juric set about making one adequate to the task. Although its extreme aspect ratio makes reproducing it a challenge (it's more than six and a half times taller than it is wide), Gott and Juric's map remains such an extraordinary document that it's presented here despite the miniaturizing effect caused by page height limitations. Citing Saul Steinberg's famous *New Yorker* cover illustration depicting a Manhattan resident's parochial view of the world—in which the buildings of Ninth Avenue are far larger and more important than the rest of the United States, the distant Pacific, or tiny China and Japan—Gott and Juric wanted to depict important objects close to the Earth as well as vast large-scale structures in the far distance. Their choice of a logarithmic scale, in which units of measurement increase exponentially, allowed them to squeeze everything onto one map. In plotting source data from the Sloan Digital Sky Survey, using their logarithmic projection technique, Gott and Juric discovered a giant wall of galaxies—here visible as the thickest blue line in the mass of galaxies on the upper right of the detail view. At 1.38 billion light-years long—about one-sixtieth the diameter of the visible universe—their Sloan Great Wall was the largest single structure yet seen. In discovering the structure while making their map, they reversed centuries of cartographic tradition, in which discoveries are made first and then charted later.

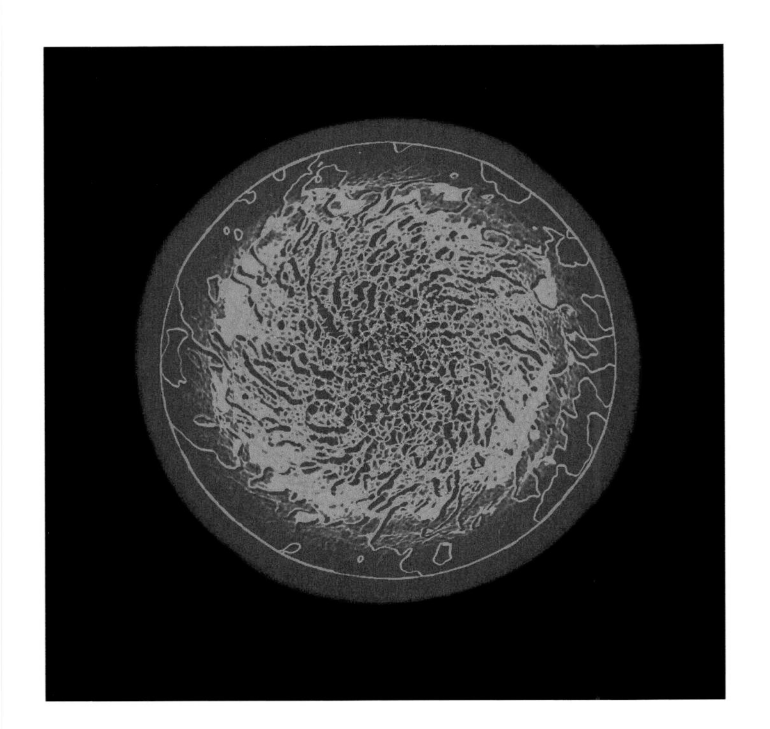

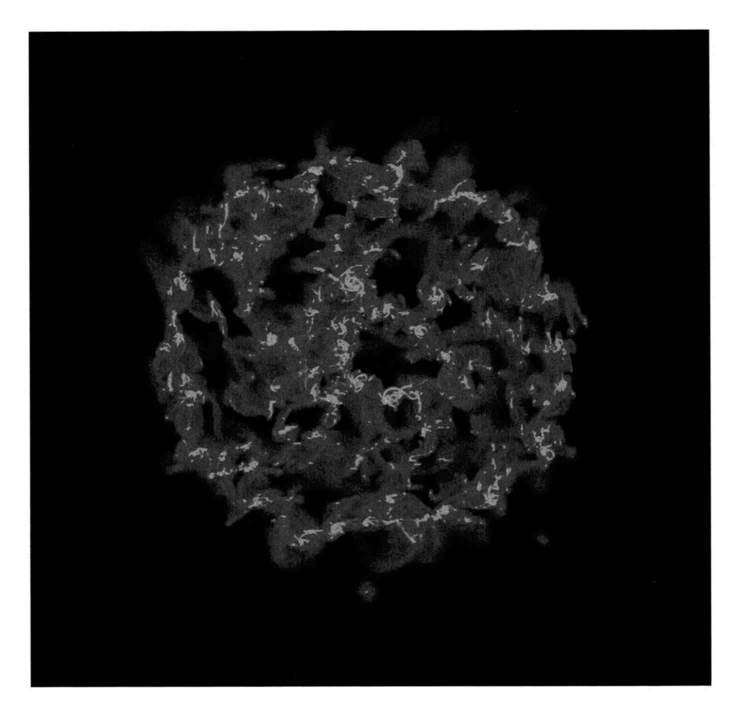

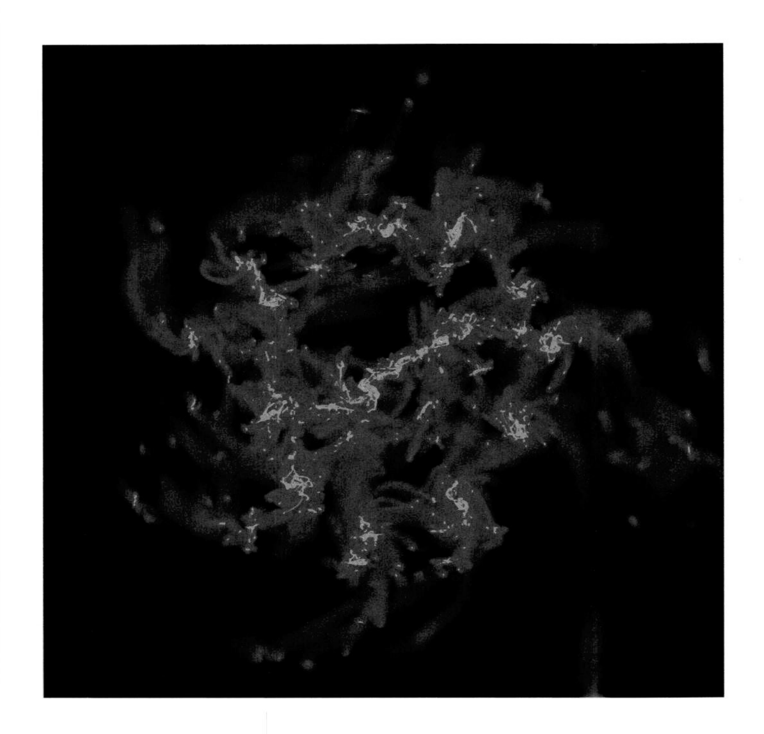

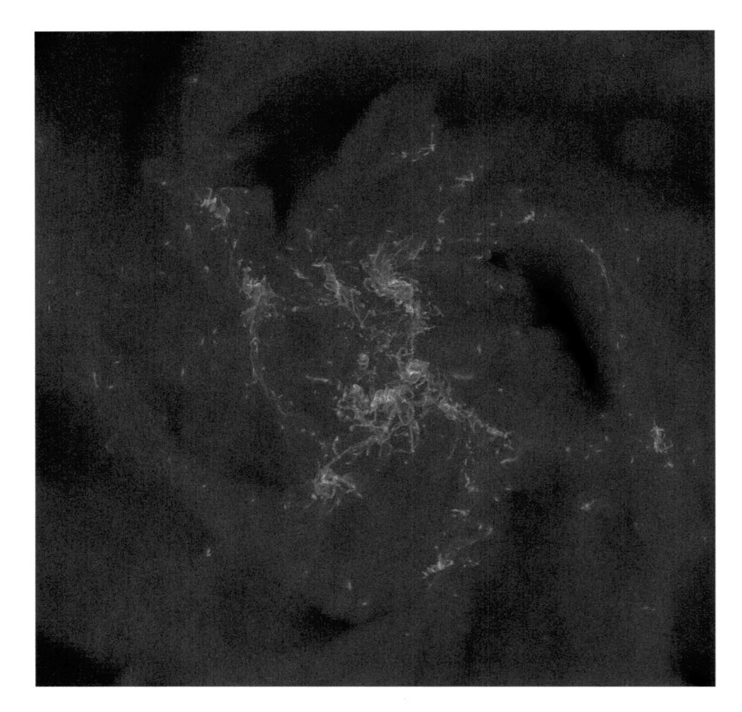

• 2010:

Four stages in the formation of a spiral galaxy are simulated in these supercomputer images by rendering baryonic gas density. Baryonic matter, which includes any kind of atoms or subatomic particles, is what makes up the ordinary material of the universe. The other hypothetical substance of the universe, dark matter, has been theorized by cosmologists as constituting the invisible mass that's having a gravitational effect on the visible material. Such simulations are used to understand the structure of the interstellar medium, the properties of molecular clouds, and the history of star formation in galaxies similar to the Milky Way, with a particular emphasis on the role of turbulence. These images required seven hundred processors of the Titane supercomputer at the CCRT computing center outside of Paris. **Top left:** A gas disc has coalesced from small progenitor clumps. **Top right:** The rotating galaxy is already showing incipient signs of structure. **Bottom left:** As it rotates, tendrils of gas extend outward due to centrifugal force. **Bottom right:** The simulated galaxy now exhibits spiral arms, a type of structure visible in galaxies throughout the universe.

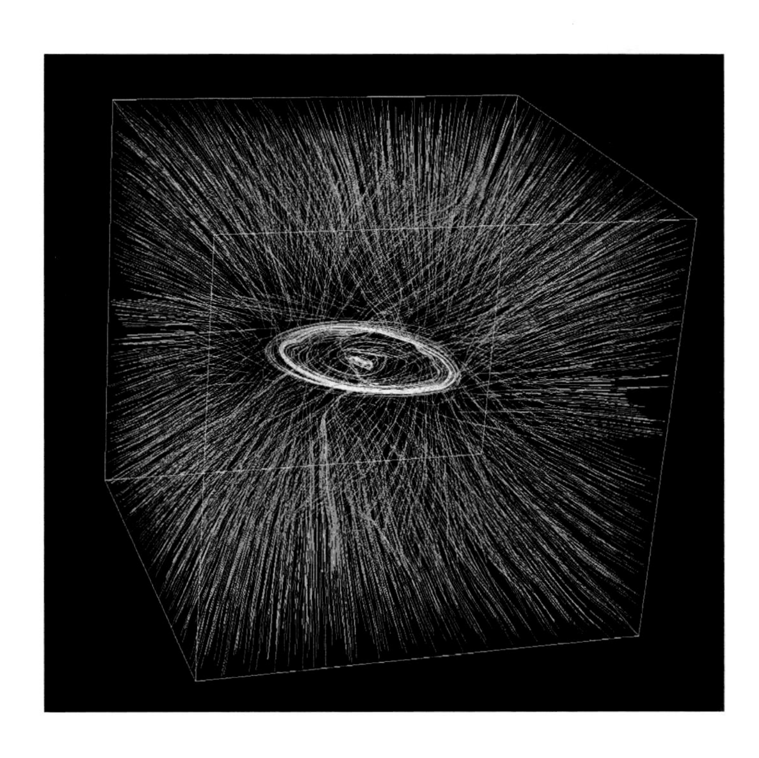

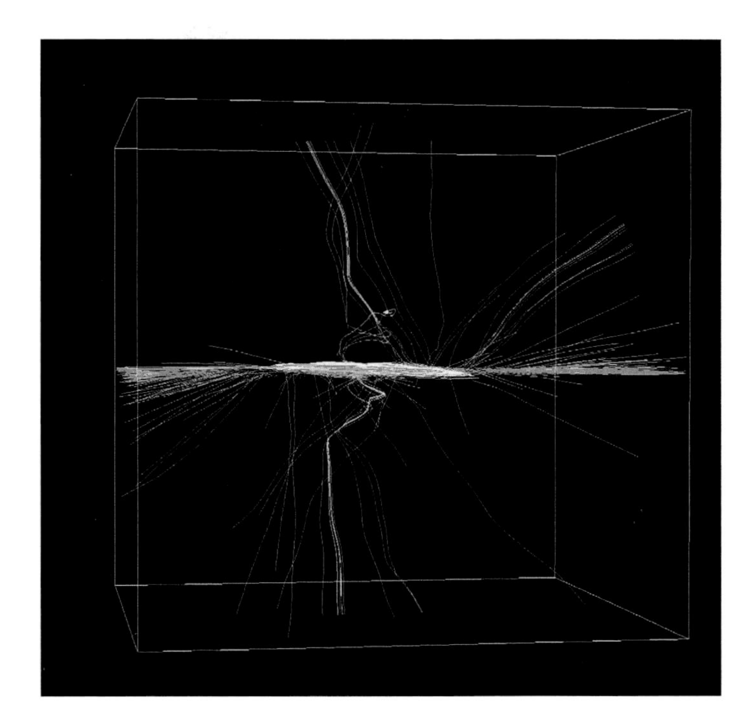

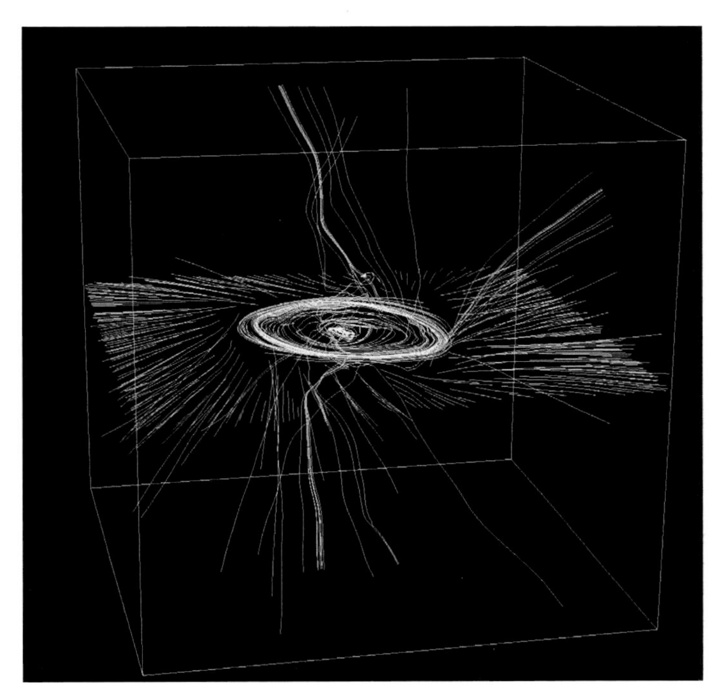

In these images, French astrophysicists are using supercomputer simulations to study star formation in gaseous clumps within high-redshift galaxies. The spidery white lines are "velocity streamlines," a way of plotting field flows in three-dimensional space, to map the kinetic dynamics of baryonic matter flows in a cube surrounding the simulated spiral galaxy (**top left**), and then on a plane within that cube (**top right and bottom left**). **Bottom right:** The last image combines the velocity streamlines with colorized depictions of the gas density in the simulated spiral (in red) and in the surrounding extragalactic space (in blue).

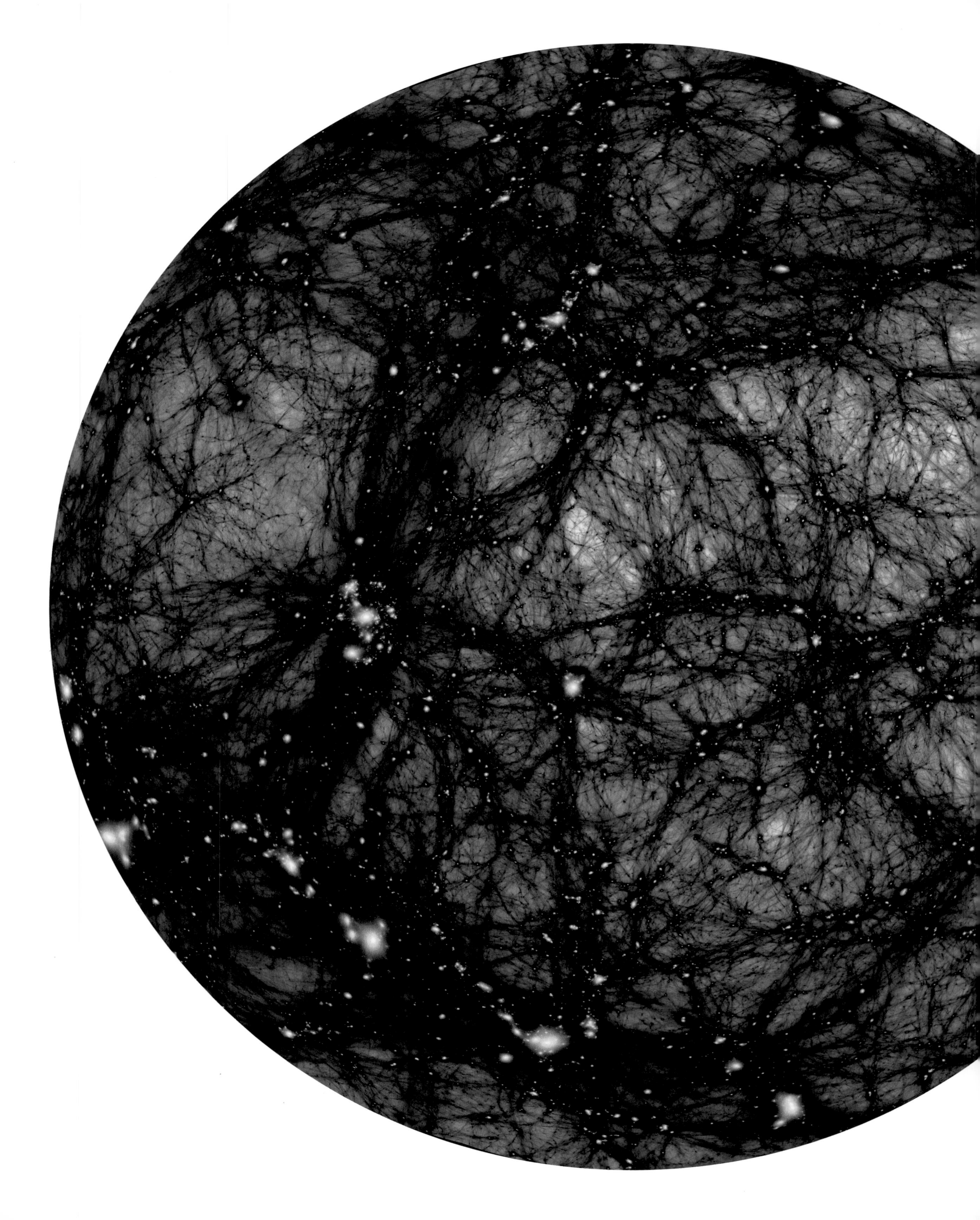

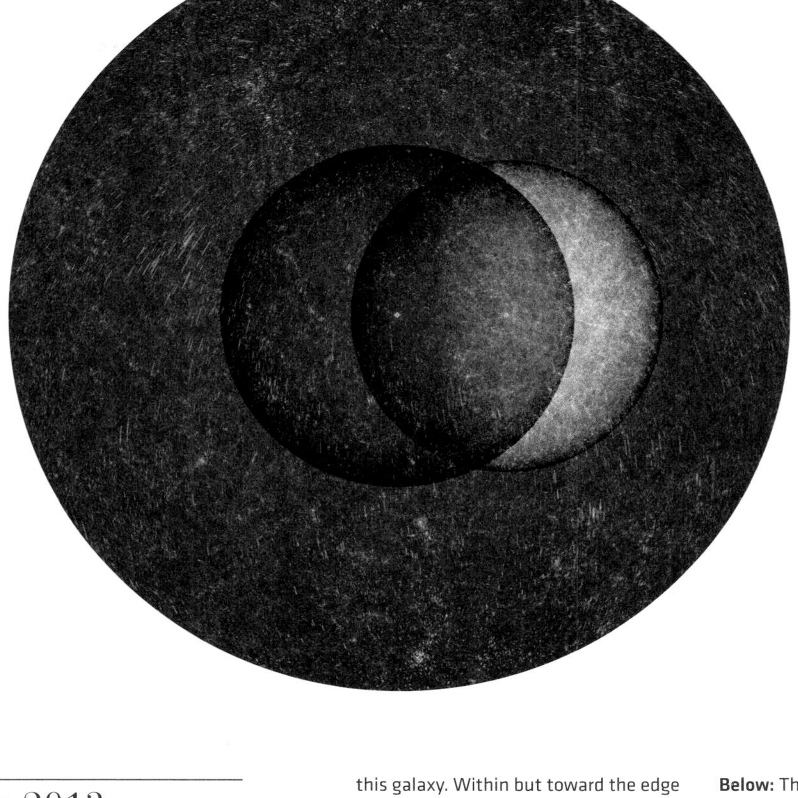

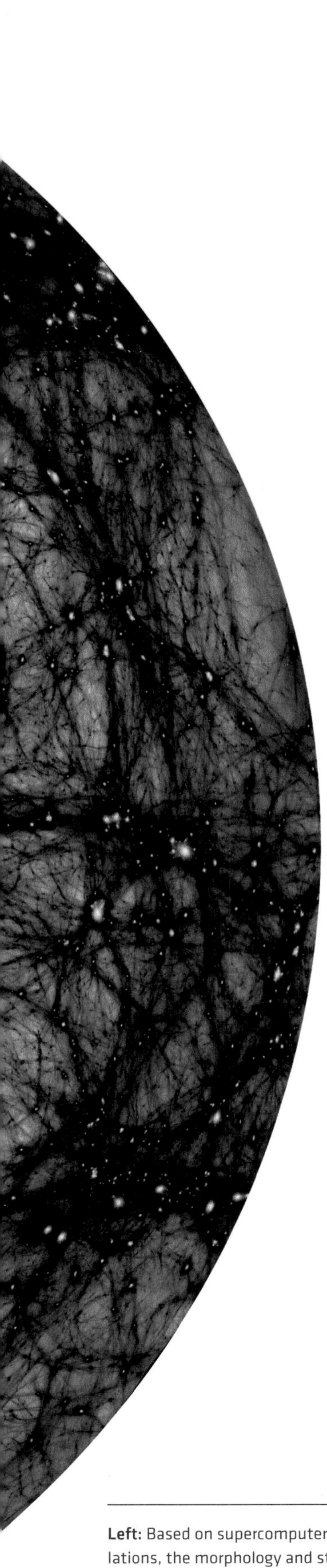

• 2013:

Above: Every individual location in the universe is at the center of its own bubble of the observable universe, as these two stills from the Hayden Planetarium show *Dark Universe* illustrate. The central bubble here simulates the entire observable universe as seen from this galaxy. Within but toward the edge of that bubble on the right is a galaxy at the center of another bubble. The yellow areas in that second bubble define areas visible from the second galaxy but that are invisible to us. (By the same token, areas on the left of the central bubble are invisible from the center of the one on the right.)

Below: Therefore, the universe is bigger—probably a great deal bigger—than the part visible from Earth. Each bubble here represents the part visible from any given point. The actual size of the universe is unknown, and may be unknowable. The radius of a single bubble of observable space is about 46 billion light-years.

Left: Based on supercomputer simulations, the morphology and structure of the universe at exceedingly large scales can be seen in this still from the Hayden Planetarium show *Dark Universe*, directed by Carter Emmart. Here galaxy clusters are arranged along weblike filaments of dark matter that extend between nodes of particularly high mass concentrations. The bright knots represent clusters of thousands of galaxies. Vast voids can also be seen between the denser areas. The light-year distance of the diameter represented here is about four hundred megaparsecs. With a parsec constituting 3.26 million light-years, that's more than 1.3 billion light-years across—or one-tenth the age of the Milky Way. To put that into perspective, the Earth is about 4.5 billion years old.

Right: In this depiction of "look back time" from the Hayden Planetarium show *Dark Universe*, galaxy clusters extend out in twin lobes from a central Earth, with the empty areas being artifacts of the masking caused by an obscuring Milky Way. (In reality, we have every reason to assume that a spongy, foamlike haze of more than 150 billion galaxies extends in every direction without substantial gaps.) The Earth is positioned somewhere at the center, and the outer rim of the picture represents the Big Bang singularity and the dawn of time itself. The bright knots represent clusters of thousands of galaxies. Vast voids can also be seen between the denser areas. *Dark Universe* was directed by Carter Emmart.

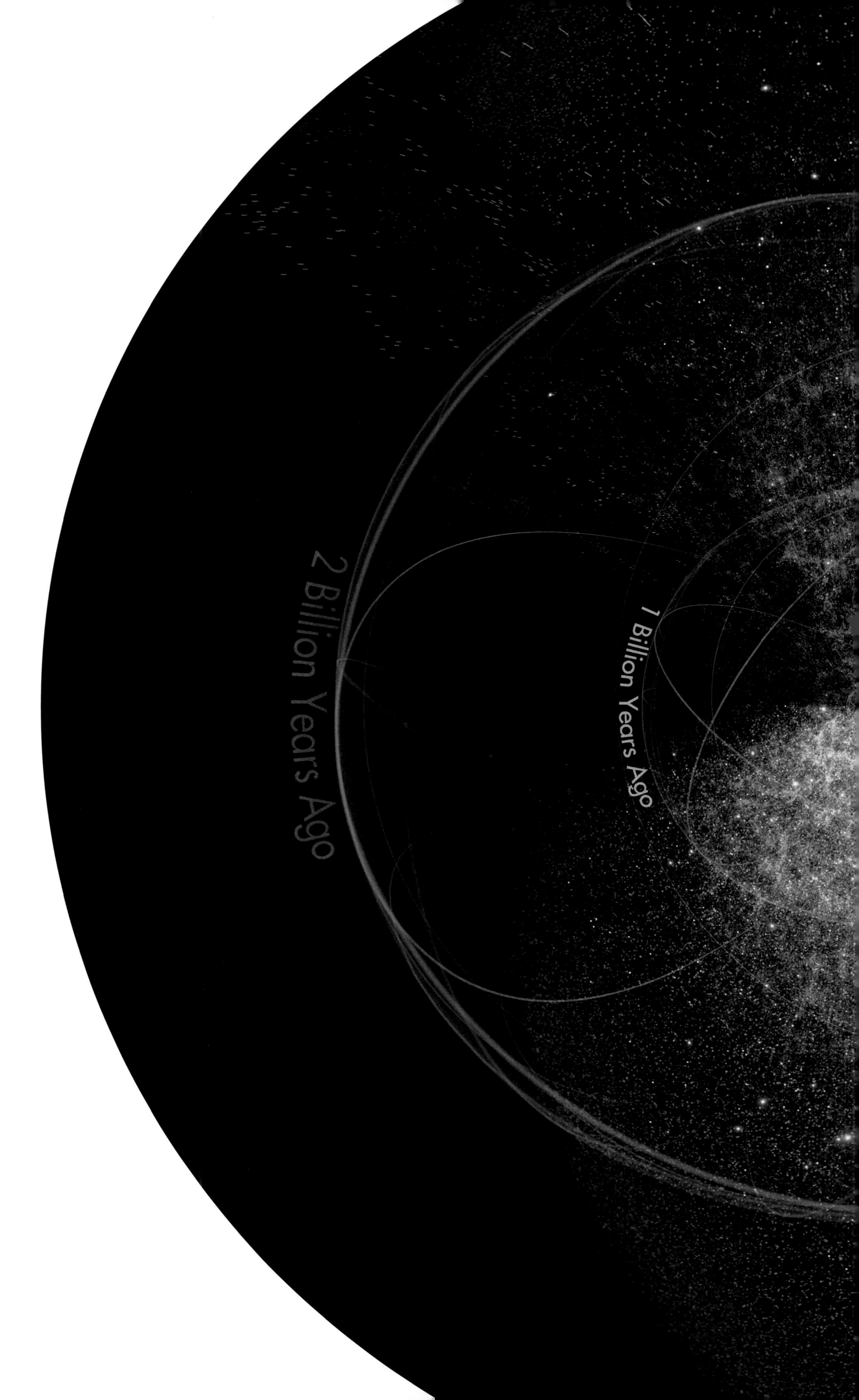

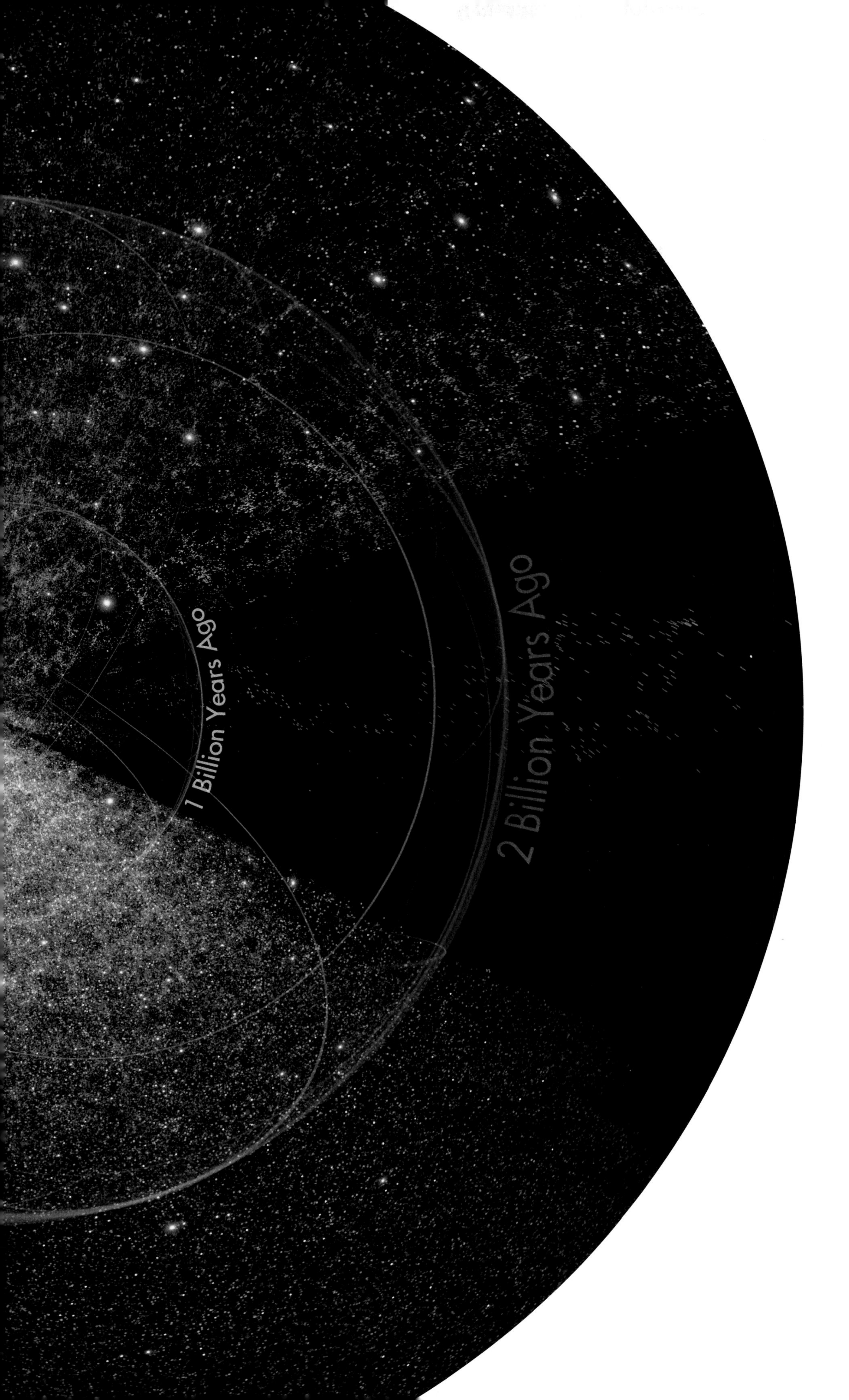

• 2014:

Overleaf: This staggering supercomputer visualization depicts for the first time gravitational flow lines knitting together a galaxy-spangled expanse of space-time well over 500 million light-years in diameter. Approximately thirty thousand galaxies are represented, with the red and black lines each associated with two distinct basins of gravitational attraction. As part of the Local Group of galaxies, which in turn is one component of the Virgo Supercluster, the Milky Way belongs to the black flow lines. These drain toward what has been called the Great Attractor, near the Norma Cluster, or directly above the green crescent shape at the center. The flows in red are associated with the Perseus-Pisces filament, our nearest neighboring large-scale structure, which is directly above the upper "Y" in the central nexus of red lines. In 2014, astronomer R. Brent Tully and his colleagues gave the new name "Laniakea Supercluster" to the full domain of the flows in black. A vast filamentary arch seems to connect the Laniakea and Perseus-Pisces structures; it surrounds the Local Void. The colors of the galaxy dots represent major structural components, with blue being galaxies that are part of the Perseus-Pisces filament; purple those belonging to the Pavo-Indus filament; green those in the historical Local Supercluster; orange the Great Attractor region; magenta the Antlia Wall and Fornax/Eridanus cloud; and gray being everything else. For the first time with this research, Tully and his collaborators are seeing the full outlines of the vast region of attraction that the Milky Way is a part of. However, it's important to mention that cosmic expansion is in fact the dominant force in the universe, so everything here is actually flying apart. The field lines are what can be discerned when that expansion is taken out.

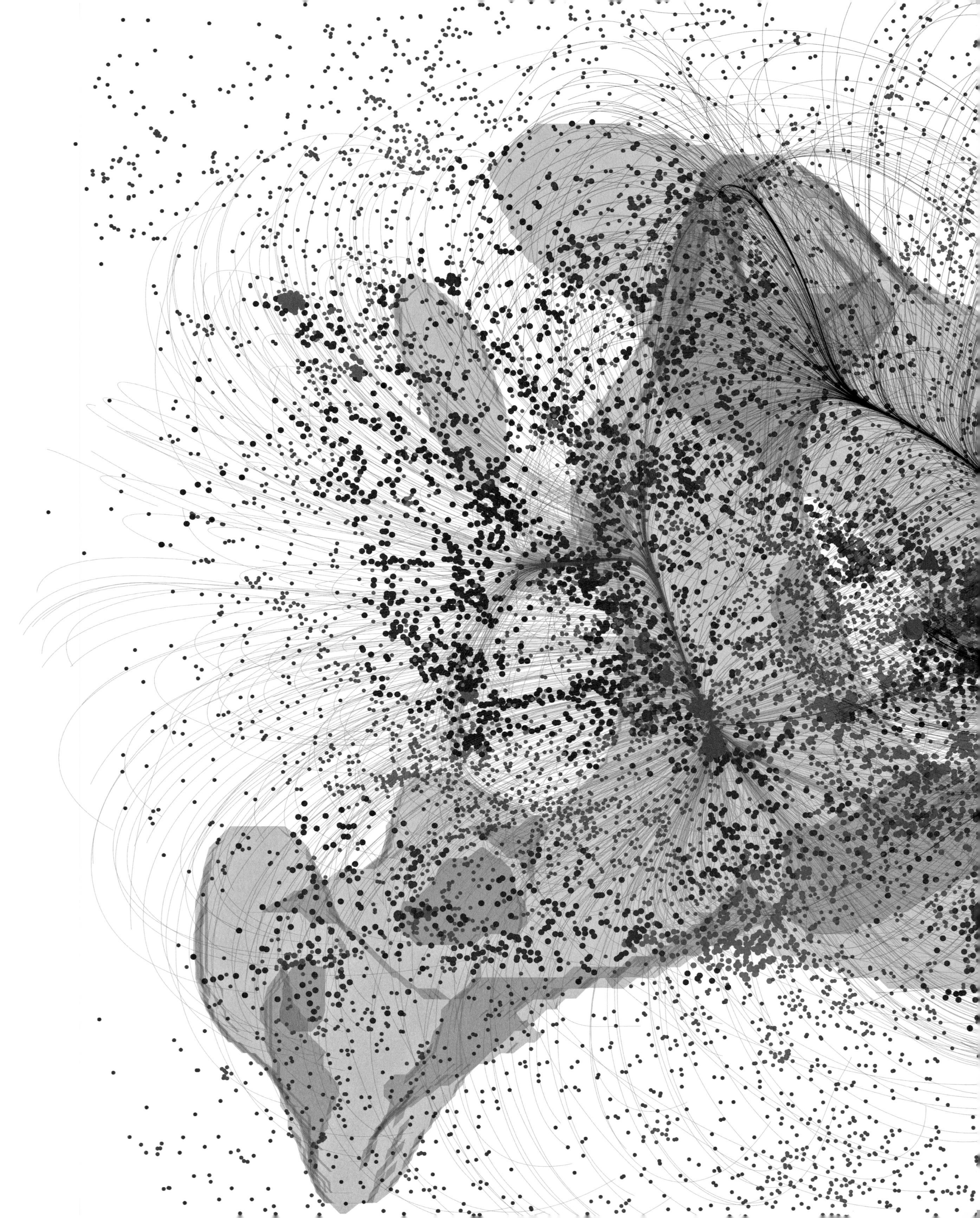

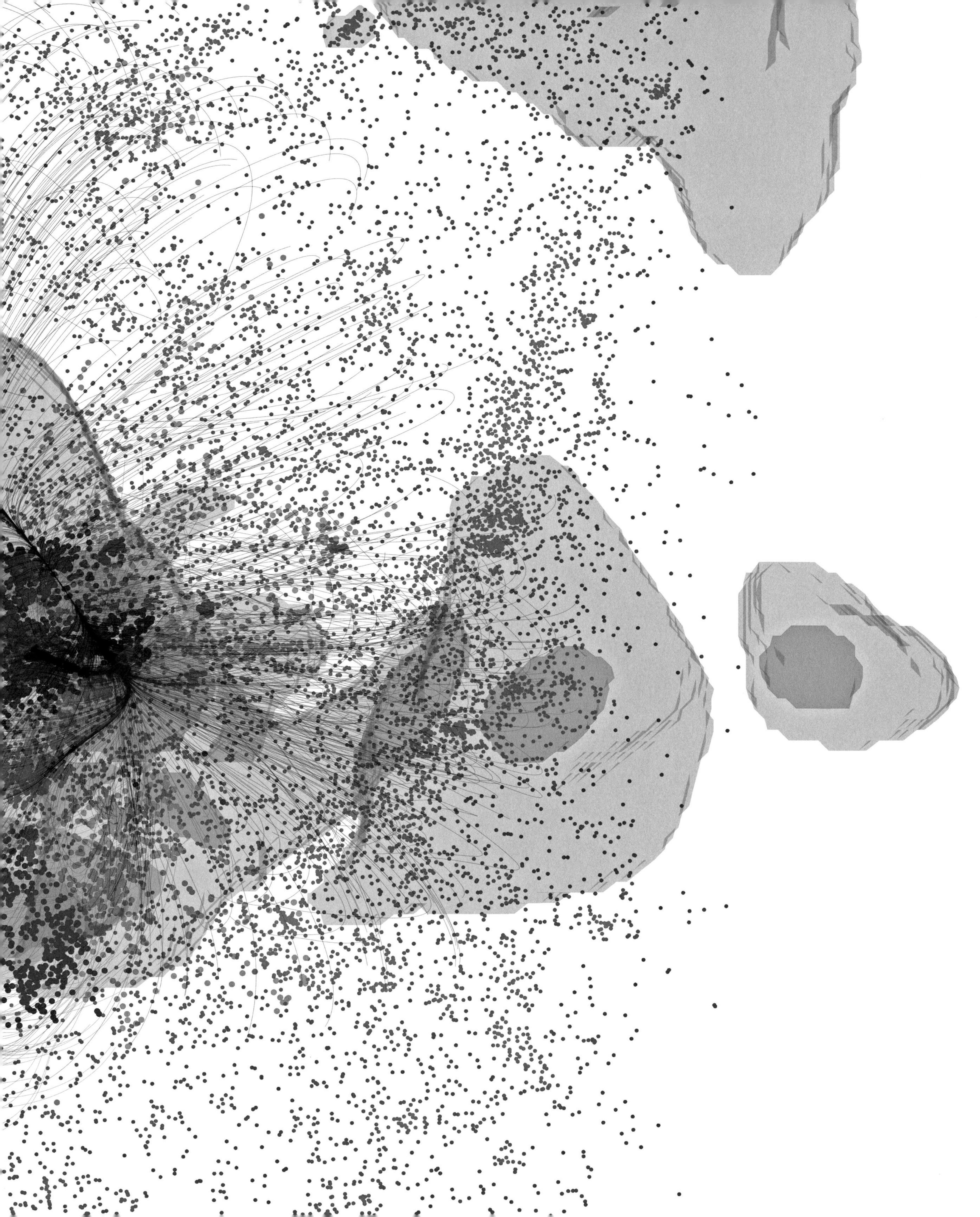

CHAPTER 6

Planets and Moons

And as if this was not enough,
you spin without a ticket in the carousel of the planets,
and along with it, dodging the fare, in the blizzard of galaxies,
through eras so astounding,
that nothing here on Earth can even twitch on time.

—Wislawa Szymborska, *Here*

UNTIL THE ARRIVAL OF THE TELESCOPE AT THE TURN of the seventeenth century, the five planets known since ancient times—Mercury, Venus, Mars, Jupiter, and Saturn—lived up to their Ancient Greek etymological roots: They were "wandering stars," or simply "wanderers"—points of light differentiated from the so-called fixed stars mainly because of their motions. They wandered up and down the band of the zodiac, and they wandered in an easterly direction against the stars, only occasionally backtracking for relatively brief periods of western, retrograde motion.

Their itinerant nature was the main source of their fascination, and starting in Babylonian times, records were kept of planetary motions. The Babylonians inherited their wandering-star preoccupation from the Sumerians, who associated leading gods in their polytheistic pantheon with the planets, the sun, and the moon.

This connection between celestial objects and capricious deities created a perceived necessity to predict what they would do next. Knowledge of astronomy also enabled a more reliable calendar, which was critical to the cycles of planting and harvesting integral to Babylonian civilization. All this fostered a class of astronomer-priests—essentially calendar keepers doubling as soothsayers. This powerful scholarly caste was believed able to predict the future based on interpreting celestial motion. Although their belief system, which is at the fountainhead of contemporary astrology, has no relevance to contemporary science, the techniques developed by the ancient Babylonians to track and predict celestial motion lie like a ziggurat foundation at the base of modern astronomy and mathematics.

The Babylonian Astronomical Diaries are the lasting cuneiform legacy of an effort to systematize celestial phenomena and give it a mathematical basis. They extend across six centuries, constituting an unbroken systematic record of planetary, lunar, and solar motion. If we exclude the Nebra Sky Disc discussed in chapter 3, the oldest fragment of recorded astronomy in existence is a Babylonian cuneiform tablet containing twenty-one years' worth of data on the rising and setting times of planet Venus. It's the first evidence that the motions of the planets were understood to be periodic, and though the tablet itself dates from the seventh century B.C., the information it contains is thought to come from the mid-seventeenth century.

The Venus tablet belongs to a series of seventy or so tablets called the *Enuma anu enlil*, which were the basis of regular astrological advisories provided to the king of the Neo-Assyrian Empire. It's easy to look at the omen-divination system underpinning the work of the Babylonian astronomer-priests and make the mistake of dismissing it as merely a kind of superstitious quackery. In fact, a close association existed between mathematics, astronomy, and astrology for most of history. As mentioned in the introduction, some of the towering figures of the Scientific Revolution spent well over half of their time involved in astrology and alchemy, including Galileo, Kepler, and Newton. In many ways, the horoscopic belief system served as a kind of motor driving the empirical data collection.

With the conquest of Mesopotamia by Alexander the Great in 331 B.C., Babylonian astronomical knowledge was systematically assimilated by the Greeks. The names we use for the five planets known to the ancients come from Roman mythology, but they're translations from the Greek, so both the generic term *planet* and most individual planetary names date back to the Hellenistic period. The Greeks classed astronomy as a branch of mathematics and continued the rigorous Babylonian mathematical approach to divining planetary motion, although they added the geocentric, multisphere Aristotelian cosmological model.

With Galileo's use of the telescope starting in the winter of 1609–10, the planets were exposed to scrutiny by the technologically assisted human eye for the first time, and wandering stars were transformed into worlds. When Galileo first observed Jupiter on the night of January 7, what immediately caught his attention was the presence of what he took to be three stars arranged in a line along the planet's equator. At first, he assumed this to be a chance alignment, but on subsequent nights he saw not only that they had tagged along with Jupiter as the planet traveled on its then retrograde westward course against background stars—but that a fourth had joined them as well. As they continued to appear and disappear on subsequent evenings, he realized he'd discovered the first extraterrestrial planetary moons ever seen. (See page 181.) Jupiter's four major moons are now rightfully called the Galilean satellites.

Saturn's rings, which weren't well resolved by Galileo's thirty-power telescope when he first observed the planet on July 15, 1610, proved confounding to the astronomer. At first, he thought they were subsidiary planets at Saturn's side, creating a tightly packed triple body, though he noted that when observed under lower magnification, they assumed what appeared to be a single ovoid shape—which as a true Italian he compared to an olive. Although Galileo wanted to confirm his conclusion, he was also concerned with establishing his priority over the discovery. The solution he hit on was to send an anagram to a network of fellow astronomers: a single word of thirty-seven letters. Deciphered and broken into four words in Latin, it meant "I have seen the highest planet [Saturn] triple-bodied."

As he continued observing Saturn, however, Galileo couldn't confirm a tripartite planet. Instead, what we now know to be Saturn's rings gradually lowered to an edge-on angle as seen from Earth. From his point of view, this meant that two of Saturn's ostensibly three bodies had performed a vanishing act—leaving a single planet comparable to Jupiter. Saturn's behavior was the single most mystifying thing Galileo witnessed during his entire career as an observational astronomer, and he never managed to explain it.

It wasn't until Christiaan Huygens observed the planet between 1655 and 1656 with his new fifty-power telescope that a potential solution to Saturn's mysterious appendages started to emerge. (The Dutch astronomer also discovered Saturn's largest moon, Titan, during these observations.) Like Galileo, however, Huygens wasn't entirely sure of his conclusions, and wanted to establish his priority without potential embarrassment. In March of 1656, he one-upped Galileo with an anagram of sixty-two letters. When deciphered, they resolved themselves into nine Latin words, translated as "It is encircled by a thin, flat ring, nowhere attached, inclined to the ecliptic."

Huygens had hit on the truth, as subsequent observations of Saturn duly confirmed.

Planetary discoveries came thick and fast throughout the next two centuries, with some of them actually being planets themselves. On March 13, 1781, William Herschel announced the discovery of what would ultimately be named Uranus. It was the first new planet in history—all the others had always been visible to the naked eye—and the first discovered with a telescope.

Neptune followed less than a century later, when an observation by astronomer Johann Galle on September 23, 1846, confirmed a prediction by French mathematician Urbain Le Verrier. Credit for the discovery went to Le Verrier: He'd been informed of an unexplained perturbation in Uranus' orbit, and had used Newton's laws of universal gravitation to determine the eighth planet's likely position.

So while Uranus provided the first telescopic discovery of a new planet, Neptune was in fact discovered using celestial mechanics, and merely confirmed by telescope. Here, the reach of Isaac Newton's genius might be usefully pointed out. Both Herschel and Le Verrier rightfully received credit for their historic discoveries. But the telescope Herschel was using in March of 1781, half a century after the English physicist's death, was based on Newton's original reflector design. And the laws of gravity that permitted Le Verrier's discovery had been formulated by Sir Isaac as well.

The first decade of the nineteenth century could boast four more "planetary" discoveries in rapid succession: Vesta, Juno, Ceres, and Pallas. All were considered planets until the 1860s, when a tide of discoveries of ever-smaller objects in similar orbits demoted them to the rank of asteroids. They're the four largest objects in a ring-shaped zone of tumbling rocks between Mars and Jupiter now called the asteroid belt. All are still considered asteroids except Ceres, which, following a decision by the International Astronomical Union in 2006, is now considered a dwarf planet—the only one in the inner solar system.

While Ceres received a promotion after more than a century of asteroidal status, another solar system object previously classed as a planet was not so fortunate. Tiny Pluto, which had been discovered by astronomer Clyde Tombaugh in 1930—the only solar system planet discovery of the twentieth century—was unceremoniously stripped of its rank by the IAU in 2006, also receiving dwarf planet status. Much as had happened with the asteroid belt in the mid-nineteenth century, twenty-first-century astronomers recognized that a new class of small planetary bodies had been discovered in what we now call the Kuiper belt, after astronomer Gerard Kuiper—a vast dispersed outer district of frozen objects left over from the formation of the solar system. Icy Pluto, which is only 70 percent the size of the moon, in effect patrols the inner part of that region, and is now considered the largest known Kuiper belt object. Three other Kuiper dwarf planets have officially been recognized, though hundreds may exist.

The arrival of the space age at the end of the 1950s transformed some astronomers into planetary scientists by vaulting their telescopes toward their objects of study in the form of robotic interplanetary spacecraft. Throughout the next half century and more, a prodigious amount of information was acquired about the worlds orbiting the sun. The geological maps presented in this chapter can only hint at the richness and scope of what has been achieved by these missions. We now have at least a provisional knowledge of all the planets and major moons of the solar system—even if much more needs to be done.

Meanwhile, hundreds of new worlds were coming into view. After centuries of speculation by Giordano Bruno, Thomas Wright, and many others, the first confirmed exoplanets—worlds orbiting other stars—were discovered during the last two decades. Because the method used to recognize them initially relied on measuring their gravitational influence on their parent stars, at first a majority were massive gas giants orbiting in very close orbits. But as detection methods grew more sophisticated, these "hot Jupiters" were joined by smaller and smaller planets, many of which are assumed to be "terrestrial," with hard surfaces like Mars, Earth, Venus, and Mercury.

In 2009, NASA launched Kepler, a space telescope designed to hunt for distant worlds in transit by detecting minute variations in the brightness of stars. Kepler soon produced a bumper crop of exoplanets. As of spring 2014, about 1,800 have been confirmed, with several thousand more awaiting verification by joyously overworked astronomers. A staggering 100 billion planets are now thought to exist in the Milky Way, with uncountable trillions more certainly scattered across the universe. It seems reasonable to believe that another green world orbits among them.

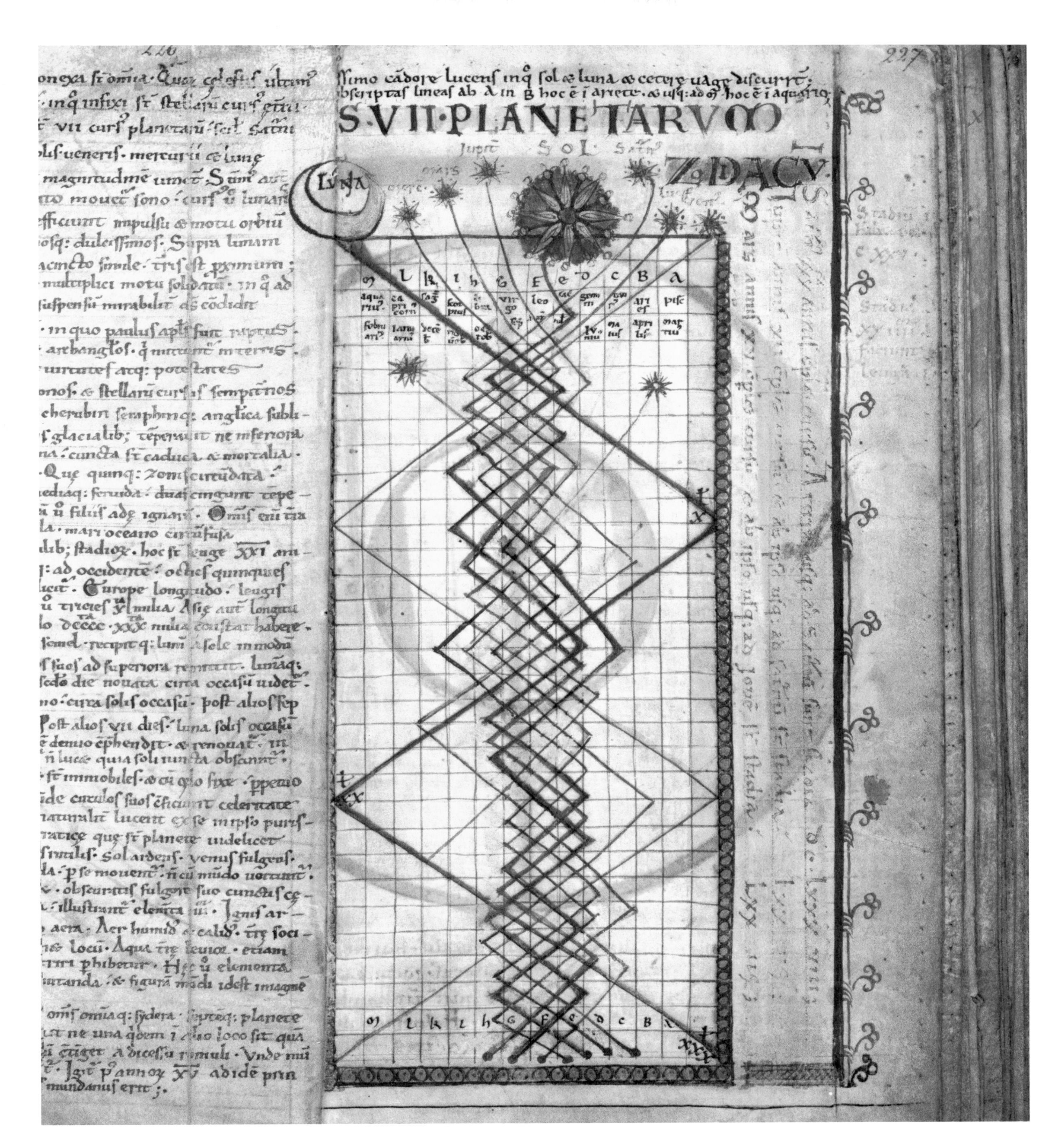

• 1121:

This extraordinary diagram representing celestial motion through time is from the medieval encyclopedia *Liber floridus* (Book of Flowers), and draws on an earlier tradition of such diagrams dating back to Pliny the Elder. Titled simply "Seven Planets," its zigzag lines trace the movements of the sun, the moon, and the five planets then known in relation to latitude lines of the zodiac. (The zodiac, a circle divided into twelve sections, centers on the ecliptic, or apparent, path of the sun across the celestial sphere during the course of a year.) Here, the sun is represented by a bright flower, the moon by a crescent, and the other planets by small stars. The star directly opposite the moon at the top of the graph—and connected to a zigzag line that's the mirror image of the moon's—represents planet Venus, with the symmetry of the two lines probably more decorative than real. Directly to the left of Venus is planet Lucifer; its presence here recalls a time when Venus was thought to be two planets, morning star Lucifer and evening star Hesperus, but by the time of this encyclopedia, the two were understood to be one celestial body, so its presence is unexplained. Also unexplained are the two star shapes that don't make it to the top of the graph. Such diagrams in hand-illuminated manuscripts can have a shocking effect when they appear. They're not unlike the sudden materialization of a Mies van der Rohe skyscraper at the center of a half-timbered medieval village—early artifacts of what architectural theorist Dalibor Vesely has called the "mathematization" of reality.

• 1444–50:

In the *Paradiso* section of Dante's *The Divine Comedy*, Dante and his guide, Beatrice, ascend beyond the moon, moving through the spheres then thought to bear the planets as they rotated around a central Earth. In these manuscript illuminations by Sienese master Giovanni di Paolo, Dante and Beatrice visit the "Heaven of Venus" (**top**), Mars (**above**), Jupiter (**facing page top**), and Saturn (**facing page bottom**). As he describes his journey through the planets, Dante reveals a knowledge of Ptolemaic astronomy, referring for example to the "third epicycle" when describing the "planet that courts the sun," or Venus. For other works by di Paolo, see pages 33, 75, 116, 144, and 256.

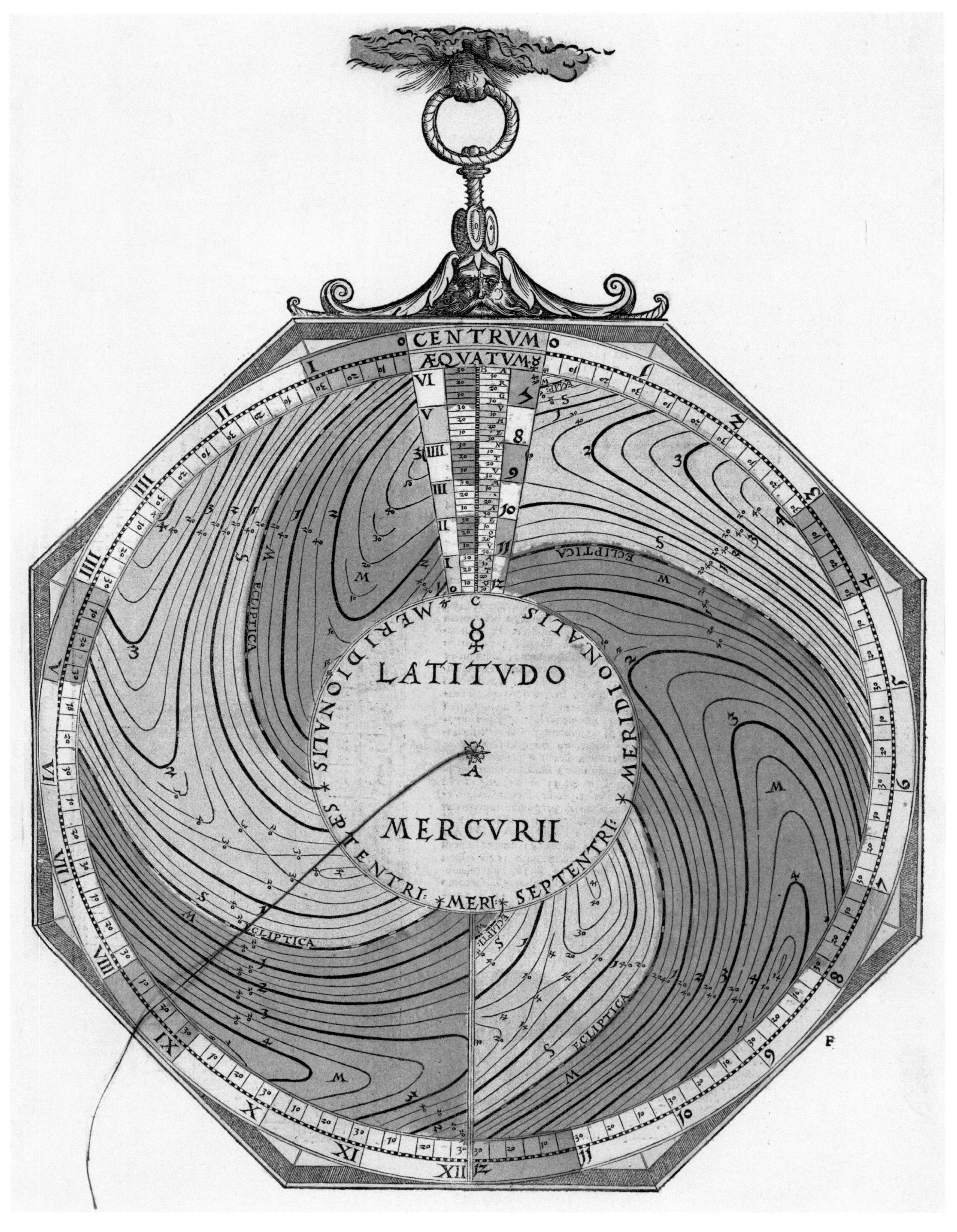

• 1540:

This disc from German printer and cosmographer Peter Apian's *Astronomicum Caesareum* (Caesar's Astronomy) allows the user to calculate the latitude of Mercury at any time during the year. By and large, the planets all move in a westward diurnal motion along with the background stars, while simultaneously gradually creeping eastward among them, until they return to something like their original position. And they largely stay within the zodiac, that imaginary band of sky eight degrees above and below the path of the sun. But the motion of all the planets is interrupted annually by brief episodes of "retrograde" westward motion, with the duration of that motion different for each. Mercury switches from an eastward to a westward direction every 116 days. When turned, the lines of the waveform visible in this volvelle allowed the user to compensate for variations in Mercury's apparent motion and position it with accuracy on the ecliptic. For more from Apian, see pages 43, 77, 119, 227, and 262.

orientales, & vna occidentalis in tali dispositione. O-

Ori. * * O * Occ.

rientalior, quæ satis exigua erat à sequenti distabat min: 4. media maior à Ioue aberat min: 7. Iuppiter ab occidentali, quæ parua erat distabat min. 4.

Die decima hora prima min: 30. Stellulæ binæ admodum exiguæ orientales ambæ in tali dispositione visæ

Ori. * *O Occ:

sunt: remotior distabat à Ioue min: 10. vicinior verò min: 0. sec. 20. erantque in eadem recta. Hora autem quarta, Stella Ioui proxima amplius non apparebat, altera quoque adeo imminuta videbatur, vt vix cerni posset, licet aer præclarus esset, & à Ioue remotior, quam antea erat, distabat, siquidem min: 12.

Die vndecima hora prima aderant ab Oriente Stellæ duæ, & vna ab occasu. Distabat occidentalis à

Ori. * * O * Occ.

Ioue min. 4. Orientalis vicinior aberat pariter à Ioue min. 4. Orientalior vero ab hac distabat min. 8. erant satis perspicuæ, & in eadem recta. Sed hora tertia

Ori. * * *O * Occ.

Stella quarta Ioui proxima ab oriente visa est, reliquis

minor

• 1610:

Previous page: On the night of January 7, 1610, a professor of mathematics at the University of Padua turned a telescope of his own design toward Jupiter—one of the "wandering stars," or planets, of the ancients. Depending on which one he was using that night, Galileo Galilei's telescope was capable of magnifying by twenty or thirty times. Either would have been enough to just make out a planetary disc, but what immediately caught Galileo's attention was the presence of what he took to be three stars arranged in a line along the planet's equator, with two on the eastern and one on the western side. At first he assumed this to be a particularly fortuitous asterism, or chance alignment. But on subsequent nights, he observed not only that they had tagged along with Jupiter as the planet traveled on its then retrograde westward course against background stars, but that a fourth had joined them as well! This page from Galileo's book *Sidereus nuncius* (Starry Messenger) details four evenings spent observing the motions of what are now known as the Galilean satellites. Along with Venus exhibiting phases like the moon, something he observed later the same year, this was the first substantial empirical evidence in support of Copernicus's theory that the sun was at the center of the solar system and the Earth moved. If Jupiter could keep its satellites while moving, Earth could as well.

• 1660:

Right: In this dramatic, seemingly Modernist plate from Andreas Cellarius's otherwise Baroque *Harmonia macrocosmica*, we see a mid-seventeenth-century (but still Ptolemaic) idea of the different sizes of various celestial objects. These start from front to back with tiny Mercury, almost invisible at the bottom of the long "thermometer" running down the center of the plate; then the moon and Venus, which are depicted as almost the same size (and practically indistinguishable within the yellow ring at the bottom); then on to Earth, the larger blue-green sphere; and Mars, right behind it in orange; and so on. The vertical scale is divided into units of one terrestrial diameter each. In the back are the assumed sizes of several stars, with the sun at the very back, as the biggest assumed celestial object of all. These size estimates range from being badly off to very wrong indeed. At the margin, one of the playful winged putti has caught a bird and put it on a string—or maybe it's the other way around. For other maps from Cellarius, see pages 47, 84, 122, 146–47, 148–49, and 232–35.

RUM
STIUM
TUDINES.

CORPORUM MAXIMI ORBICULARIS CIRCUITUS, ET MAGNITUDO

9000 MILLIARIA GERMANICA.

V. DIAMETRI TERRÆ.

MAGNITUDO STELLARUM PRIMÆ MAGNITUDINIS

MAGNITUDO, ET ORBICULARIS CIRCUMFERENTIA

SECUNDÆ MAGNITUDINIS, EANDEM FERE MAGNITUDINEM HABENTIUM

STELLARUM ORBICULARIS CIRCUITUS ET MAGNITUDO

IV DIAMETRI TERRÆ.

MAGNITUDO STELLARUM QUARTÆ MAGNITUDINIS

6000

UDO ET CORPUS STELLARUM QUINTÆ MAGNITUDINIS

III. DIAMETRI TERRÆ.

5000

MAGNITUDINIS CORPUS, ET CORPORALIS AMBITUS

TÆ MAGNITUDINIS

4000

II DIAMETRI TERRÆ.

3000

MARTIALIS MAGNITUDO, ET CIRCUMFERENTIA

I. DIAMETER TERRÆ.

CORPORIS TERRENI CIRCUMFERENTIA, ET MAGNITUDO

DIAMETER TERRÆ.

1000

van Loon f.

Apud GERARDUM VALK, et PETRUM SCHENK, Amstelædami.

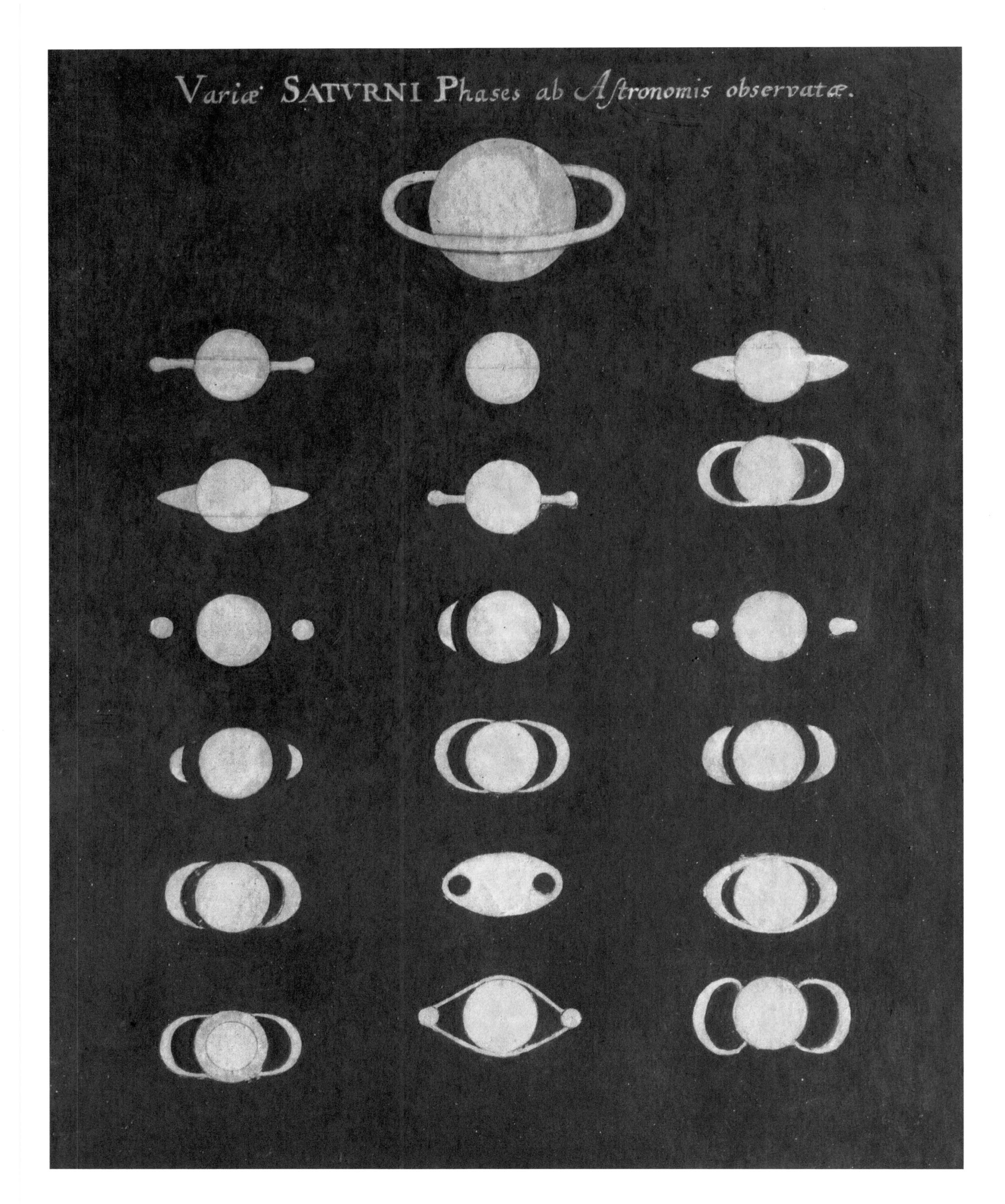

• 1693–98:
Women did in fact sometimes find a route to becoming astronomers in the seventeenth century—albeit very rarely—with German astronomer-artist Maria Clara Eimmart being one of them. The daughter of a Nuremberg artist and amateur astronomer, Eimmart based these depictions of Saturn's mysterious shape-shifting on a 1659 engraving by Dutch astronomer Christiaan Huygens. (She also did many paintings based on her own observations.) Using a more powerful telescope than his predecessors, Huygens had been able to confirm his hypothesis that Saturn's mysterious appendages, which had confounded Galileo and subsequent astronomers, were in fact "a thin flat ring, nowhere touching." The plate in Huygens's book *Systema Saturnium* (The Saturn System), which Eimmart's painting is partly based on, presented observations of Saturn by astronomers prior to Huygens, including Galileo. At the top, Eimmart adds a more accurate depiction of Saturn and its rings—one that for some reason is not as true to the actual planet as several plates by Huygens from the same book, however. Still, this unique painting contains the results of observations by ten or more astronomers—a kind of sequential astronomical palimpsest. For another painting by Eimmart, see page 87.

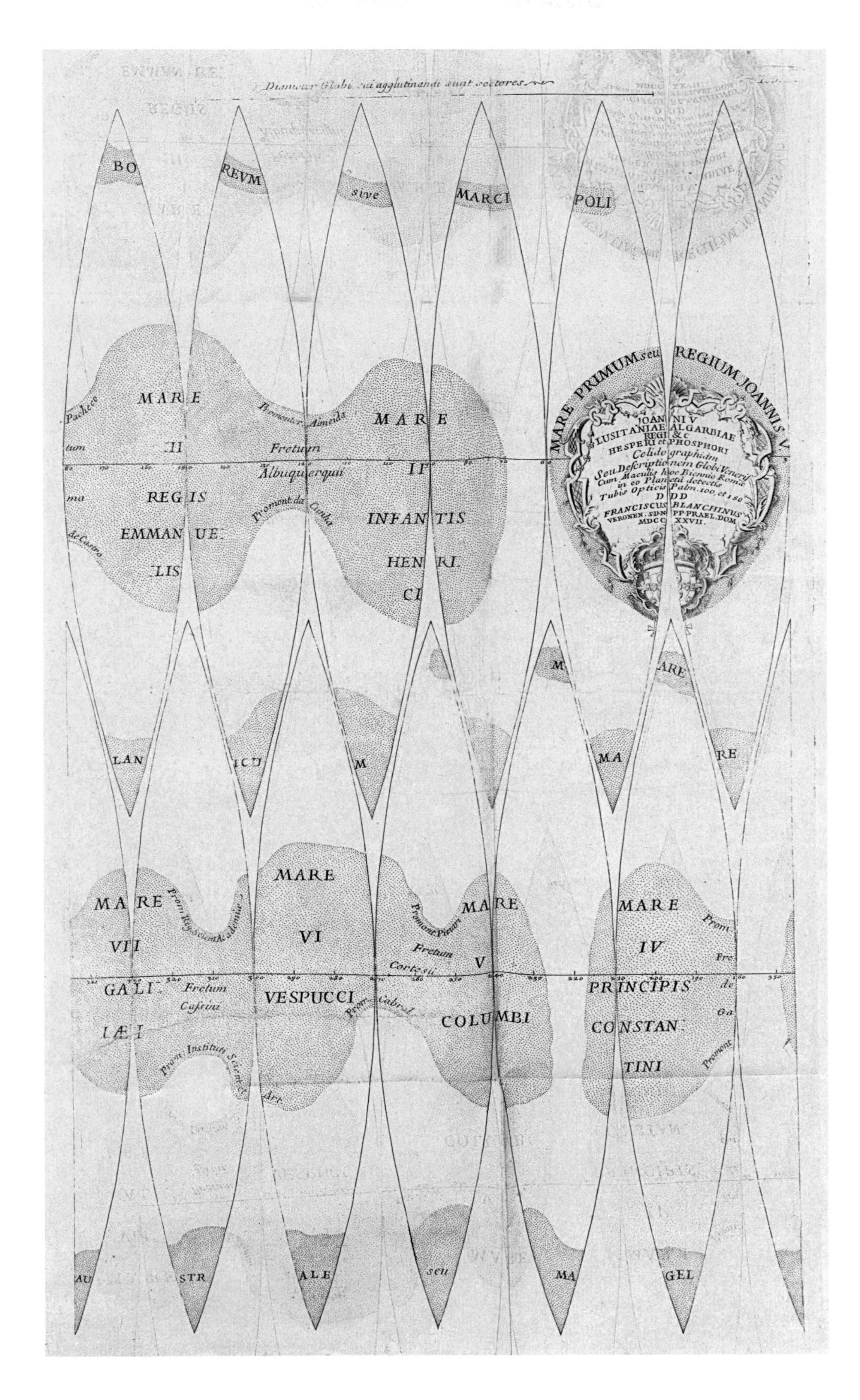

• 1728:

Although the many details of the planets could be glimpsed through eighteenth-century telescopes, details among the solar system's inner or "terrestrial" worlds—Mercury, Venus, and Mars—were exceedingly hard to make out. These map gores of Venus were made by Italian scientist and astronomer Francesco Bianchini, best known for working at the service of three popes to create a more accurate calendar. Using a clumsy "aerial telescope" (a tubeless refractor in which the objective lens can be mounted up to two hundred feet from the eyepiece), Bianchini saw what he believed to be permanent dark patches on the surface of Venus, which he thought were seas and named as such, as visible here. He even calculated the rotation rate of the planet based on them. In fact, Venus is utterly dry and permanently swathed in clouds, so there is no way that he could have seen the surface, let alone seas there, though he may have seen gray patches in the dense Venusian atmosphere. Bianchini's map gores were in fact made into a globe, now in the University of Bologna's Astronomy Museum.

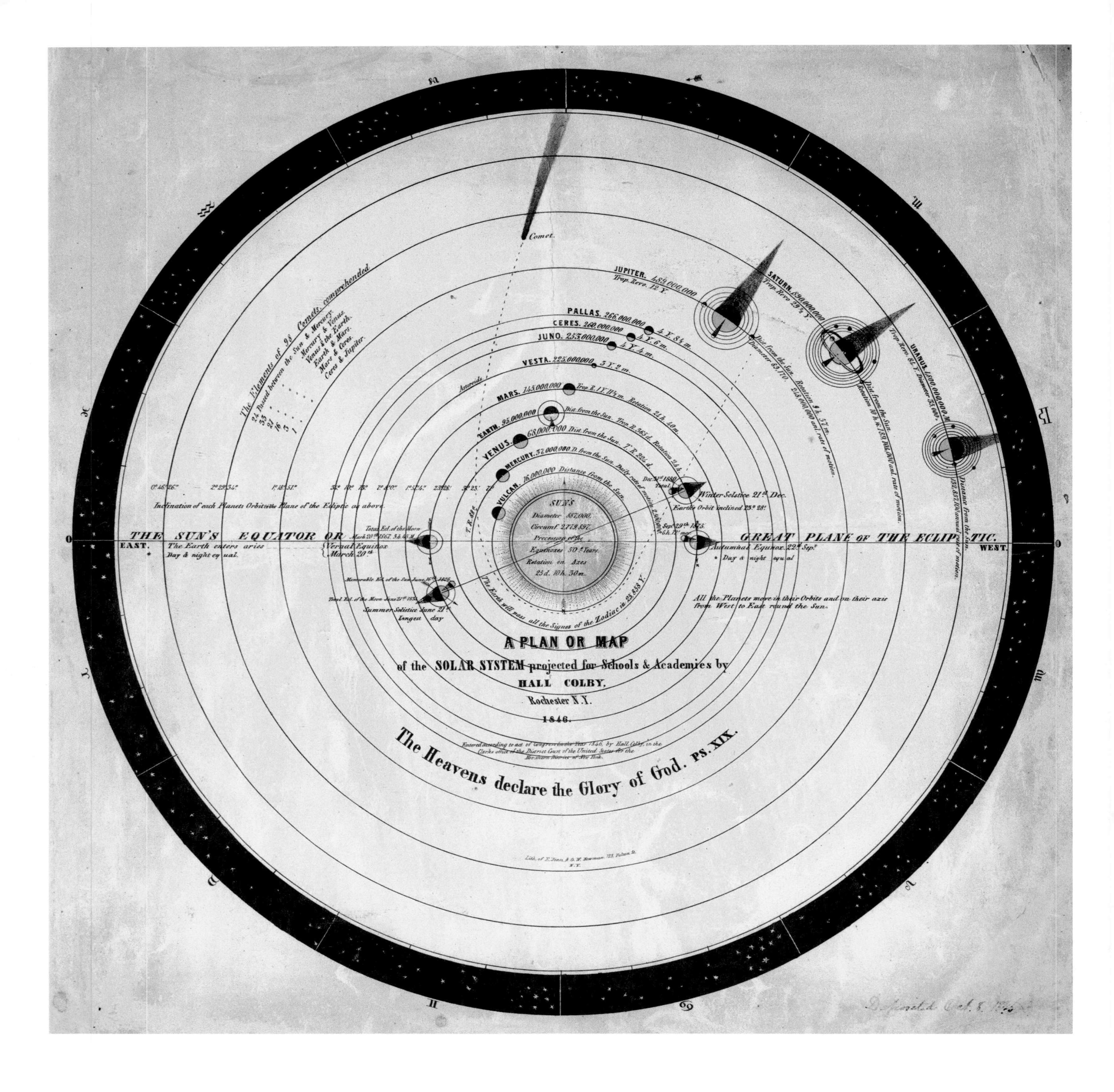

• 1846:

This map of the solar system by one Hall Colby is notable for one absence and several presences. While Colby's map does portray planet Uranus and five of its moons, that's not particularly notable, because Uranus had been discovered by John Herschel sixty-five years earlier. Among the interesting presences, then, are four "planets" visible here between the orbits of Mars and Jupiter: Vesta, Juno, Ceres, and Pallas. All were discovered in the first decade of the nineteenth century, and all were considered planets until the 1860s, when a tide of discoveries of ever-smaller objects in similar orbits demoted them to the rank of mere asteroids. (The four largest objects in the asteroid belt, all are still considered asteroids except Ceres, which is now a dwarf planet, the only one in the inner solar system. Pluto, missing here, has also been classified as a dwarf planet since 2006.) Another notable presence, if you look closely at the detail view of Colby's map to the right, is that of a planet inside the orbit of Mercury: Vulcan. Although it may sound like it comes from *Star Trek*, Vulcan first entered the language of popular culture when its existence within this solar system was predicted by French mathematician Urbain Le Verrier in 1843. Le Verrier was so sure he named it, and promoted it widely in the hopes that astronomers would confirm his "discovery." They tried for decades but came up short. As for the notable *absence* in Hall Colby's map, it's our current seventh of eight planets: Neptune. That's because Neptune wasn't discovered until September 24, 1846—months after this map, which was intended for schools, was printed. But what's truly interesting is that Neptune was discovered due to the prediction of a French mathematician: one Urbain Le Verrier.

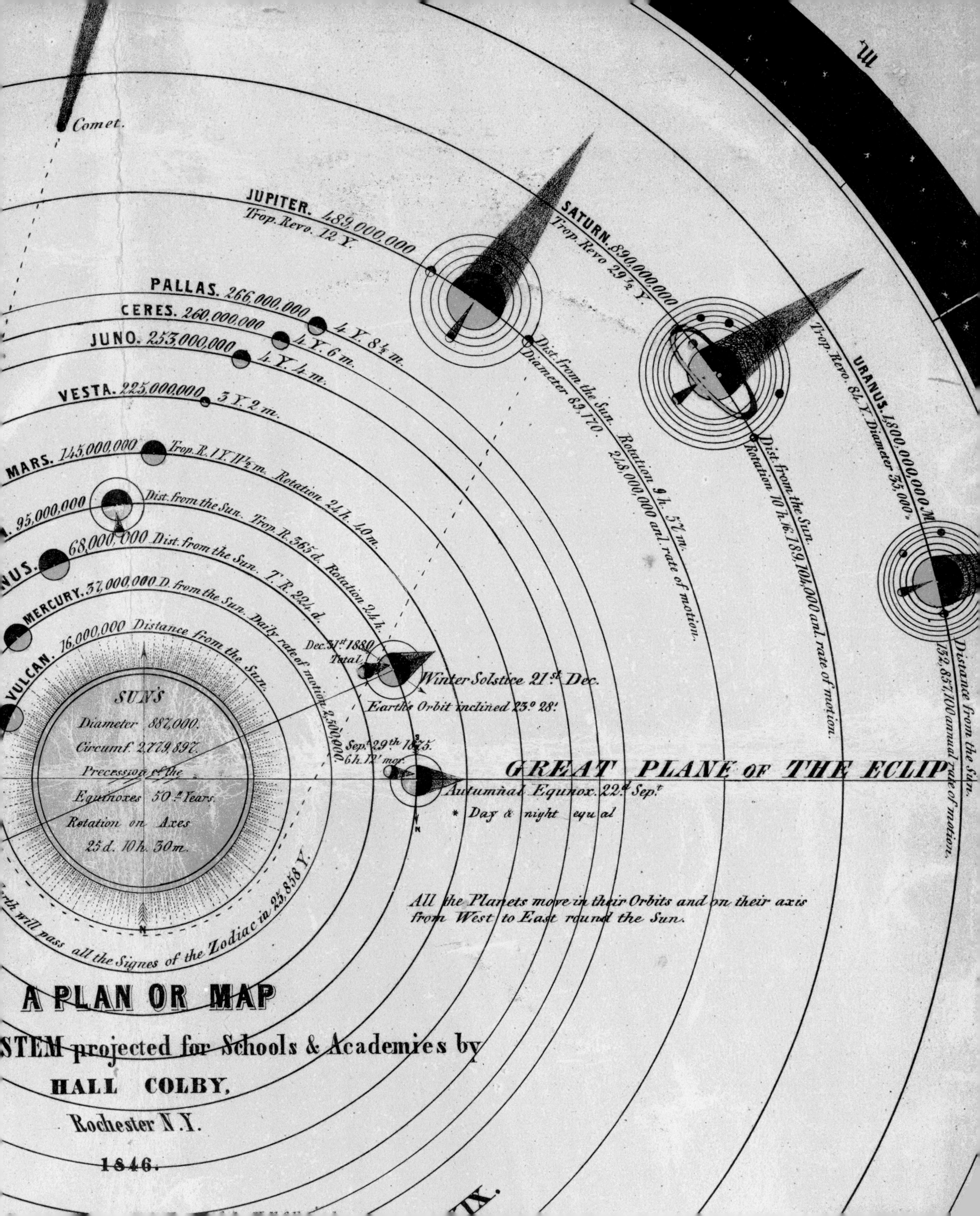
Comet.
JUPITER. 489,000,000
Trop. Revo. 12 Y.
SATURN. 890,000,000
Trop. Revo 29½ Y.
PALLAS. 266,000,000
4 Y. 8½ m.
CERES. 260,000,000
4 Y. 6 m.
JUNO. 253,000,000
4 Y. 4 m.
VESTA. 225,000,000
3 Y. 2 m.
MARS. 145,000,000
Trop. R. 1 Y. 11½ m. Rotation 24 h. 40 m.
95,000,000
Dist. from the Sun. Trop. R. 365 d. Rotation 24 h.
NUS. 68,000,000 Dist. from the Sun. T. R. 224 d.
MERCURY. 37,000,000 D. from the Sun. Daily rate of motion 2,500,000.
VULCAN. 16,000,000 Distance from the Sun.
Dist. from the Sun.
Diameter 89,170.
Rotation 9 h. 56 m.
248,000,000 anl. rate of motion.
Trop. Revo. 84 Y. Diameter 35,000 m.
URANUS. 1,800,000,000 M.
Dist. from the Sun.
Rotation 10 h. 16.
189,104,000 anl. rate of motion.
Distance from the Sun.
152,857,100 annual rate of motion.
Dec. 31st 1880
Total
Winter Solstice 21st Dec.
Earth's Orbit inclined 23° 28'.
Sept. 29th 1875.
6 h. 12' mor.
GREAT PLANE OF THE ECLIP
Autumnal Equnox. 22d Sept.
* Day & night equal
SUN'S
Diameter 887,000.
Circumf. 2,779,897.
Precession of the
Equinoxes 50 1/4 Years.
Rotation on Axes
25 d. 10 h. 30 m.
rth will pass all the Signes of the Zodiac in 25,858 Y.
All the Planets move in their Orbits and on their axis
from West to East round the Sun.
A PLAN OR MAP
STEM projected for Schools & Academies by
HALL COLBY,
Rochester N.Y.
1846.

PLATE VIII.
Copyright 1881 by Charles Scribners Sons.
THE PLANET MARS.
Observed September 3, 1877, at 11h. 55m. P.M.
E. L. Trouvelot

• 1881:

Between 1872 and circa 1880, artist-astronomer Étienne Trouvelot had access to two of the best telescopes in the United States, the fifteen-inch Great Refractor at the Harvard College Observatory and the even greater twenty-six-inch refractor at the U.S. Naval Observatory in Washington, D.C. He used his time on them to produce a series of superb illustrations of planets and other celestial objects. In 1881, Charles Scribner's Sons released a limited-edition collection of chromolithograph reproductions of his best illustrations. The images on this and the next three pages come from that series. **Above:** Despite his use of the best telescopes in the country, Mars was (and remains) notoriously difficult to see in any detail from Earth. Although the planet does have a dark patch extending north toward the equator called Syrtis Major, which is likely reflected here, Trouvelot's depiction of Mars is beautiful but largely unsupported by later evidence. (All of his planet portraits are south side up—a convention of the time.)

PLATE IX.

E. L. Trouvelot.

THE PLANET JUPITER.

Observed November 1, 1880, at 9 h. 30 m. P.M.

Trouvelot's Jupiter is far more accurate than his rendition of Mars. Here, two of the planet's large Galilean satellites are depicted in transit, with their shadows visible on Jupiter's face. The planet's vast Great Red Spot, an anticyclonic storm system several times the size of Earth, is visible as well, as are Jupiter's belted clouds. While the size of the Great Red Spot may seem exaggerated here, it in fact matches other observations made at around the same time (and even photos taken in the mid-twentieth century). While the spot is somewhat smaller now, it's still the largest such phenomenon in the solar system.

Trouvelot's rendering of Saturn is extraordinarily accurate, and even captures such details as the so-called spokes in Saturn's rings. These subtle features had been seen by only a handful of observers—and largely dismissed as figments of their imagination—before the twin Voyager missions flew by Saturn in 1980–81 and confirmed their existence. They are thought to be due to electrostatic charges in the ring particles.

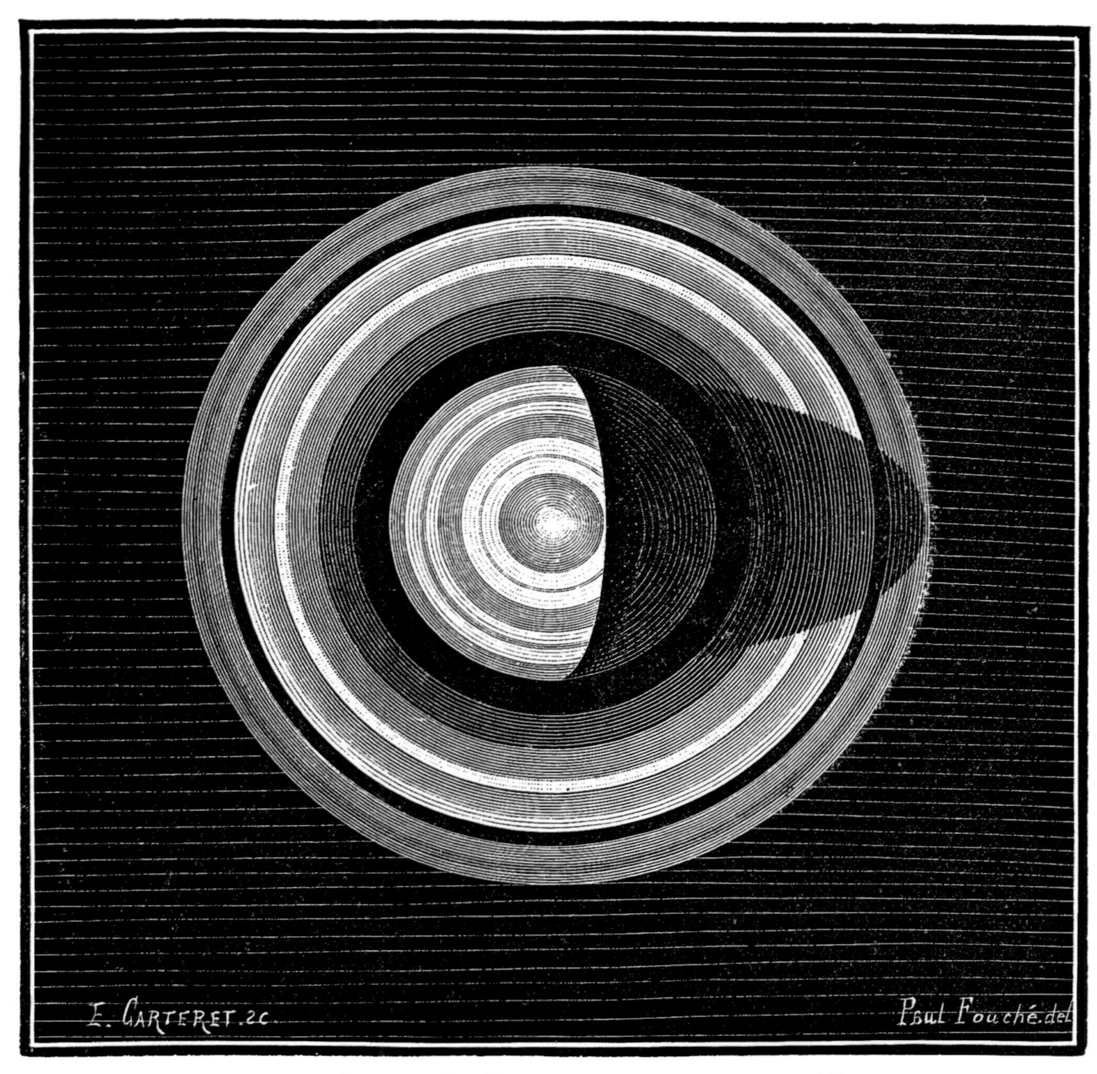

FIG. 190.—SATURN'S RINGS SEEN FROM THE FRONT

• 1894:

Above: This print portraying Saturn as seen from high above one of its poles is both impressively realized and prescient. It would be more than a century before a spacecraft, NASA's Cassini orbiter, would allow us to see the planet from this perspective; there's nothing inaccurate about this illustration. (From French astronomer Camille Flammarion's *Astronomie populaire*.)

• 1888:

Facing page: When Italian astronomer Giovanni Schiaparelli observed Mars during a particularly close opposition in 1877, he thought he glimpsed a web of lines on its surface. He called them *canali*, which is Italian for "channels," and set about giving them names, as is visible on these maps. Subsequent observations from Milan's Brera observatory in 1882 and 1888 seemed to confirm their presence. Schiaparelli never claimed these features were artificial. We now know them to be the result of a kind of cognitive optical illusion caused by the mind's need to connect poorly discerned features.

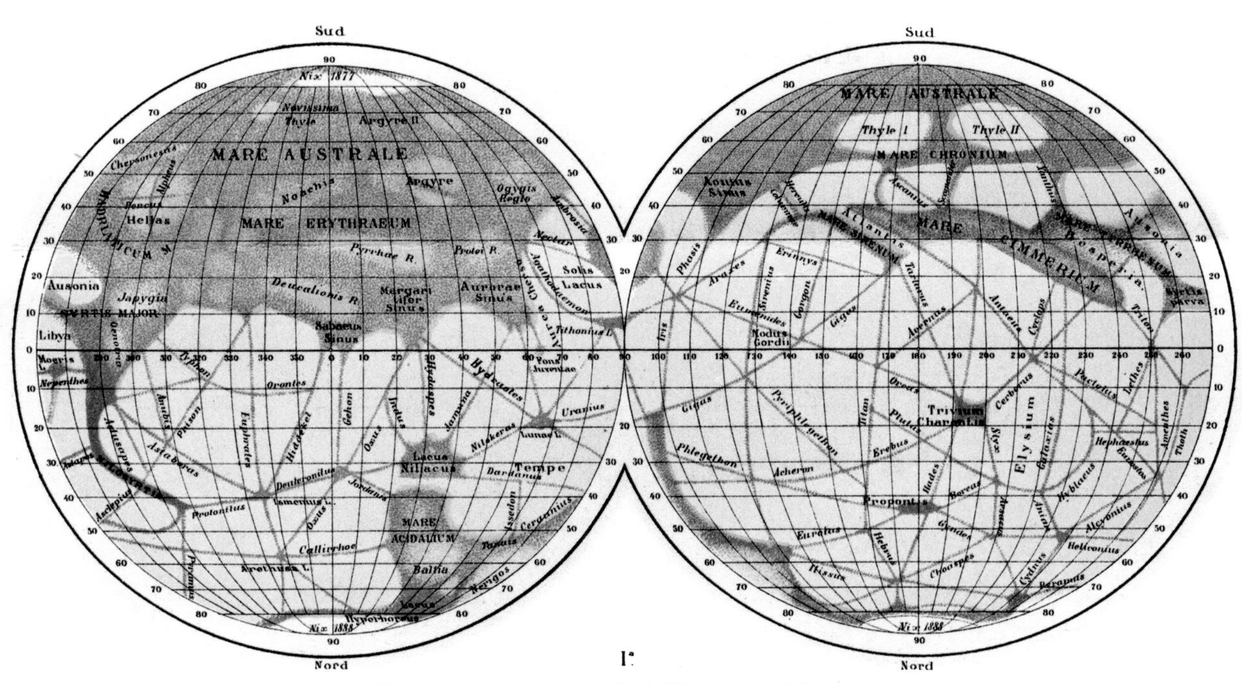

I

Carta generale del Pianeta Marte

secondo le osservazioni fatte a Milano
dal 1877 al presente

(N.B. Le linee o strisce oscure che solcano i continenti sono in questa carta presentate nel loro stato semplice cioè come appaiono quando non sono geminate.)

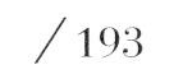

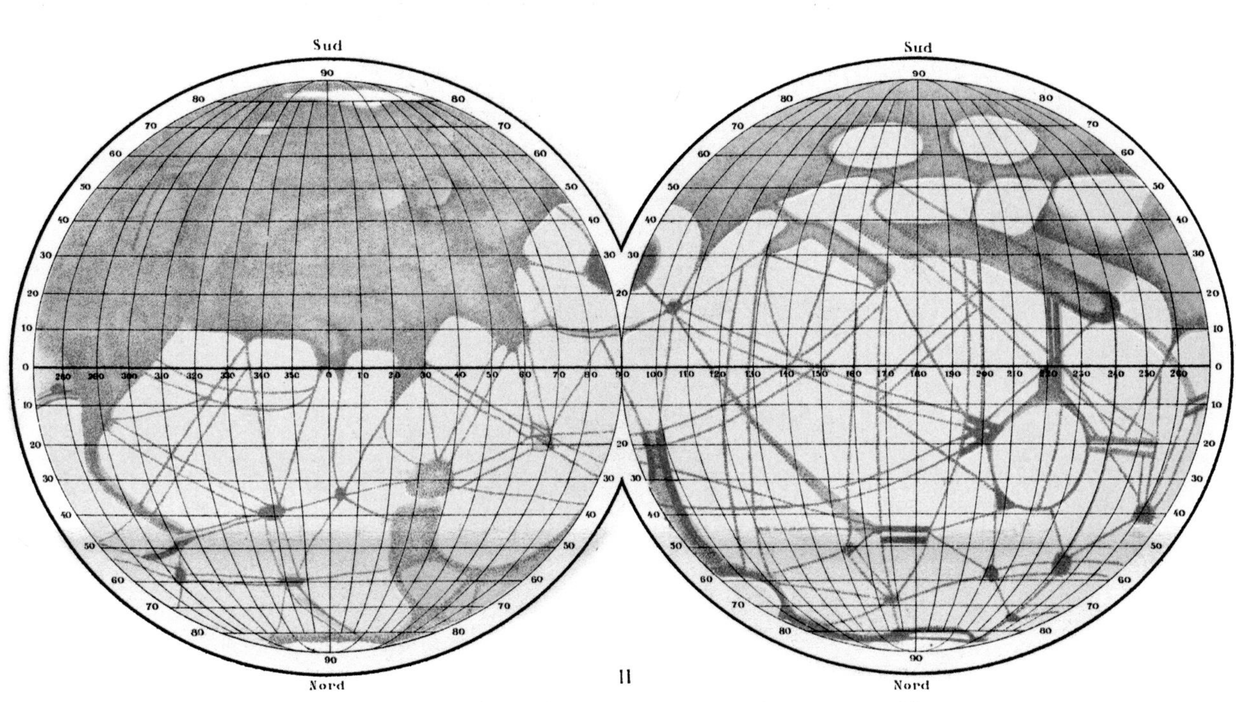

II

Le geminazioni delle linee oscure del pianeta Marte

quali furono osservate a Milano principalmente
nel 1882 e nel 1888

Lit. della Casa Edit. Dott. Francesco Vallardi
Proprietà Letteraria

G. Schiaparelli dis.

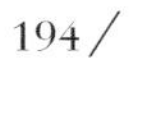

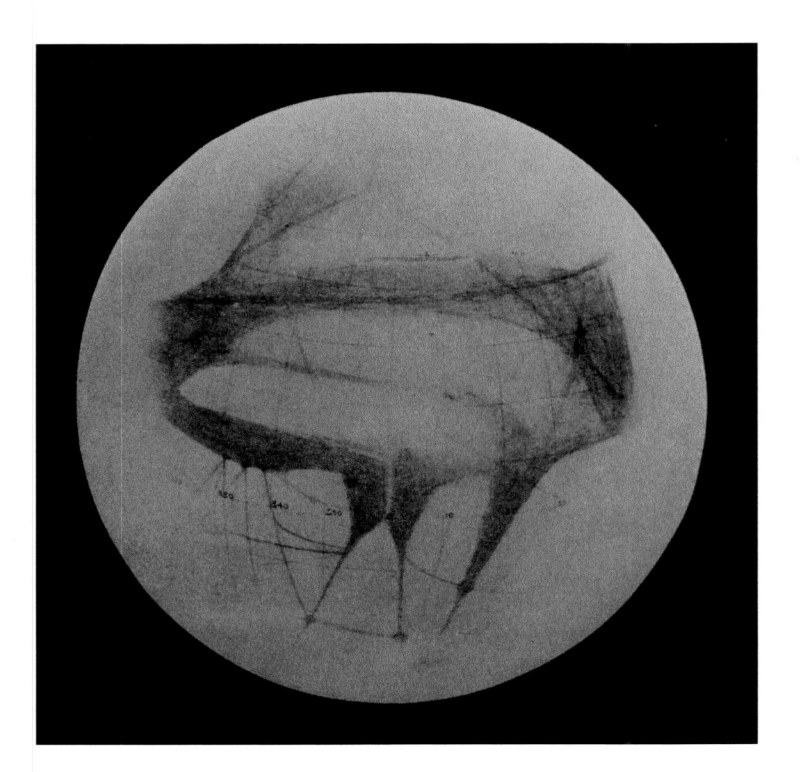

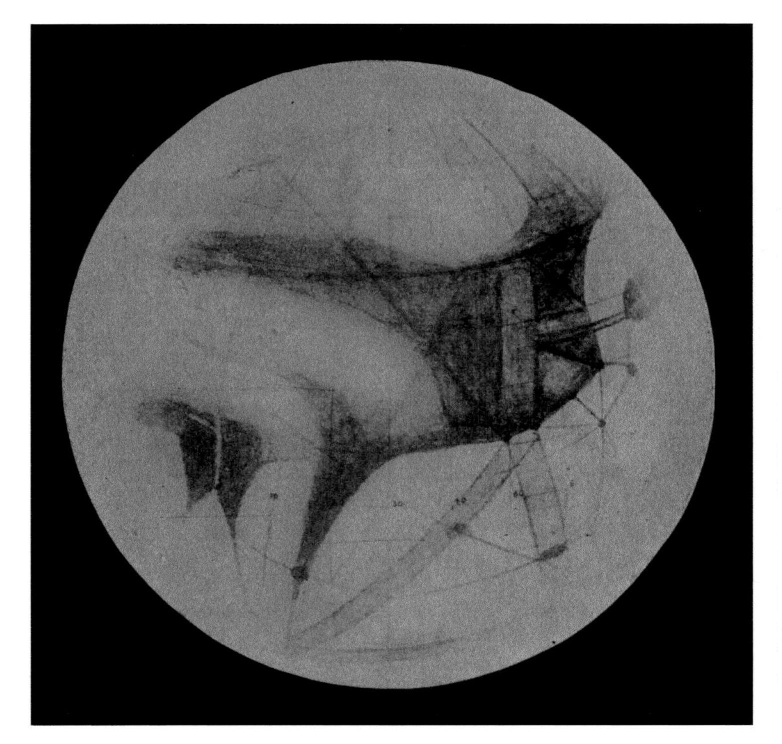

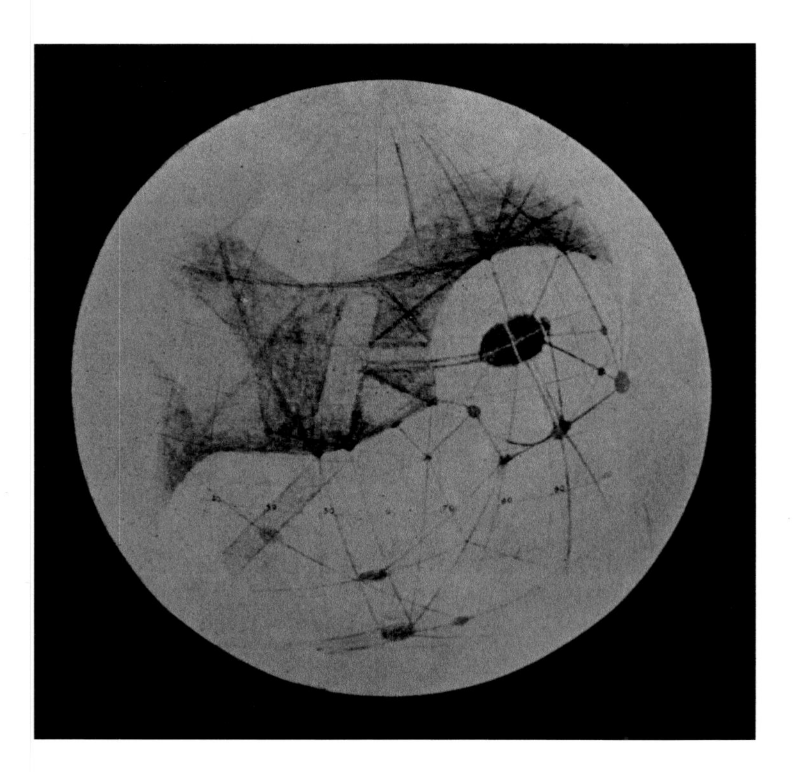

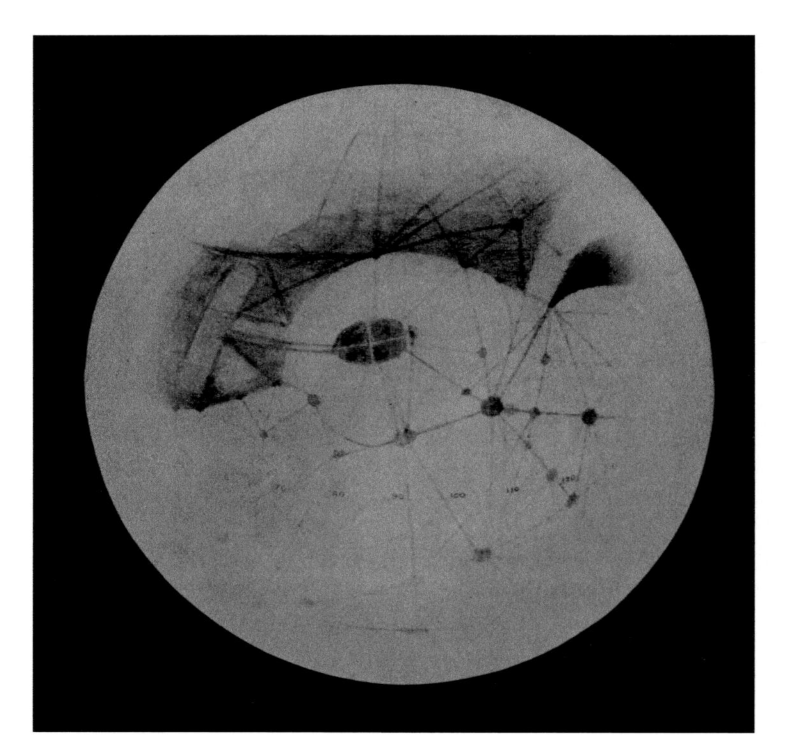

• 1896:

In the last decade of the nineteenth century, a Boston Brahman named Percival Lowell read a book on Mars by French astronomer Camille Flammarion, and decided that he wanted to dedicate himself to astronomy. In the book, Flammarion argued for a habitable Mars, and Lowell also became aware of Schiaparelli's descriptions of there being *canali*—a word which had been translated into English as "canals." By 1894, Lowell had built an observatory in Flagstaff, Arizona, to see them for himself—reportedly the first time an observatory had been located in a specific place based on good viewing conditions—and he soon became an enthusiastic proponent of the idea that a Martian civilization had built an extensive canal network. These illustrations are from *Mars*, the first of three books by Lowell popularizing the notion that his ersatz canals indicated the presence of intelligent life on the Red Planet.

• 1944:

Along with French illustrator and astronomer Lucien Rudaux, American painter Chesley Bonestell pioneered a genre of speculative solar system landscapes sometimes called "space art." Bonestell's *Saturn as Seen from Titan*, first published in *Life* magazine, remains the best-known example of the genre. From 1952 to 1954, Bonestell teamed up with Wernher von Braun, designer of the Nazi V-2 rocket and, later, the Saturn V rocket of NASA's Apollo program that took men to the moon, to produce an influential series of illustrated articles for *Collier's* magazine. Titled "Man Will Conquer Space Soon!" it paved the way for the actual space age that followed. We now know that Titan's atmosphere is so thick, a view like this would be impossible, which takes nothing away from the power of Bonestell's achievement.

• 1963:

Another illustrator who became fascinated with space exploration was Czech artist Ludek Pesek. This view by Pesek of Saturn's rings as seen from high in its atmosphere is from Josef Sadil's book *The Moon and Planets*. One of the planet's icy inner satellites, probably Mimas, is visible to the upper left. On the right Saturn's curving shadow cuts across the rings, which are made of innumerable fragments of ice and rock and are the most extensive planetary rings in the solar system. The upper of the two gaps in the rings is the Cassini division, named after its discoverer in 1675, Italian-French astronomer Giovanni Cassini. When this scene was painted, it was not known if Saturn possesses a solid surface under its atmosphere. It is now thought to have a thick layer of liquid hydrogen under gaseous hydrogen. The upper clouds, however, which we see here, are made of ammonia crystals. Saturn likely does have a solid iron and nickel core, deep under all the liquid hydrogen.

• 1965:

On July 15, 1965, U.S. space probe Mariner 4 flew by Mars at a closest approach of about six thousand miles, transmitting twenty-one images—the first close-up pictures of another planet. Because computers of the time were so slow, scientists at NASA's Jet Propulsion Lab in Pasadena, California, knew that it would be many hours before they would see any print from the stream of numbers constituting Mariner's image data. Impatient to see Mars, JPL engineer Richard Grumm and several colleagues printed out individual strips of image data, assembled them on a backing board, and ran out to the local art supply store to buy pastel crayons. Because the numbers on each strip indicated the brightness values of that part of the image, Grumm was able to color the strips in close approximation to what Mariner was actually seeing of Mars. The result was a landmark not only in the graphic representation of space, but also in the early history of digital photography. Grumm's initials can be seen in the lower right-hand side of the picture. The darker brown patch also on the lower right is space, with the limb of the planet represented by the wavy red-brown line. Neither Mariner 4 nor any subsequent spacecraft found any evidence of canals.

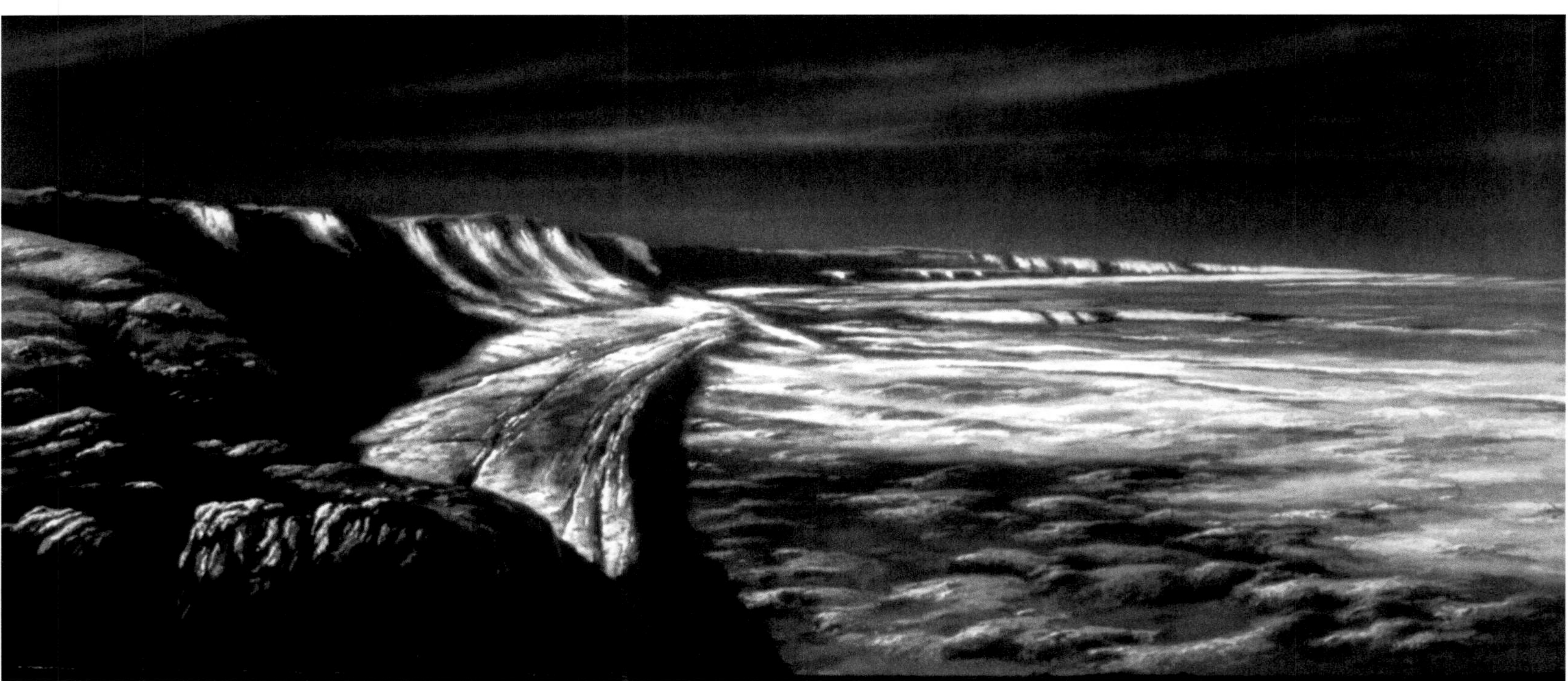

• Early 1970s:

Although Mariner 4 and a subsequent pair of flyby missions revealed a highly cratered surface similar to the moon's, 1971's Mariner 9 actually entered the orbit of Mars and had time for a detailed reconnaissance of the planet. Starting in January 1972, it began revealing the tallest volcanoes and the biggest canyon in the solar system. Working for *National Geographic* and also the Smithsonian Institution, Ludek Pesek was asked to illustrate some of those discoveries, and produced a series of approximately thirty Mars landscapes. Although in 1976 the two Viking landers revealed that the Martian sky was a muted pink-orange in hue, rather than Pesek's tenuous blue, in other respects his paintings hold up remarkably well. **Above:** Mars as seen from Phobos, one of the planet's two tiny moons. **Facing page top:** The walls of the Valles Marineris canyon system, which is as wide as the continental United States and was named after its spacecraft discoverer. **Facing page bottom:** A view of the immense caldera of the Olympus Mons volcano, which is approximately fifty miles across and two miles deep.

• 1984:

Between 1974 and 1975 the Mariner 10 spacecraft conducted the first reconnaissance of Mercury, the innermost planet. Because the same face of the planet was lit for each of the spacecraft's three fly-bys, just under half of Mercury's heavily cratered surface was photographed, with this geological map of the southern Michelangelo Quadrangle based on that data. Four ancient multi-ringed basins are clearly evident. These and most of the innumerable smaller craters are assumed to be the result of the same heavy bombardment phase of solar system history that scarred our own moon. Between the craters, comparatively smooth undulating plains can be seen. In this map, the yellows, greens, blues, and browns signify different types of cratered terrain; the maroon, red, and tan the smooth to intermediate plains material; and the purple and olive green the underlying basins.

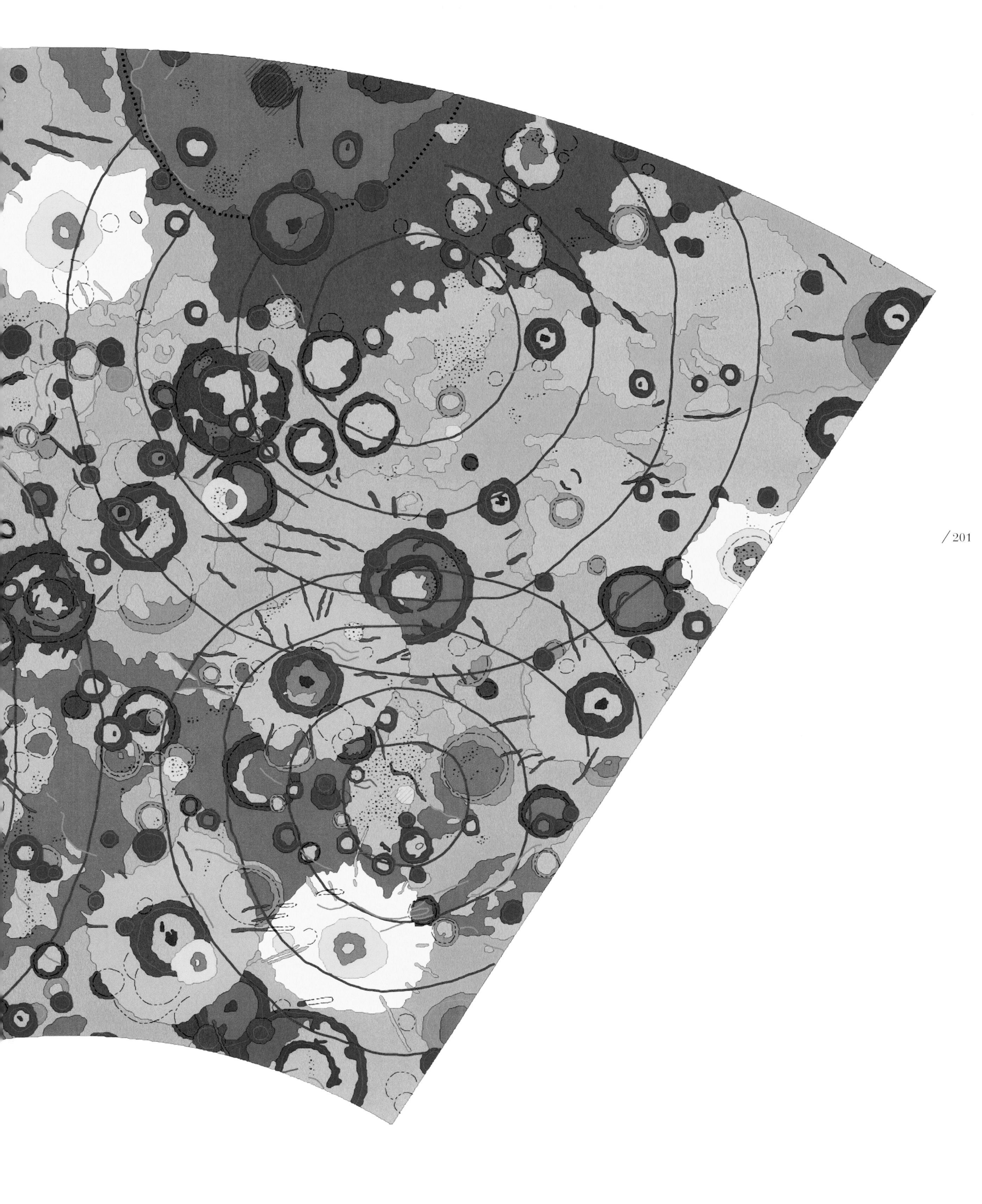

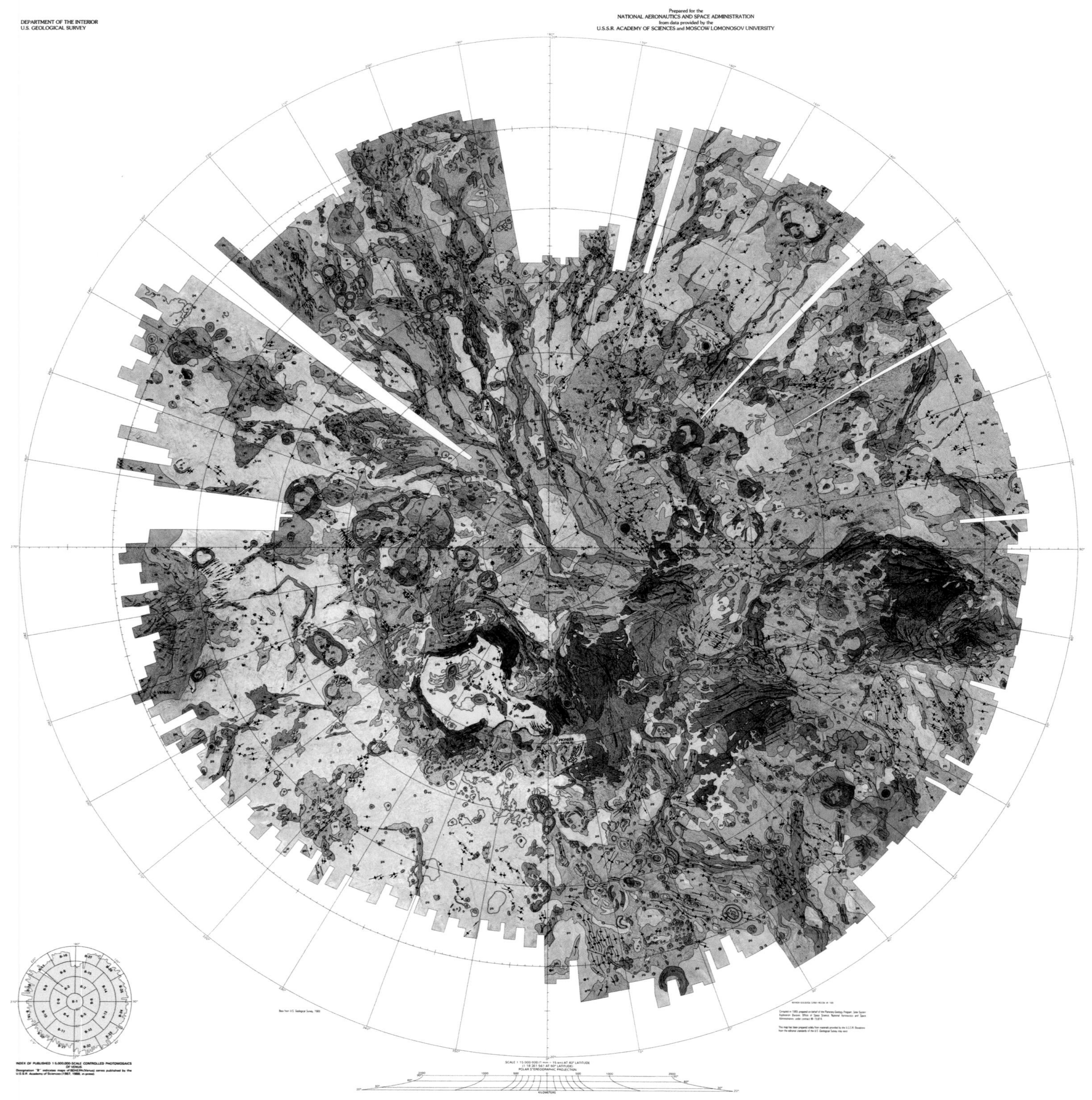

GEOMORPHIC/GEOLOGIC MAP OF PART OF THE NORTHERN HEMISPHERE OF VENUS
By
A.L. Sukhanov, A.A. Pronin, G.A. Burba, A.M. Nikishin, V.P. Kryuchkov, A.T. Basilevsky,
M.S. Markov, R.O. Kuzmin, N.N. Bobina, V.P. Shashkina, E.N. Slyuta, and I.M. Chernaya
V 15M 90/0 G
1989

• 1989:

Between 1961 and 1984, the Soviet Union sent an astonishing thirteen Venera spacecraft to investigate Venus; ten of them landed on its surface. Among the last in the series, the 1983 Venera 15 and 16 missions carried radar imaging equipment, permitting the first mapping of a surface obscured by the planet's dense atmosphere. As its attribution to a list of twelve Russian scientists makes clear, this U.S. Geological Survey geological map of the Venusian northern hemisphere is based on a series of twenty-six maps published after those missions by the USSR Academy of Sciences in 1987 and 1988. In this map, the red to orange areas are volcanic terrain, the various shades of green signify rough terrain, and the blues are ridges.

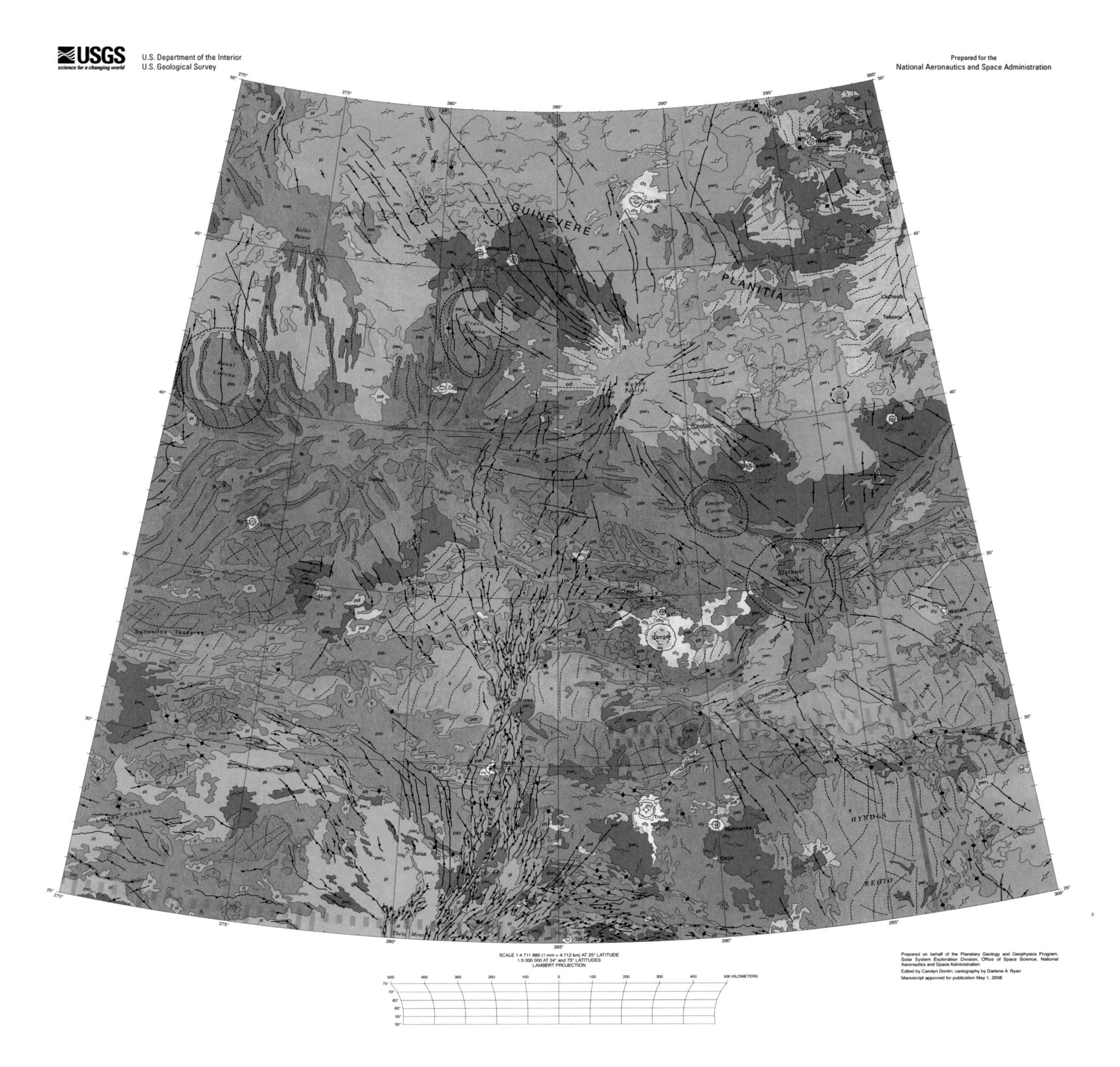

• 1989:

The map on the facing page was made in preparation for an American radar orbiter, Magellan, which mapped Venus from 1990 to 1992. The above geological map is based on Magellan data and charts the Beta Regio quadrangle of the planet, a 15,500-mile-wide area in the northern midlatitudes cut by the deep tectonic valley of Devana Chasma, visible here as the irregular light-maroon patchwork extending up from the bottom of the map. The Beta Regio rise is so prominent on the Venusian surface that it was even seen in long-range radar studies conducted from Earth in 1965 and 1978. In this map, the blues and light greens signify plains material; the red and maroon, faults such as Devana Chasma; the olive green, intensely fractured terrain; and the yellow, impact crater material. The oval areas with "Corona" designations are some of the immense volcanic features that define the Venusian surface.

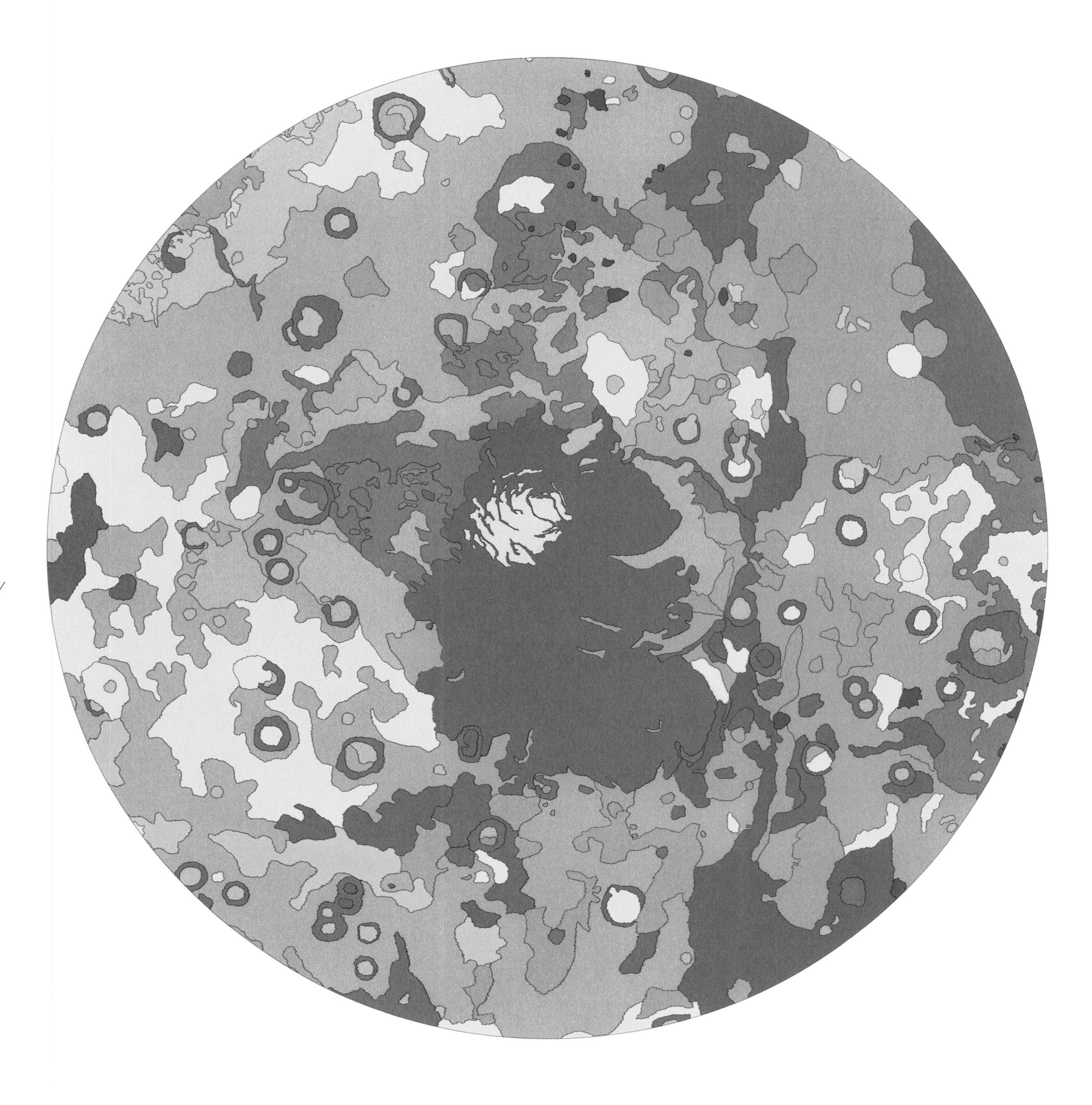

• 1987:

NASA's highly successful pair of Viking Mars missions of the mid-1970s deposited soft-landers on the surface and left orbiters to map the planet for years to come. This geological map of Planum Australe, the Martian south polar region, is based on Viking Orbiter data. Here, the orange sections are smooth plains material, the purple are ravaged ancient highlands, and the blue represents polar deposits and such eolian (windblown) materials as dunes. Polar ice can be seen as well: the irregular white patch within the blue. The southern polar cap of Mars is comprised of water and carbon dioxide ice, and is thought to be about two miles thick, depending on the season. In a geographical quirk, it's just under one hundred miles north of the South Pole.

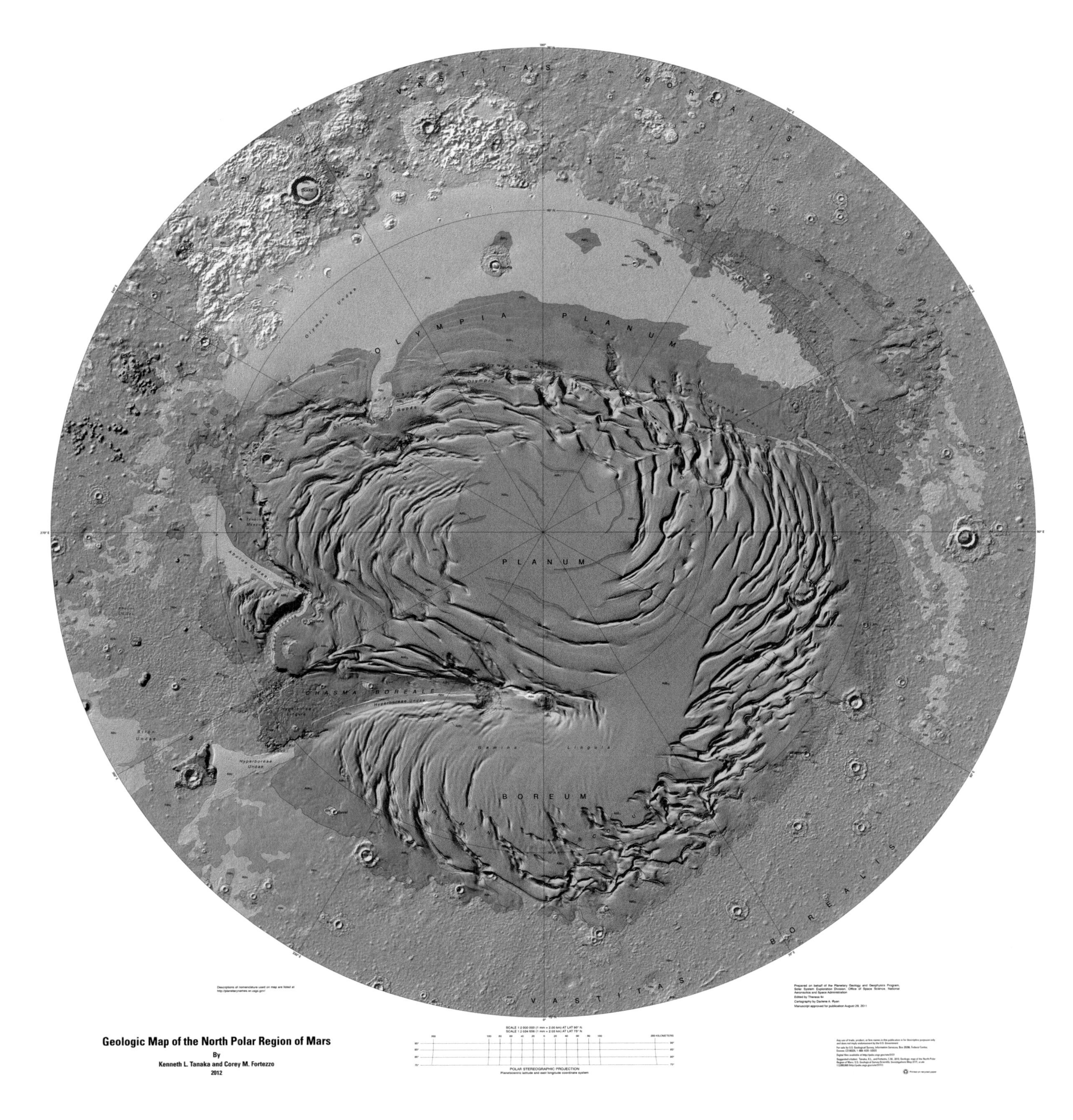

• 2012:

This much more recent map of Planum Boreum, the north polar region of Mars, benefits from a fire hose of data from such contemporary missions as the Reconnaissance Orbiter, Global Surveyor, and Odyssey missions. The northern polar cap is far larger than the southern one, encompassing an area about 1.5 times the size of Texas. It rests on the vast lowland plain that dominates the Martian northern hemisphere, and is bisected by Chasma Boreale, an immense canyon approximately 350 miles long and 60 miles wide—far larger than the Grand Canyon. In this map, the blue-green areas are moderately cratered plains and outflow channels, the metallic blue is mostly pure water ice, and the gray area above the polar cap is a vast sand sea. The dull-orange areas constitute crater material.

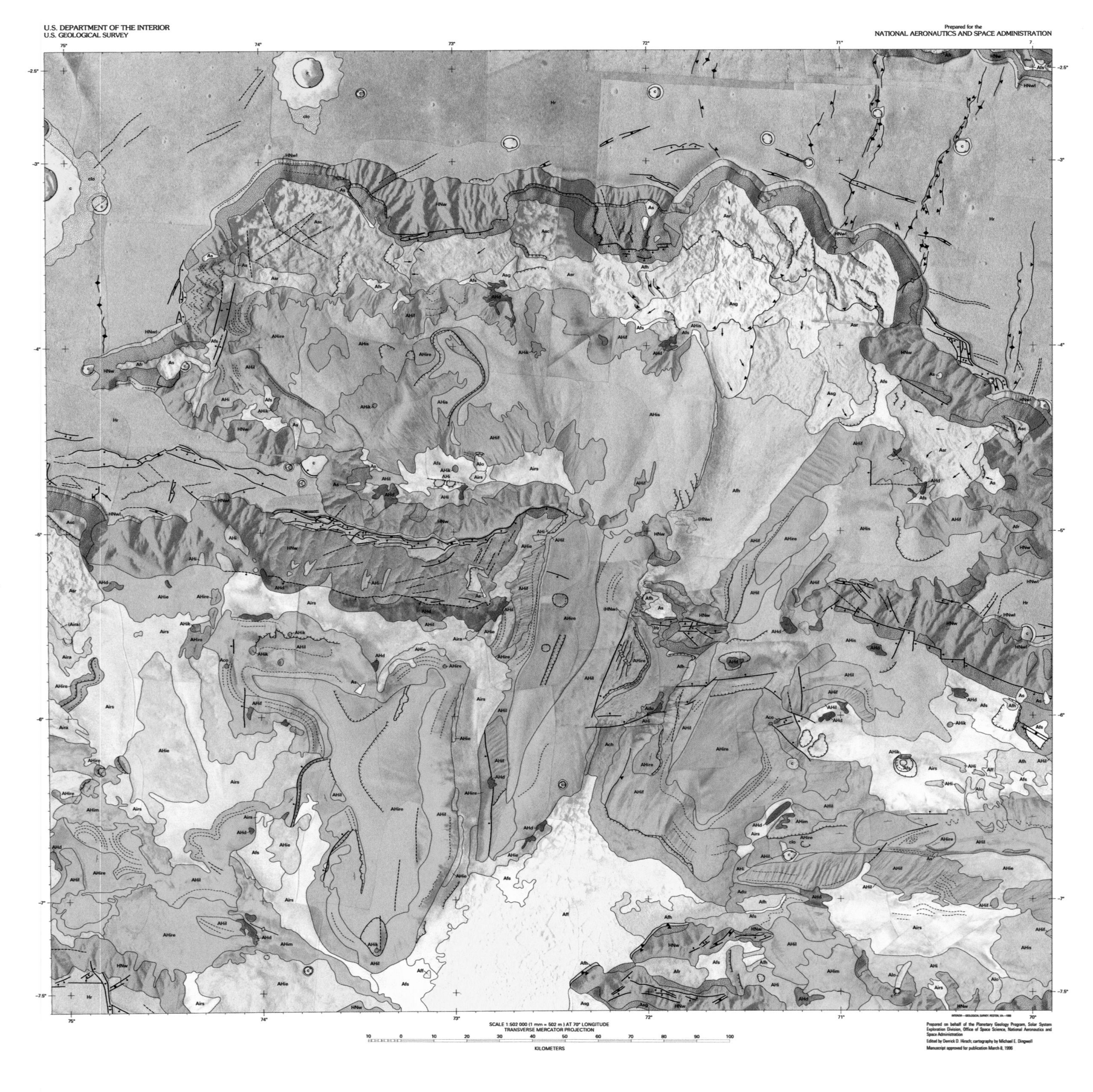

• 1999:

This geological map of the Ophir and Candor chasmas, vast formations in the central part of Valles Marineris, charts some of the most spectacularly rugged landscapes ever seen. Here, plateaus give way to steep-sided walls, a landscape likely produced by a graben process in which valleys are bordered by vast scarp walls. (The word *chasma* signifies such a steep-walled depression in an extraterrestrial context.) Ophir, the canyon at the top, is about two hundred miles wide from end to end, while Candor, at the bottom, is more than five hundred miles wide, with much of it unseen here. In this map, the purple areas are high plateaus, the blue signifies fluted interior mesas, the pale yellow is relatively smooth trough floors, and the red indicates dark areas associated with landslides. Dark beige indicates rock walls.

The Western Hemisphere of Mars

• 2005:

Freelance planetary cartographer Ralph Aeschliman mapped planets at the U.S. Geological Survey for eleven years before moving on to do his own work based on data from NASA missions. This highly detailed map of the western hemisphere of Mars features many of the planet's most striking features, from its vast shield volcanoes (including the highest mountain in the solar system, Olympus Mons, above the equator to the left) to its 2,500-mile-wide Valles Marineris canyon system, below the equator and right of center.

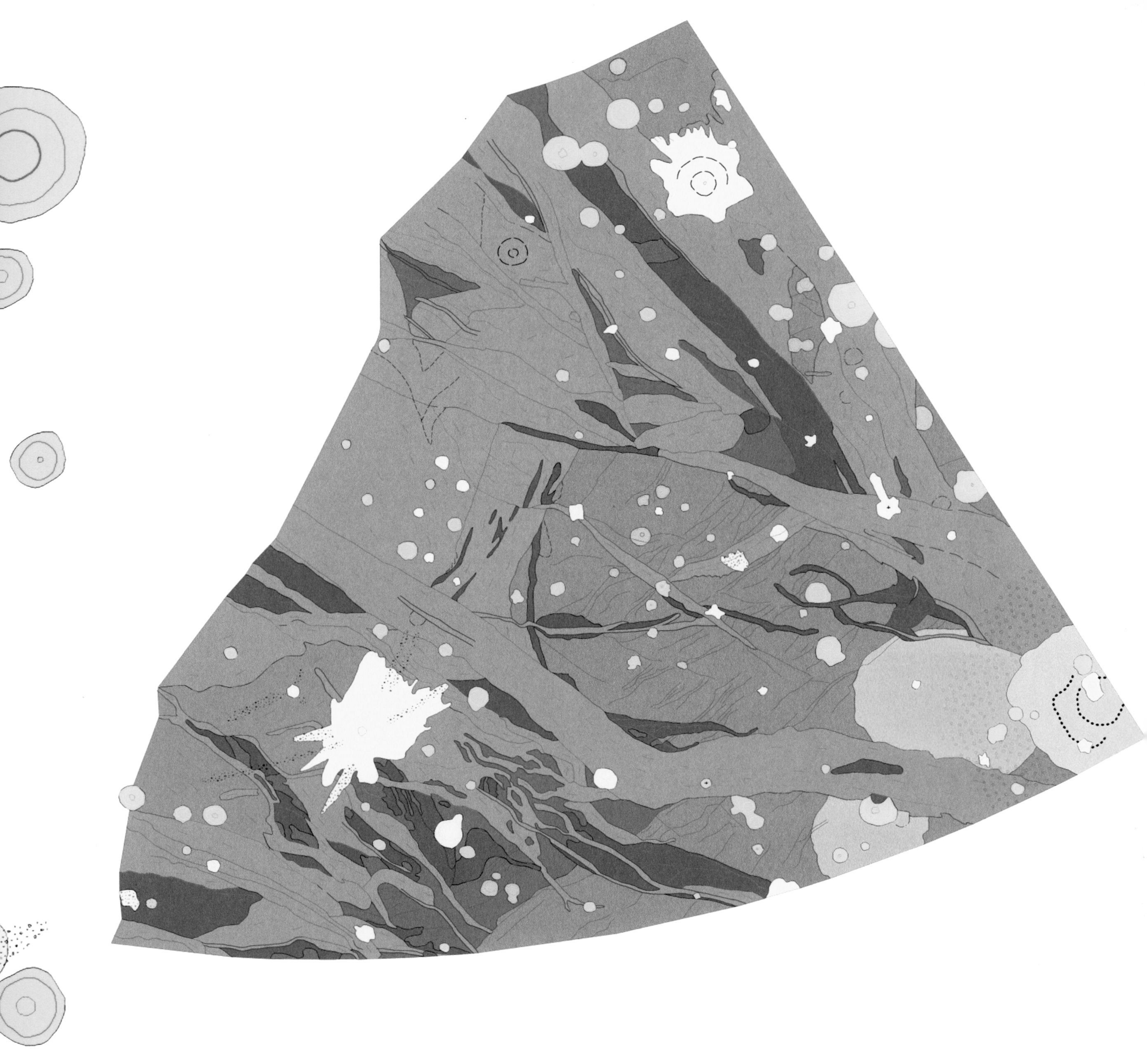

• 1989–1992:

These two geological maps of regions on Jupiter's moon Ganymede, which has a surface made predominantly of rock-hard water ice, are perhaps the most striking yet created for an extraterrestrial body. At 3,273 miles in diameter, Ganymede is the largest moon in the solar system and slightly larger than planet Mercury, though not as massive. An icy satellite subject to powerful gravitational stresses due to continuous interactions with its three large sister moons and massive Jupiter, Ganymede's surface exhibits an ancient cratered landscape interrupted by lighter grooves of terrain disrupted by global tectonic forces. **Above:** Furrows, troughs, and linear grooves characterize the topography of the Philus Sulcus quadrangle of Ganymede, opposite the Jupiter-facing side. In this 1989 map, blues and greens signify the moon's lighter, disrupted materials; maroon and olive green, its darker materials; and various shades of yellow, crater material. **Facing page:** The previous color scheme also applies to this 1992 map of Ganymede's Memphis Facula quadrangle, with red signifying narrow, widely spaced furrows. This region is also opposite the Jupiter-facing hemisphere, toward the moon's leading side as it orbits the planet. In this map, Ganymede's complex layered palimpsest terrain is particularly evident.

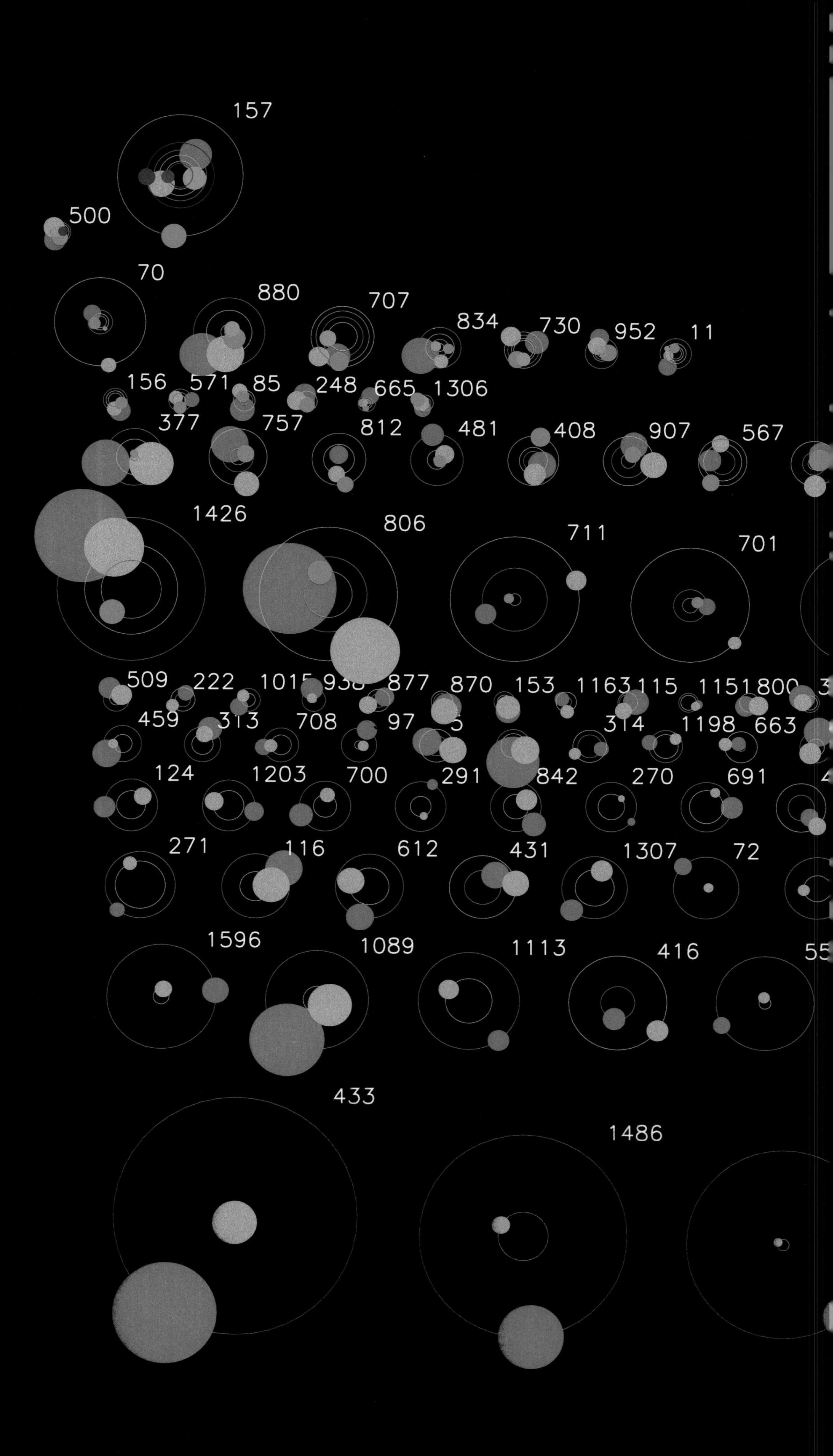

• 2011:

In 2009, NASA launched Kepler, an orbiting space telescope designed specifically to hunt for distant planets by detecting minute variations in the brightness of stars in its field of view. By May 2013, when the space telescope became disabled, it had found about 3,500 or so candidate planets. In 2011, Daniel Fabrycky of the Kepler science team put together an animated "orrery," or simulation of a mechanical model of a solar system, presenting all the multiple-planet systems discovered by Kepler until February 2 of that year. The hotter the color, the larger the planet relative to other planets in its system, with cooler colors indicating smaller planets (red to yellow to green to cyan to blue to gray). This still image from Fabrycky can only hint at the effect of seeing all of Kepler's worlds spinning in his orrery, but it's easy to find online.

The Kepler Orrery

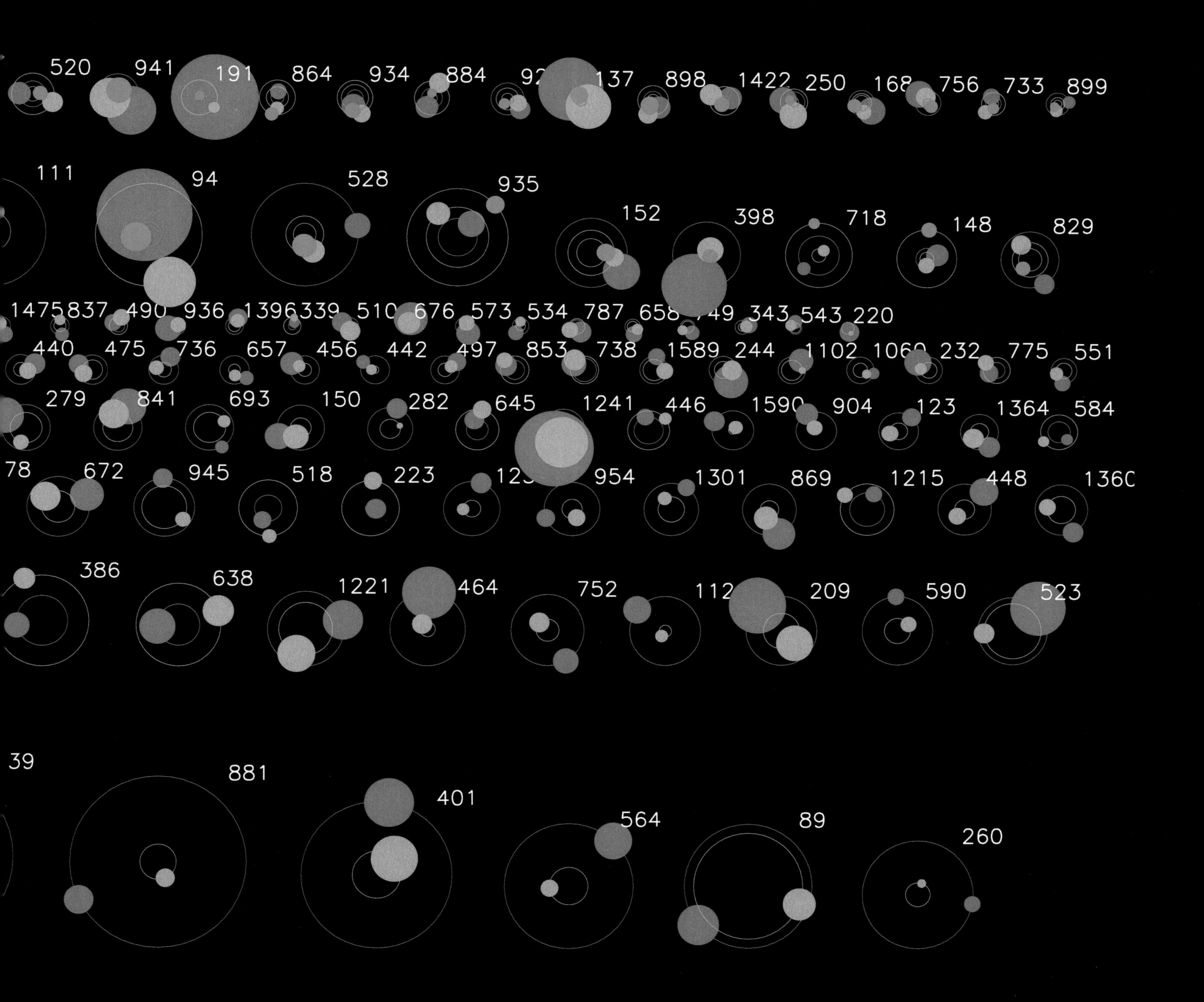

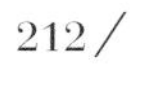

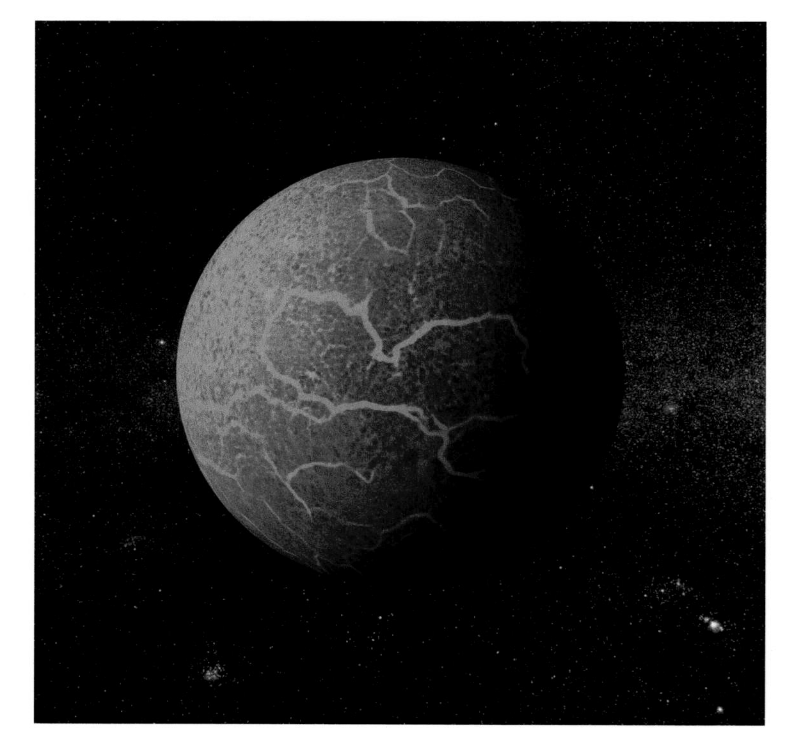

• 2011:

Planetary scientist Abel Mendez Torres fed data concerning the newly discovered planets into a digital visualizations tool at the Planetary Habitability Laboratory at the University of Puerto Rico in Arecibo. Called the Scientific Exoplanets Renderer, it produced highly plausible renderings of what some of them might look like. **Top left:** "Cold Subterran": an icy, Mars-size planet. **Top right:** "Hot Jovian": a Jupiter-size hot planet. **Above left:** "Warm terran": a habitable exoplanet about the size of Earth. **Above right:** "Hot Subterran": a hot Mars-size world.

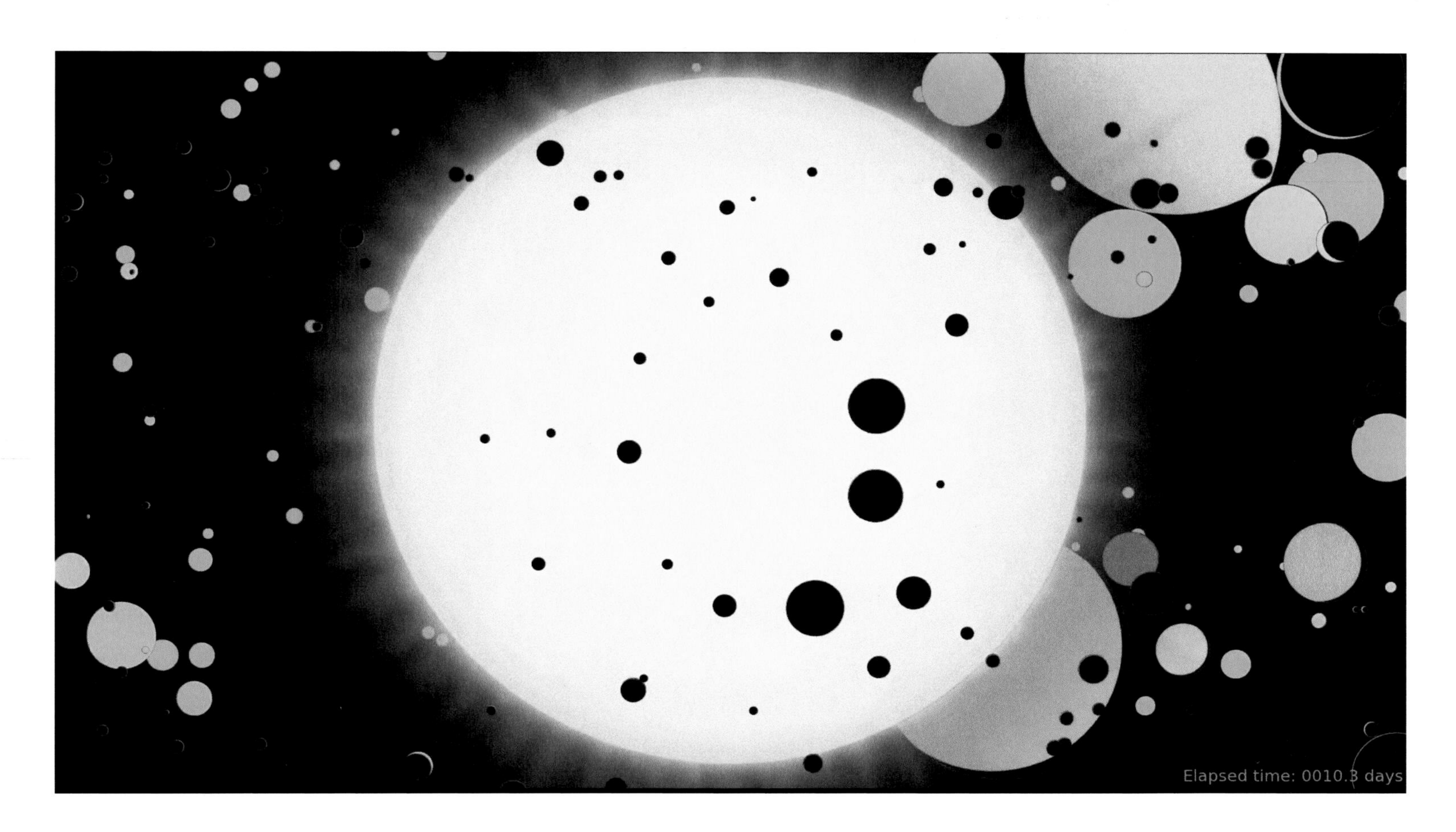

• 2013:

Alex Parker, an astronomer and planetary scientist at the Harvard-Smithsonian Center for Astrophysics, hit on another way of illustrating the bounteous crop of Kepler exoplanets: He created an animated film. Titled *Worlds: The Kepler Planet Candidates*, it presents 2,299 candidate worlds as seen orbiting a single star. (In fact, those planets orbit around a total of 1,770 stars.) The result is a whirling blizzard of planets, underlining the magnitude of Kepler's accomplishment. (The colors indicate their estimated temperatures.) As with Daniel Fabrycky's orrery, the stills presented here can only hint at the effect of seeing the film, but it's easy to find online.

CHAPTER 

Constellations, the Zodiac, and the Milky Way

The sky is colorless and set
With Constellations pale as milk

—Guillaume Apollinaire, *Alcools*

There's something moving about the constellations, and it's not their endless westerly roll enforced by Earth's rotation. These stellar asterisms, or coincidental alignments of stars, have functioned as clothes hangers on which to drape the forms of mythological figures since the dawn of recorded history. What we have with the constellations is a kind of stellar Rorschach test onto which the deepest collective human concerns and stories have been projected. If their utilitarian function is as a mnemonic device—a kind of easily remembered celestial coordinate system—their mix of animal, human, and mythopoetic forms also provides a revealing communal scratch pad laid across the universe. Anyone tempted to argue that we humans don't feel at home in the cosmos should consider the comfort level behind inscribing the animals, herdsmen, charioteers, sea monsters, dragons, cups, and crowns of Earth on the domed fabric of space.

Of course, another way to look at the constellations is that they're the way we have found to humanize the cold light of the stars—to bring the celestial sphere home to the human one. As with the bison and bears inscribed on the rock walls of Chauvet and Lascaux, they're artifacts of the human need to see recognizable patterns within nature and to fix them in time, by pinning them to the heavens or painting them on the wall. And the constellations have a definite cartographical significance as well. They enable celestial objects to be located within recognizable territories on the sky's starry dome, functioning as a kind of combination scaffolding and aide-mémoire.

Many of the eighty-eight constellations recognized in the Western world today are in the southern skies, and were unknown to classical astronomers. But nearly all of Ptolemy's forty-eight constellations are still there as well. These descended from Babylonian sources, which derived them in turn from their Sumerian predecessors.

Most cultures have constellations of one kind or another. The Hindu system shares many similar figures with the one originating in Mesopotamia. The Chinese have constellations entirely unconnected to those of the Fertile Crescent, as do the Hawaiian, Polynesian, and Australian aboriginal cultures. The Hawaiian constellation "The Canoe-Bailer of Makali'i" rises like a cup in the east, encompassing what we call Orion and Taurus, and then pours its contents into the sea as it sets in the west. The indigenous tribes of the Atacama Desert saw figures such as llamas in the heavens by discerning forms not as defined by the stars themselves, but rather due to the irregular dark areas of the Milky Way occluded by interstellar gas and dust. Aboriginal groups also saw animistic shapes in galactic negative space—a giant emu, for example, defined by the Coalsack Nebula and neighboring tendrils of obscuring interstellar material.

While we tend to visualize them as chalk stick-figures on a blank slate, the modern constellation system is more like a map of regions crowded together with no gaps in between, divided into eighty-eight areas of the celestial sphere that together cover the entire sky. So, while we think of Gemini as the asterism known as the twins, or Castor and Pollux, if you look on the constellations map you'll find them eerily twinned for eternity within an irregular boxy border vaguely reminiscent of a generic midwestern state. Gemini borders on Taurus and Orion to the east, and Cancer to the west, with the line of the ecliptic running like an interstate directly through them.

Although one might think that the Milky Way would be the superhighway, rather than the invisible path followed by the sun and the planets, in fact the ecliptic is far more important. It was divided into twelve zodiacal signs associated with constellations during the first half millennium B.C., with some of its signs—Gemini and Cancer among them—known to come from even older Bronze Age sources. The twelve-sign zodiac created a celestial longitude system with divisions based on the lunar calendar, and was used to determine planetary positions through time. The Babylonian zodiac permeates the Hebrew Bible as well; some scholars link the organizing principles of the twelve tribes of Israel to its signs, which if true provides an intriguing connection between what we now know as astrology and Old Testament verities.

Concerning the alleged influence of the stars on human destiny, although Ptolemy's *Almagest* was rendered largely irrelevant with the arrival of Copernican heliocentrism, his companion book *Tetrabiblos* (Four Books) codified Babylonian zodiacal signs and astrological ideas, and remains at the core of contemporary astrology. It systematized a belief system in which celestial cycles are thought to impact the atmosphere, affecting its heating, cooling, drying, and humidity. "Most events of a general nature draw their causes from the enveloping heavens," Ptolemy wrote.

The oldest known circular (or planispheric) depiction of the classical constellations was found on a temple ceiling in Egypt dating back to the first century B.C. When Napoleon invaded Egypt in 1798, he brought along artist-archaeologist Vivant Denon, who spotted an intriguing bas-relief on the ceiling of a temple in Dendera (see page 217). It's now understood to be a copy of the Mesopotamian zodiac, and contains constellations in forms familiar to later Greek and Roman zodiacs, including Taurus, Libra, Scorpio, and Capricorn, as well as others in previously unknown Mesopotamian-Egyptian incarnations. Like so many precious antiquities, Denon's find was soon expropriated, and is now in the Louvre.

Star mapping in graphic form within manuscripts followed Ptolemy's *Almagest* star catalog by almost a thousand years, with one of the first attempts to present the constellations graphically being Persian astronomer Abd al-Rahman al-Sufi's A.D. 964 *Book of Fixed Stars*. Al-Sufi translated and corrected the *Almagest*, revising Ptolemy's magnitude indicators. He also started the tradition of depicting each constellation both as seen from Earth and in mirror image. The latter is supposed to represent its appearance when seen from *outside* the outermost celestial sphere, the star-bearing globe that was assumed to rotate beyond Saturn. Hundreds of manuscript copies of the *Book of Fixed Stars* were made in succeeding centuries. The book contains the first recorded mention of a celestial object external to the Milky Way—what we now know to be the Andromeda Galaxy.

In Europe, star mapping dates back to illustrations in various manuscript copies of Ancient Greek poet Aratus's book *Phaenomena*, which described the Greek constellation in some detail. (In this, Aratus preserved the work of Greek astronomer Eudoxus of Cnidus, who wrote two otherwise long-lost works on the constellations circa 370 B.C.) Some early medieval copies of the book contained planispheres not unlike the Dendera temple ceiling, as visible on page 218. Celestial cartography really came into its own, however, at the end of the Renaissance, and extended to the end of the seventeenth century, a period when letterpress books were increasingly augmented by prints made using copper and steel plates, which were frequently hand-colored.

In 1515, German Renaissance artist Albrecht Dürer produced the first printed star maps. His two planispheres, one of the northern and one of the southern constellations, were both depicted from an "outside-in" perspective. Although based on Ptolemy, Dürer's maps incorporated information from other astronomers, including al-Sufi. His depiction of the southern sky is particularly interesting because it's so underpopulated. (See page 226.) Its gaps simply reflect a lack of information; the Age of Exploration was just starting, and the constellations of the southern hemisphere were not yet described. Less than a century later, after numerous South Seas expeditions, twelve new southern constellations would be added by German celestial cartographer Johann Bayer in his landmark *Uranometria* star atlas.

Probably the most striking European stellar maps can be found in Dutch-German cartographer Andreas Cellarius's 1660 magnum opus *Harmonia macrocosmica* (Cosmic Harmony). This lavish demonstration of the printer's art contained twenty-nine double-page plates depicting the rival proposed designs of the cosmos, as well as exceedingly innovative depictions of the constellations. In particular, four plates presented an entirely new and inclusive way of looking at the universe. While in many star atlases the constellation figures were reversed, none depicted those reversed constellations with anything behind them. In *Harmonia macrocosmica*, however, the unknown engraver employed by Cellarius's Amsterdam printer had an insight: If Earth is at the center of the starry sphere, and it is depicted as though seen from the outside, then Earth should be visible behind its crystalline wall as well. As a result, in the plates reproduced on pages 233–35, we see the northern and southern skies in front of their respective hemispheres of Earth—an epic merger of celestial and terrestrial cartography.

Despite his grand treatment of the sun at the center of the planetary system in the early part of the atlas, Cellarius managed to avoid its inclusion in the index of books banned by the Catholic Church by including the rival system of Tycho Brahe as well as the classical Ptolemaic geocentric model in *Harmonia macrocosmica*. It couldn't have hurt that he also included two double-page planispheres in which the classical pagan constellations had been replaced with a teeming panoply of biblical figures. These were based on a Christianized constellation scheme first presented by German lawyer Julius Schiller in 1627, as visible on page 232. Schiller's biblical constellations never caught on, however, and remain an odd footnote in the history of celestial mapping.

The opaque Milky Way was a curiously unimportant feature in the classical constellations, which are all based on resolvable stars. Even before Cellarius, however, English astronomer Thomas Digges had discarded the concept of a fixed stellar sphere, proposing instead that they were scattered in a vast space beyond the solar system (though he never grasped that they might all be parts of a single galaxy). (See page 228.) Within a few decades of his 1576 publication of a Copernican solar system depicted with stars salted across the page on the outside, Galileo's telescope revealed that the Milky Way was comprised of innumerable individual stars. And as touched on in the introduction and in chapter 5, by 1750 English astronomer Thomas Wright had intuited that the galaxy could be shaped as a flattened disc.

Although Wright depicted two possible galactic forms in his 1750 book, they were not based on systematic observational evidence. In 1785, one of the great observational astronomers of all time, William Herschel, tried to use a technique he called "star gauging" to measure the Milky Way's shape. Herschel's results fell short, however, in part because his method assumed that stars are distributed uniformly, which is not the case, and also because he assumed he could see all the stars above a certain magnitude within a given distance—which was also not so, due to the galaxy's obscuring gas and dust. In fairness to Herschel, many particulars concerning the Milky Way's structure remain in dispute to this day, although we now know its general form: it is a flattened disc with spiral arms. For Herschel's Milky Way, see page 238.

By the mid-twentieth century, star mapping was achieving new levels of sophistication, largely because of the meticulous work of Czech astronomer Antonin Bečvář. Working in the observatory he had founded during the Second World War in Skalnate Pleso, in the high Tatra mountains, Bečvář compiled a vast index of deep-sky objects, including stars, galaxies, nebulae, and interstellar dust clouds. In 1948, he published the first edition of his *Skalnate Pleso Atlas of the Heavens*, which contained stars down to a magnitude of 6.25, or very dim indeed. It became an instant success on the international market, with most observatories and many thousands of amateur astronomers acquiring copies. Bečvář had set a new standard, and his sixteen hand-drawn charts became the basis for follow-on celestial atlases.

Anyone examining Bečvář's scrupulous star charts who is already familiar with previous celestial maps will soon come to an intriguing realization. While the constellation figures of antiquity had appeared in thousands of incarnations across the centuries, and were virtually synonymous with the heavens for more than three thousand years, in the *Skalnate Pleso Atlas* they're nowhere to be seen.

• 50 B.C.:

When Napoleon Bonaparte invaded Egypt, he brought along artist-archaeologist Vivant Denon specifically to carry out cultural research. Among other discoveries, Denon spotted an intriguing bas-relief on the ceiling of a temple in Dendera: He'd stumbled on the first-known depiction of the classical zodiac. The zodiac is a band of sky encompassing an area about eight degrees on either side of the sun's apparent path, divided into twelve equal segments. It functions as a celestial coordinate system, and is filled by constellations, frequently represented by such figures as Taurus the Bull and Libra, a set of scales. Although the twelve-part zodiac is known to date back to Babylonian astronomy of the first millennium B.C., its representation in circular (or planispheric) form was previously unknown. The Dendera temple ceiling relief has been interpreted as a copy of the Mesopotamian zodiac, and contains constellations in forms that are familiar to later Greek and Roman zodiacs, including Taurus, Libra, Scorpio, and Capricorn, as well as others in previously unknown Mesopotamian-Egyptian incarnations. Based on a study of planetary configurations and eclipse representations on that stone ceiling, researchers have dated it to 50 B.C.: the Ptolemaic period. Denon's find was later removed to Paris and is now in the Louvre.

舊藏天象列次分野之圖

• Circa 1 B.C.– A.D. 6:

This Korean planispheric depiction of constellations was published in 1777, but is based on an engraving made on a stone column in 1395. Korean scholars who have studied it estimate in turn that the configuration for those stars dates some of these constellation figures, none of which are given human or animal forms, to between 1 B.C. and A.D. 6. That information has been handed forward, like a baton in a race, through the centuries.

• Ninth Century:

This medieval planisphere depicting the constellations is from a parchment codex of a Latin translation by Marcus Cicero of Ancient Greek poet Aratus's sole surviving major work, *Phaenomena*. (Aratus lived circa 310–240 B.C., or two centuries before the Dendera ceiling zodiac was made, but the image is from much later.) The poem gives an introduction to the constellations, and this particular copy contains images thought to have derived in turn from constellation figures in Latin writer Gaius Julius Hyginus's *Poeticon astronomicon* (Poetic Star Atlas). Hyginus lived circa 64 B.C.– A.D. 17, and could theoretically have witnessed the Dendera ceiling being made.

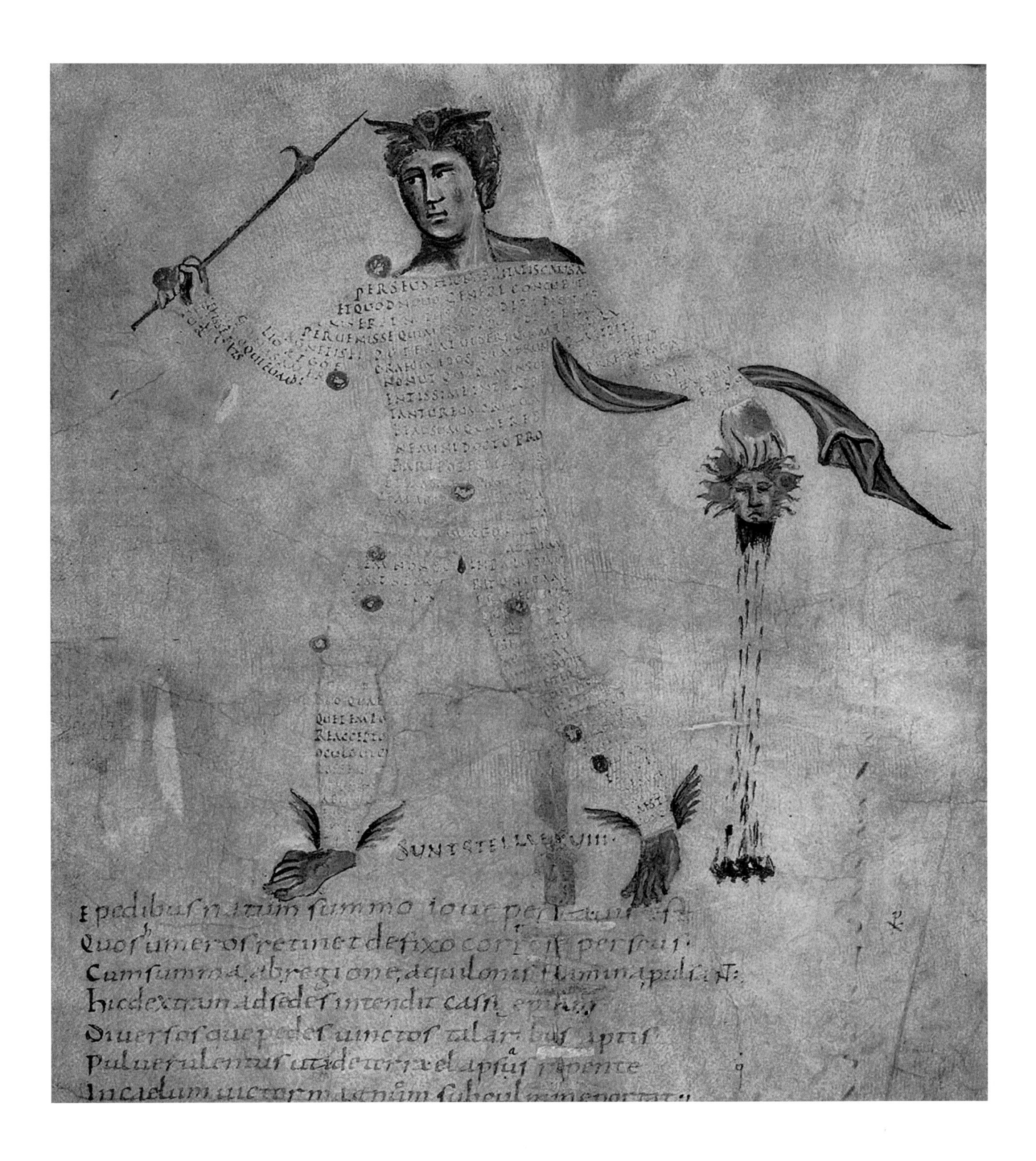

Depiction of the constellation Perseus from *Phaenomena*. Studded by the stars of his eponymous constellation, the Greek hero wields Medusa's oddly shrunken head. The stars just provide a stick figure; in fact Perseus's body is entirely defined by Latin letters—the writing of Hyginus. Below it is a second level of text: Cicero's translation of Aratus. There's a strangely contemporary, hypertextual quality to this thousand-year-old manuscript.

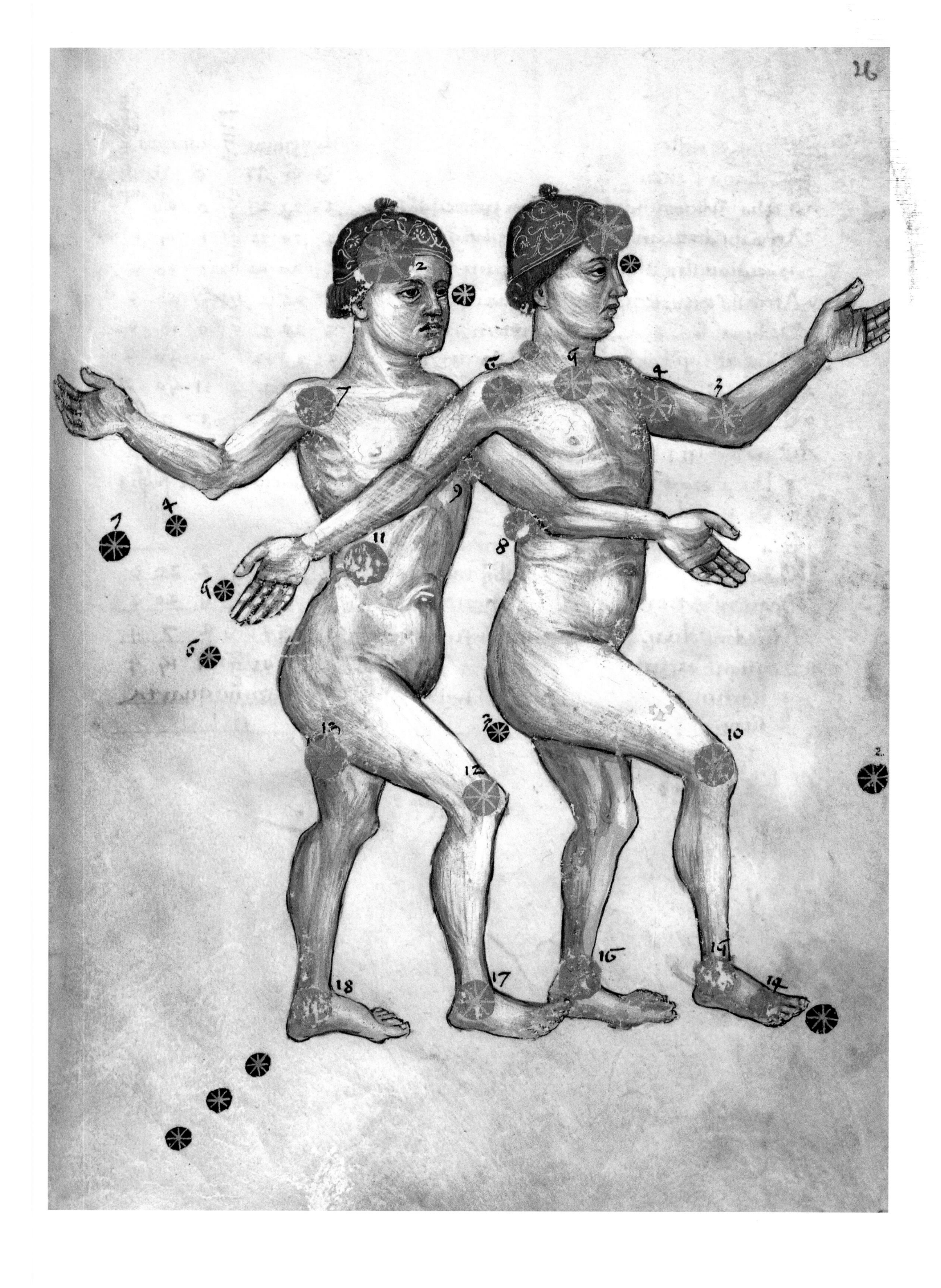
26

• 1428:

The role of the Middle East in the history of astronomy can scarcely be exaggerated. Arab and Persian astronomers did far more than preserve the work of the Ancient Greeks while Europe went through centuries of medieval indifference. In 964, Persian astronomer Abd al-Rahman al-Sufi wrote his heavily illustrated, Arabic text *Book of Fixed Stars*. It combined his own astronomical research and that of the Arab and Persian worlds with Greek-Alexandrine astronomer Claudius Ptolemy's treatise the *Almagest*. It also contains the first recorded mention of two nearby galaxies: Andromeda and the Large Magellanic Cloud. This bizarre depiction of Gemini, the twins, is from the same fifteenth-century edition of al-Sufi's book. The twin bright stars of Castor and Pollux can be seen at each head; in Greek mythology, when his twin brother, Castor, was killed, Pollux asked Zeus to share his immortality and keep them together forever. Zeus then placed them in the constellation Gemini for eternity.

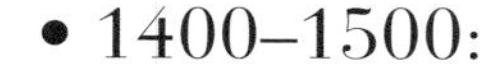

• 1400–1500:

Italian poet Francesco Petrarch is credited with sparking the Renaissance, partly by discovering and promulgating Roman philosopher Cicero's letters, and partly with his own works. These masterful illuminations are from the first French manuscript of Petrarch's *Triumphs*, probably commissioned for King Louis XII. They illustrate the "Triumph of Time," and the sun's progress across the zodiac and background stars as it steals human accomplishments away. Although the zodiac is defined by the sun's path, it's extremely rare to see the sun and stars presented together, as here, let alone with zodiacal figures as well. Below, two triumphs are literalized. (A "triumph" in Ancient Rome was a parade for a victorious general on returning to the capital—sometimes with elephants, as here.) A central figure on both pages is Petrarch's muse, Laura, a married woman who he pined for productively throughout his life.

le triūphe du temps

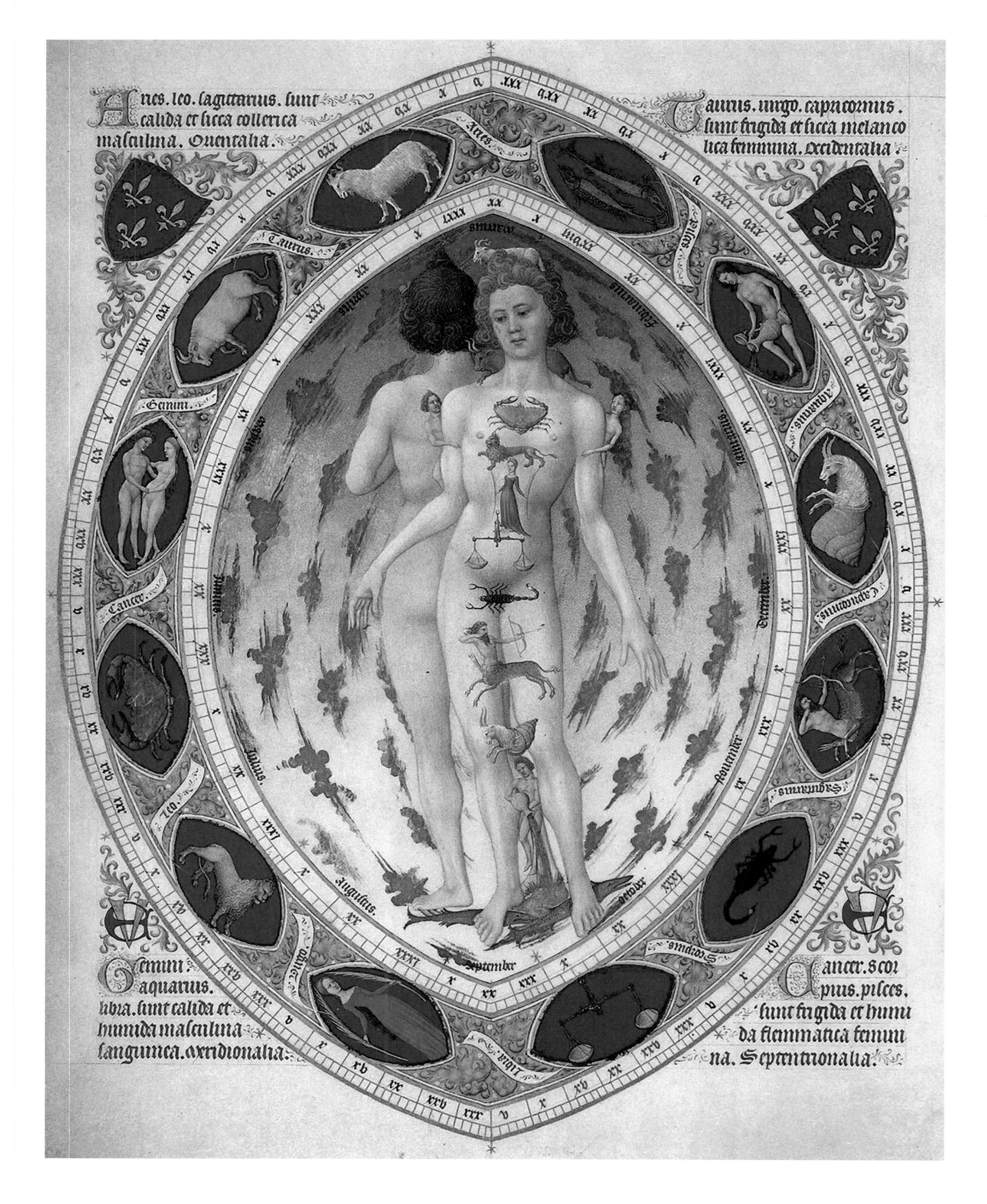

• 1412–16:

Manuscript illumination of an Anatomical Zodiac Man, or *homo signorum*, from the famous book of hours, *Très riches heures du Duc de Berry* (The Very Rich Hours of the Duke of Berry). For much of history the signs of the zodiac were thought to govern the health of parts of the body. Here, astrological signs can be seen from the head (Ares) to the feet (Pisces; he's standing on fish!). Latin inscriptions in each corner detail the supposed properties of these signs. This illumination can be understood as belonging to medical astrology, in which decisions on advanced medical treatments such as bloodletting were conducted with reference to the stars.

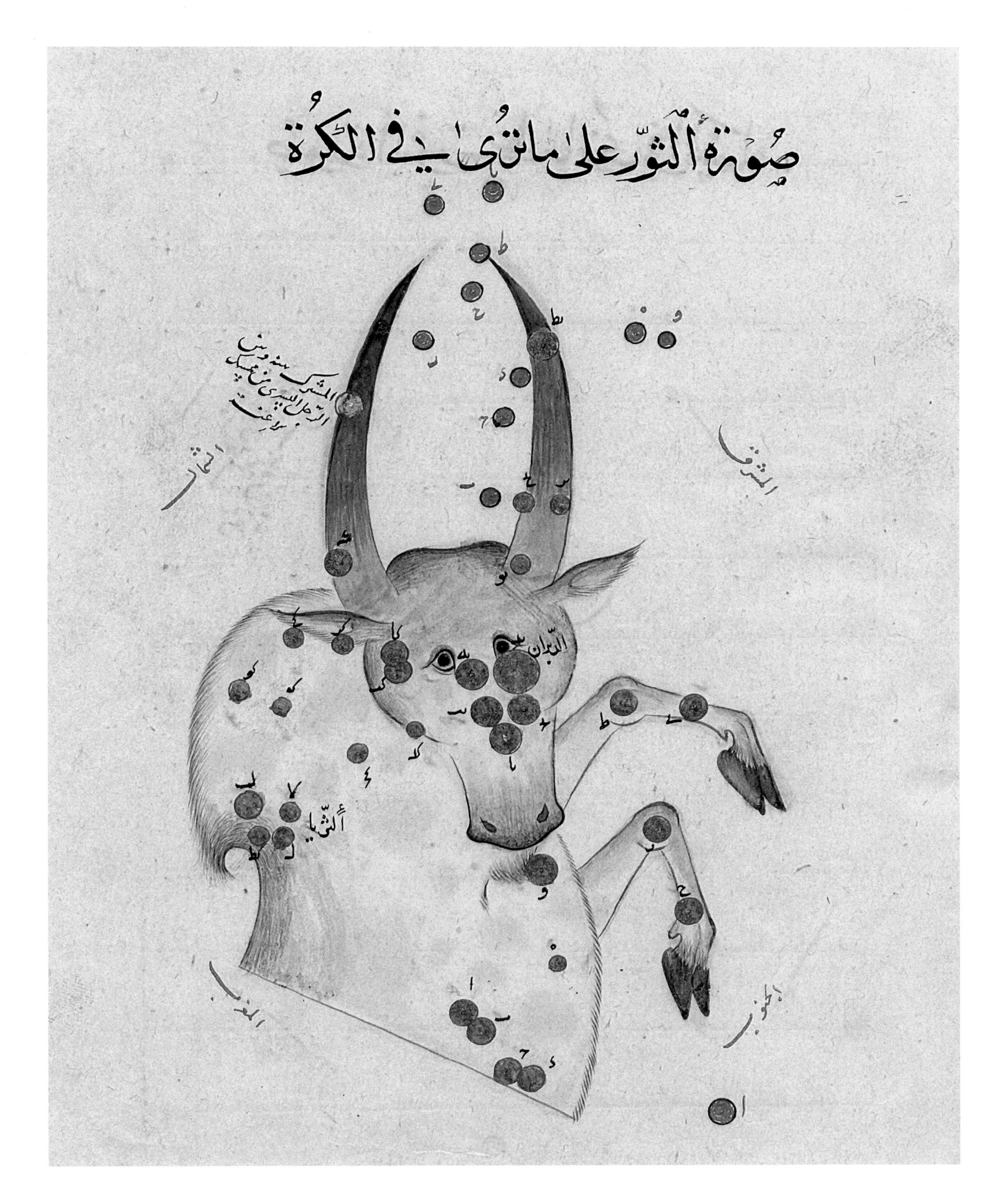

• 1436:
In this illumination of Taurus the Bull from an updated and annotated edition by Ulugh Beg of Abd al-Rahman al-Sufi's *Book of Fixed Stars*, the gold stars represent those that Ptolemy listed as internal to a constellation, and the red ones are those near that constellation. As with the al-Sufi illuminations on page 220, a star's size represents its magnitude; one of the Persian astronomer's major contributions was in fact his tabulation of star brightness. This particular illumination contains a star catalogued by Ptolemy that can't be corroborated by actual observation. Ulugh Beg indicates its presence by circling it in red (it's in the left horn).

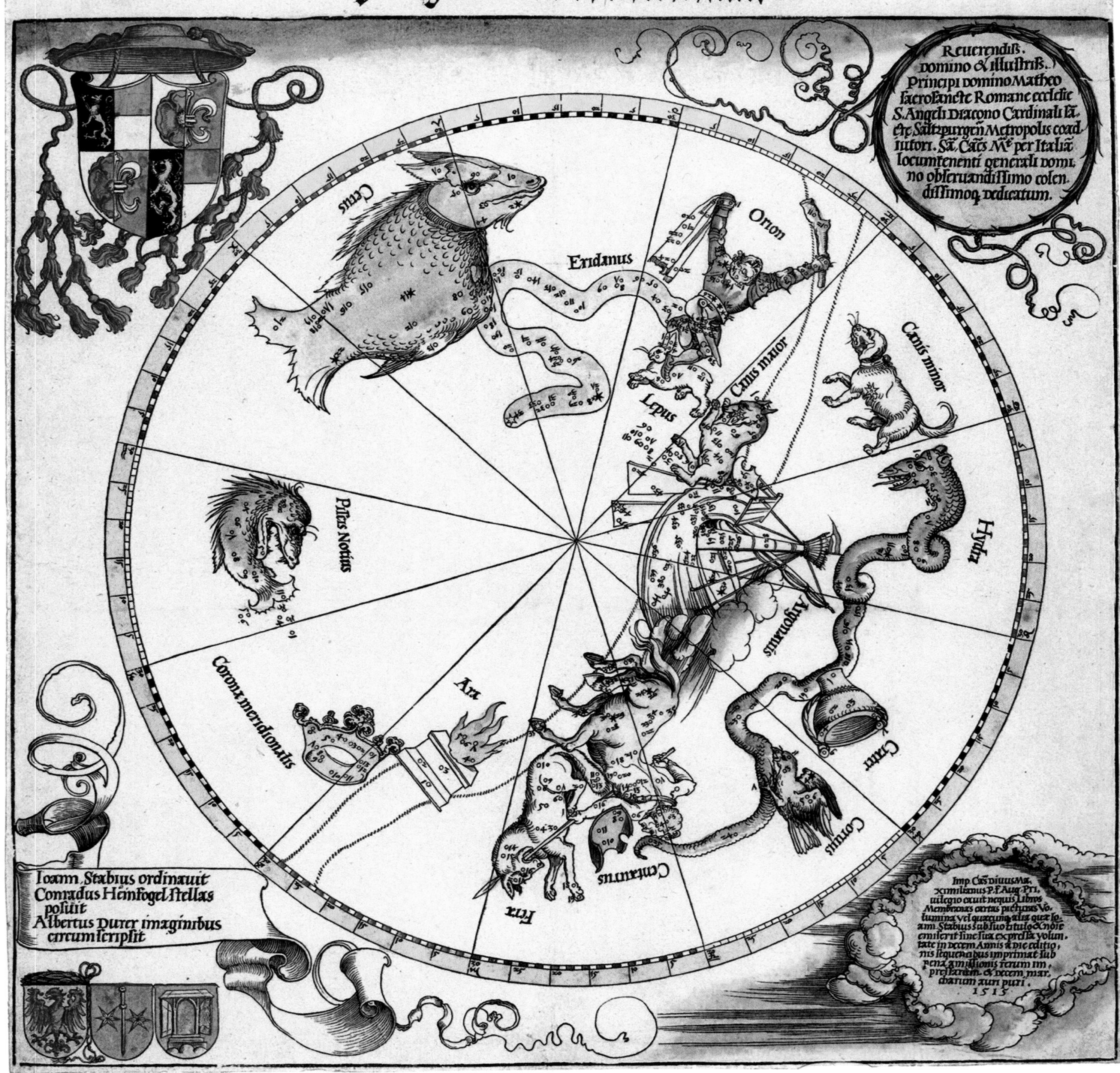

• 1515:

The intersecting lines in this polar stereographic projection, or planisphere, of the southern constellations by German Renaissance artist Albrecht Dürer give it the distinction of being the first printed star map. An enthusiastic amateur astronomer, Dürer made this woodcut and a companion map of the northern constellations in collaboration with two professional astronomers, Johannes Stabius and Konrad Heinfogel, who devised its coordinate system and provided its star positions. Here, the constellations are seen as though from outside the celestial sphere, reversing their orientations as seen from Earth (a so-called convex projection). Although based on Ptolemy, Dürer's map incorporated information from other astronomers, including al-Sufi. Empty areas of the map reflect a lack of information concerning the southern sky; the Age of Exploration was just starting.

• 1540:

This volvelle containing a planisphere of the constellations visible from the northern latitudes by German printer, mathematician, and cosmographer Peter Apian's *Astronomicum Caesareum* (Caesar's Astronomy) borrows directly from Dürer's map of the same (the companion to the one on the facing page). The first volvelle in Apian's book-instrument, it's in many ways the conductor and key to the succeeding volvelles. This paper dial is meant to turn one full revolution every thirty-six thousand years, which was the rate of precession according to Ptolemy; in fact, because by Apian's time this was known to be incorrect, the oval directly beside Cetus's fishy tail on the planisphere compensates by providing an auxiliary scale. (Precession, which is the movement of the axis of a spinning body around another axis due to torque, gradually causes the terrestrial poles to trace the shape of a cone over a long period of time; as a result, the position of the pole star shifts over thousands of years.) Each of the tabs visible at the edge of the volvelle represent a planet; they can be set to compensate for the Earth's incremental precession. After being correctly set, the positions established with this volvelle are meant to be exported to successive volvelles in the book. For more from Apian, see pages 43, 77, 119, 180, and 262.

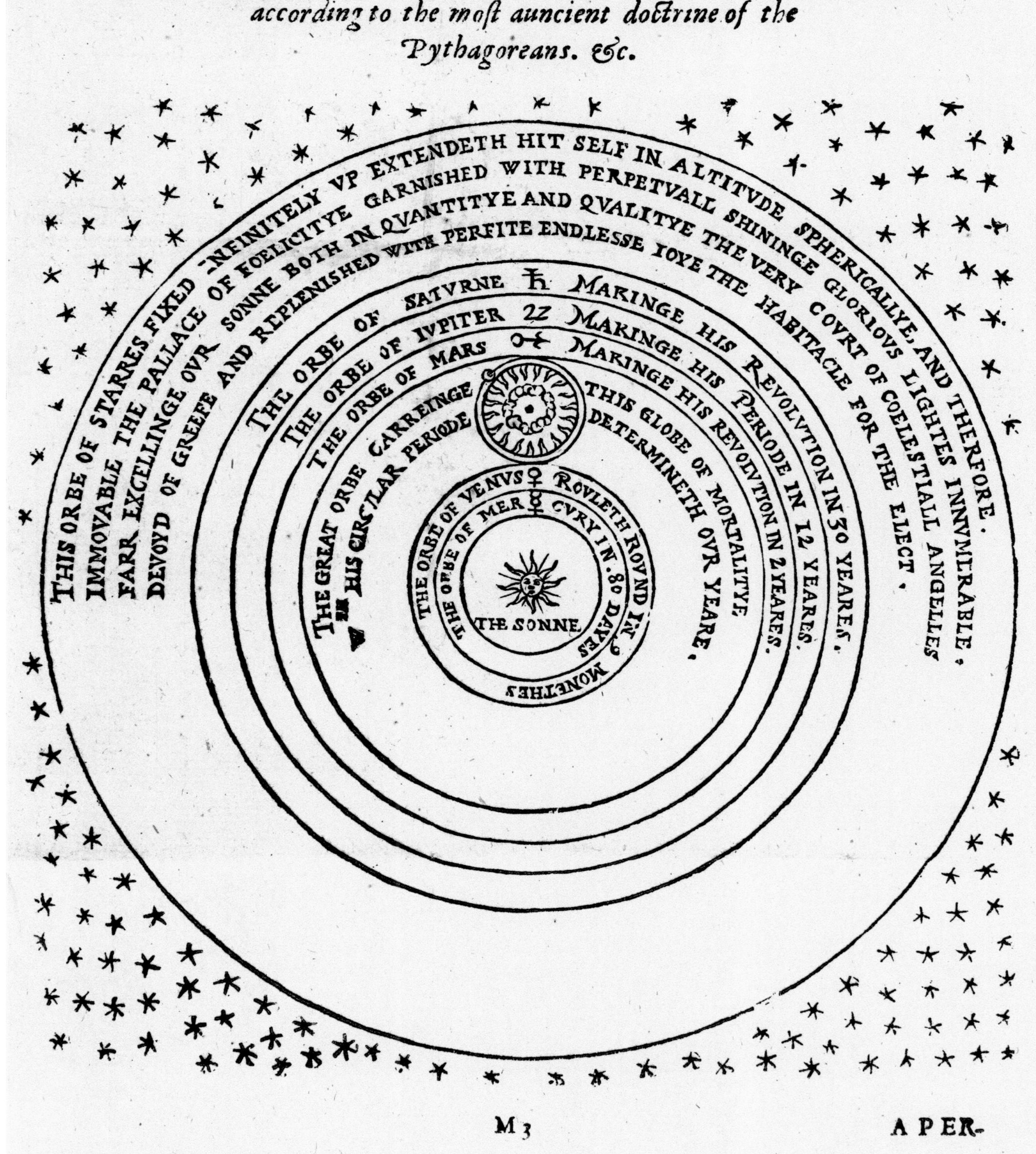

• 1576:

This print from a book by English astronomer Thomas Digges was the first to discard the notion of a fixed sphere of stars. Digges wrote a long appendix to his father's almanac, *A Perfit Description of the Caelestiall Orbes*. It contained the first published presentation in England of Copernicus's heliocentric universe, but also implicitly posited an endless universe. In this print, the sun rules its planets, but evidently has no authority over the stars.

• 1603:

German cartographer Johann Bayer's celestial atlas *Uranometria* is considered a big advancement on prior star maps, in part because it contained a system whereby stellar magnitudes are designated by Greek and Roman letters, and in part because its grid system allowed their positions to be measured with a high degree of accuracy. The maps shown here are as seen from Earth, not from outside the celestial sphere. These hand-colored plates depict the constellations Auriga (**top**) and Argo Navis (**bottom**). The lower constellation, a boat containing Jason and the Argonauts, is the only one of Claudius Ptolemy's original forty-eight constellations that's no longer recognized. Ptolemy's ship was broken up by astronomers in the late eighteenth century because of its immense size, and is now Carina, Puppis, Pyxis, and Vela (for keel, poop deck, mast, and sails). As for Auriga's shepherd, he has an interesting genealogy: the stars behind him were recognized in Mesopotamian astronomy as a shepherd's crook, and later by Bedouin astronomers as a herd of goats. This eventually merged, becoming a shepherd with goats, as here. In Bayer's depiction, he and his whiplashes look weightless, as though flying through the Milky Way.

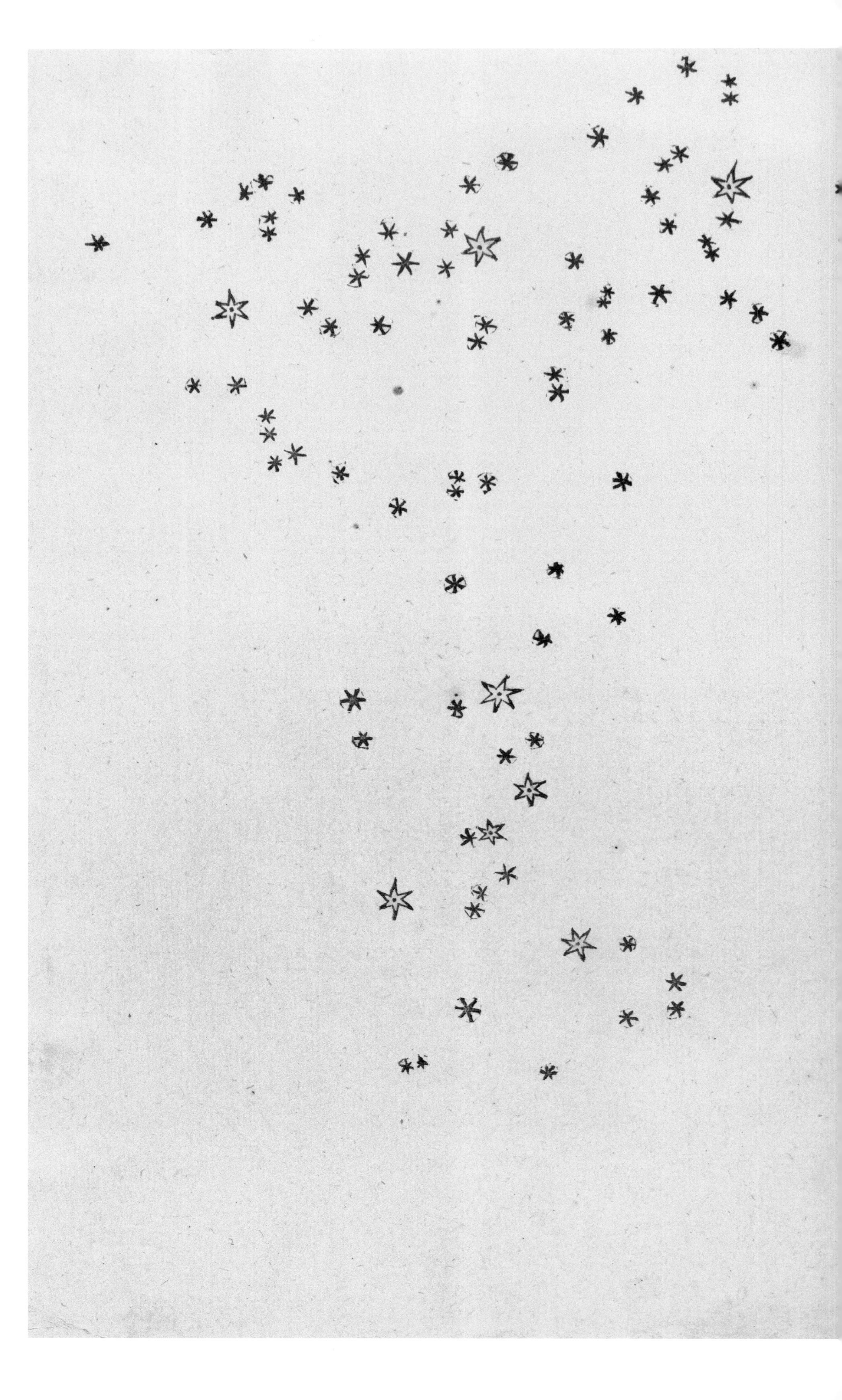

• 1610:

During his revelatory nights of observation using a telescope in the winter of 1609–10, Galileo observed the stars as well as the moon and planets. He discovered that while they didn't assume the shapes of discs, like planets, many more of them were visible than could be seen with the naked eye. (Later, he contradicted himself, asserting that he could see stellar discs—something even the most powerful telescopes can't. He was confused by the diffraction produced by his instrument.) In this depiction of the Pleiades open star cluster from *Sidereus nuncius* (Starry Messenger), a new plenitude of stars seems to burst off the pages. Only six of the Pleiades can be seen by the naked eye on a clear night from Earth; at a stroke, Galileo's telescope had multiplied that number many times. For what is thought to be a Bronze Age depiction of the same star cluster, see page 73.

PLEIADVM CONSTELLATIO.

Quòd tertio loco à nobis fuit obſeruatum, eſt ipſiuſmet LACTEI Circuli eſſentia, ſeu materies, quam Perſpicilli beneficio adeò ad ſenſum licet intueri, vt & altercationes omnes, quæ per tot ſæcula Philoſophos excrucia runt ab oculata certitudine dirimantur, noſque à verboſis diſputationibus liberemur. Eſt enim GALAXYA nihil aliud, quam innumerarum Stellarum coaceruatim conſitarum congeries; in quamcunq; enim regionem illius Perſpicillum dirigas, ſtatim Stellarum ingens frequentia ſe ſe in conſpectum profert, quarum complures ſatis magnæ, ac valde conſpicuæ videntur; ſed exiguarum multitudo prorſus inexplorabilis eſt.

At cum non tantum in GALAXYA lacteus ille candor, veluti albicantis nubis ſpectetur, ſed complures conſimilis coloris areolæ ſparſim per æthera ſubfulgeant, ſi in illarum quamlibet Specillum conuertas Stellarum conſtipatarum

cętum

• 1627:

Above: Augsburg lawyer Julius Schiller may have been inspired by his fellow citizen Johann Bayer's *Uranometria* to publish his own star atlas, something he accomplished with Bayer's help. As its title promises, *Coelum stellatum Christianum* (Christian Heavenly Stars) replaced the old pagan constellation with Old and New Testament figures, with the twelve zodiacal constellations taken over by the twelve apostles, and Jason and the Argonaut's multi-oared golden fleece–seeking Greek galley was replaced by Noah's Ark, as here. Schiller's Christian constellations never caught on, although Andreas Cellarius does present versions of them in his 1660 celestial atlas *Harmonia macrocosmica* (Cosmic Harmony).

• 1660:

Right: This plate from Andreas Cellarius's *Harmonia macrocosmica* is one of four that present a new way of looking at the celestial spheres. In Dürer's southern sky and Apian's northern one on pages 226 and 227 respectively, our vantage point is from the outside looking in—but there is no backdrop. Here, the unsigned engraver employed by Cellarius's Amsterdam printer had a rare insight: If Earth is at the center of the starry sphere, and that crystalline sphere is seen from the outside, then Earth should be visible, too. Here, we see the southern sky and southern hemisphere together—a merger of celestial and terrestrial cartography. For other maps from Cellarius, see pages 47, 84, 122, 146–47, 148–49, 182–83, and 234–35.

Overleaf: Detail of a similar view of the northern hemisphere from Cellarius's atlas, in which Earth and sky are seen together as though from outside the sphere of the stars.

VM SCENO
AUSTRALE
STELLA:
TERRÆ.
Prestant Amstelaedami apud
Petrum Schenk et Gerardum Valk
Perseus
Andromeda
Pegasus
Triangulum
Apis
Auriga
TAURUS
ARIES
PISCIS
ÆQUINOC
Cetus Bal.
Orion
OCEANUS
ÆTHIOPICUS
Lepus
Phoenix
Piscis Aust.
Grus
Indus
Toucan
Hydrus
Dorado
Columba Noachi
Fluvius Eridanus
Canis major
Canobus
Pavo
COGNITA
Apis Indica
Chameleon
Arca Noachi
28

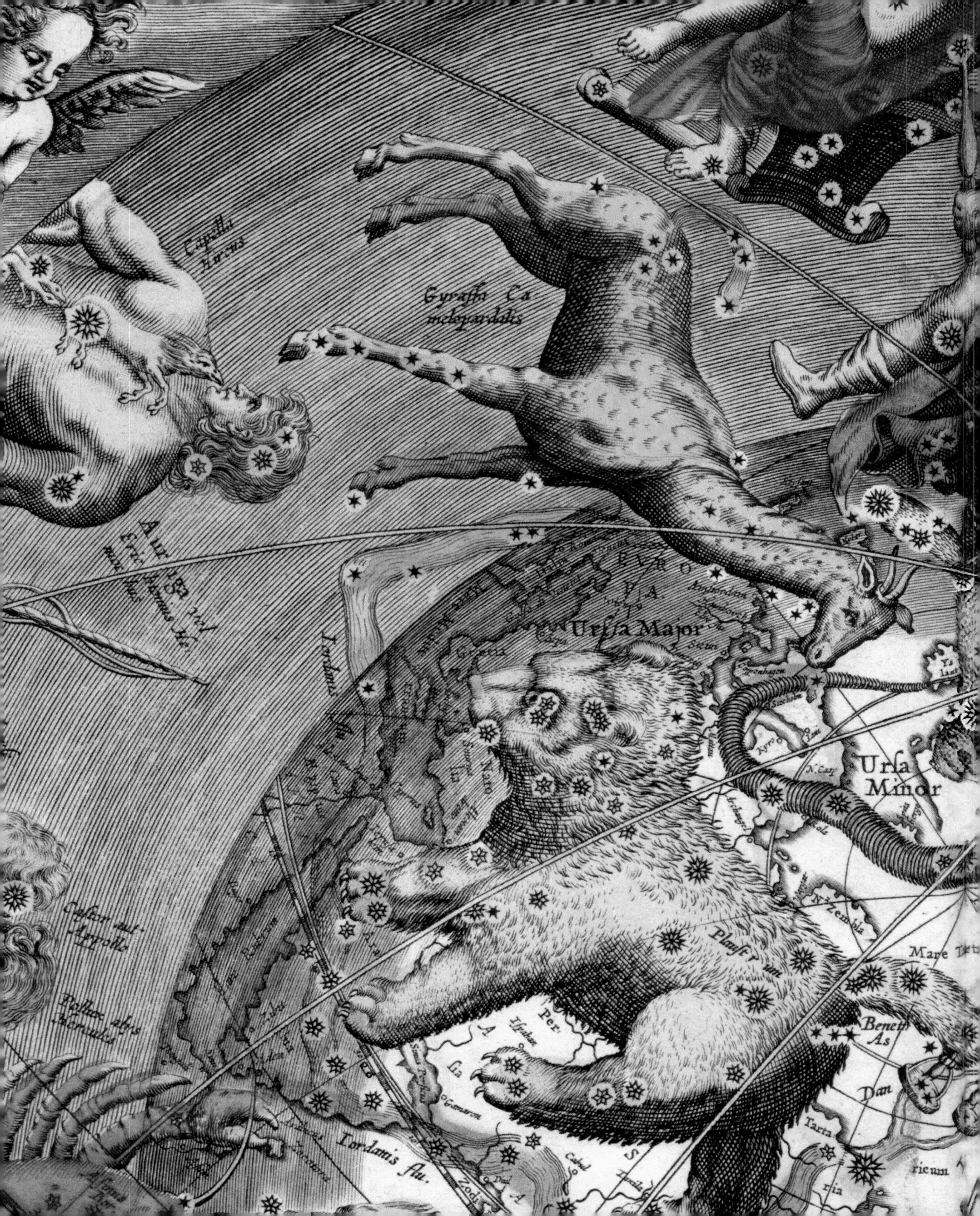

Capella Hircus
Gyraffa Camelopardalis
Ursa Major
Ursa Minor
Plaustrum
Castor aut Appollo
Pollux aut Hercules
Iordanis flu.
N. Zembla
Mare Tartaricum

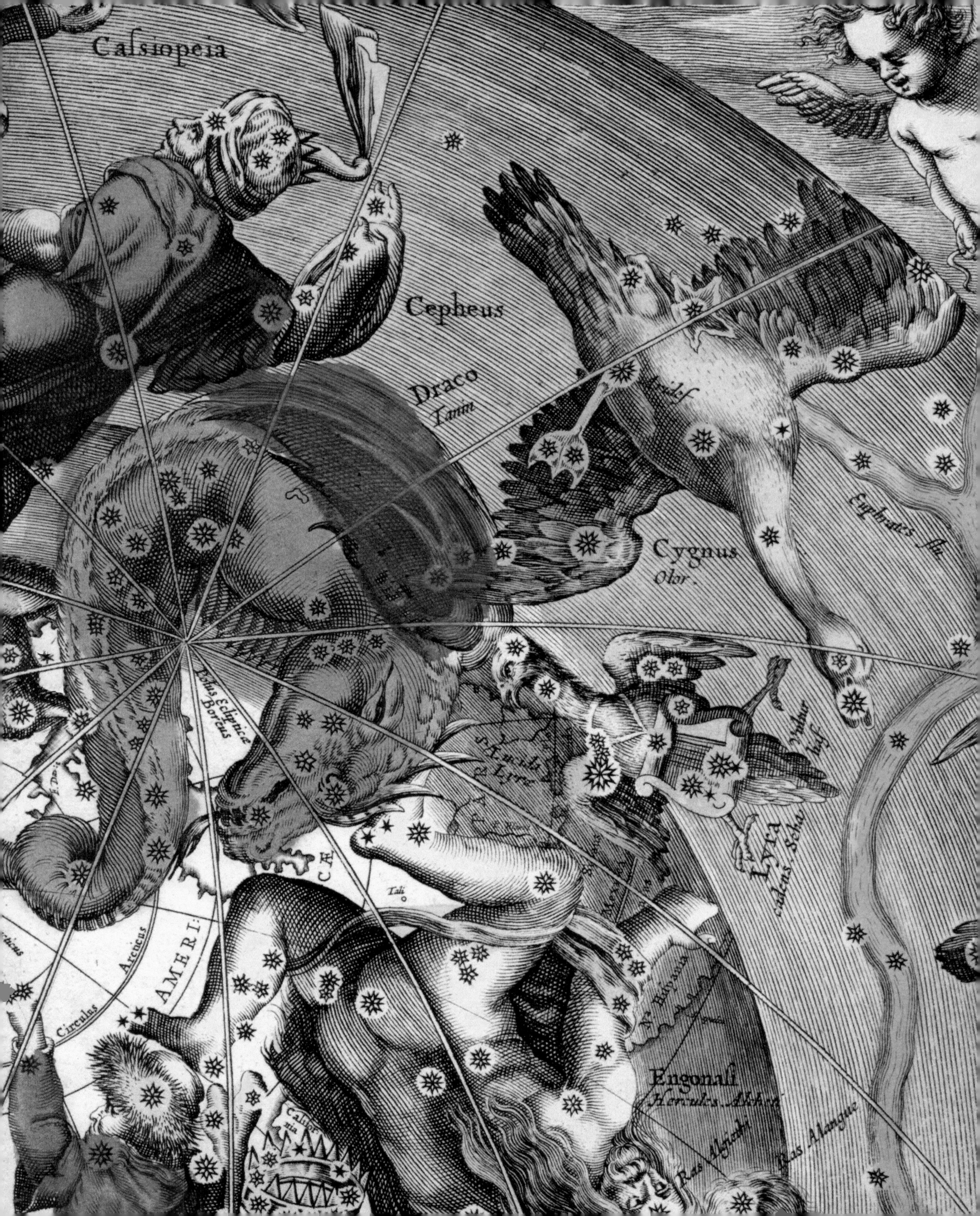

Cassiopeia
Cepheus
Draco
Tanin
Cygnus
Olor
Lyra
Engonasi
Polus Eclipticæ Borealis
AMERI
Arcticus
Circulus
Ras Alangue

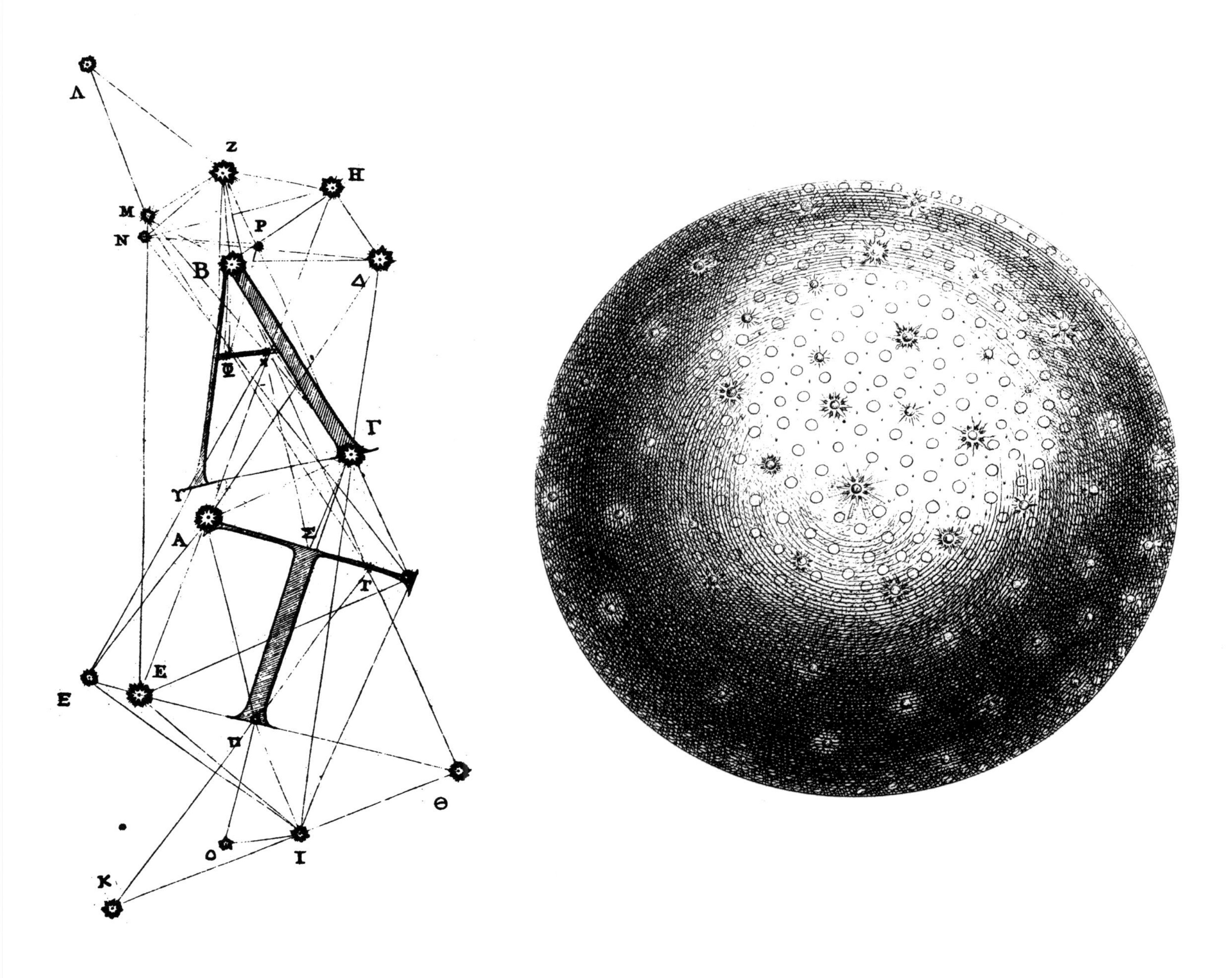

• 1750:

At the end of the eighteenth century, English astronomer Thomas Wright had taken to profitably questioning previous certainties concerning the form and structure of the universe. In his 1750 book, Wright presented his readers with a theory: *"all the Stars are, or may be, in motion . . ."* (italics per the original). **Above left:** Wright invents a new kind of constellation figure in order to test his hypothesis, taking the Pleiades star cluster as a test subject (the cluster is also seen on pages 230–31). Wright proposes that if any of the stars that form the letter A or T are found to deviate from their positions in ten or twenty years, then his theory will be vindicated. **Above right:** Wright posited that one form a vast collection of moving stars might take is "a general motion . . . round a common centre," as illustrated here. He had intuited the shape of what we now call an elliptical galaxy, many of which are almost spherical. **Facing page:** Wright proposes another organizing principle for a "kind of regular irregularity of objects," or a galaxy-size collection of individual stars. Asking the reader to imagine them "extended like a plane," he puts Earth at A, and uses the other letters to illustrate how the view from Earth would be of a "perfect zone of light" on one plane—exactly as the Milky Way appears from the vantage point of a solar system inside it. He then states that if we agree with the premise that they must all be in motion, it stands to reason that they move not in straight lines but rather "in an orbit," and that one of two ways they could do this is by "not much deviating from the same plane," as here. Although he hadn't quite envisioned the spiral arms within the flattened disc, Wright had reasoned his way to the shape of what we now call a spiral galaxy. For more images by Wright, see pages 153–54.

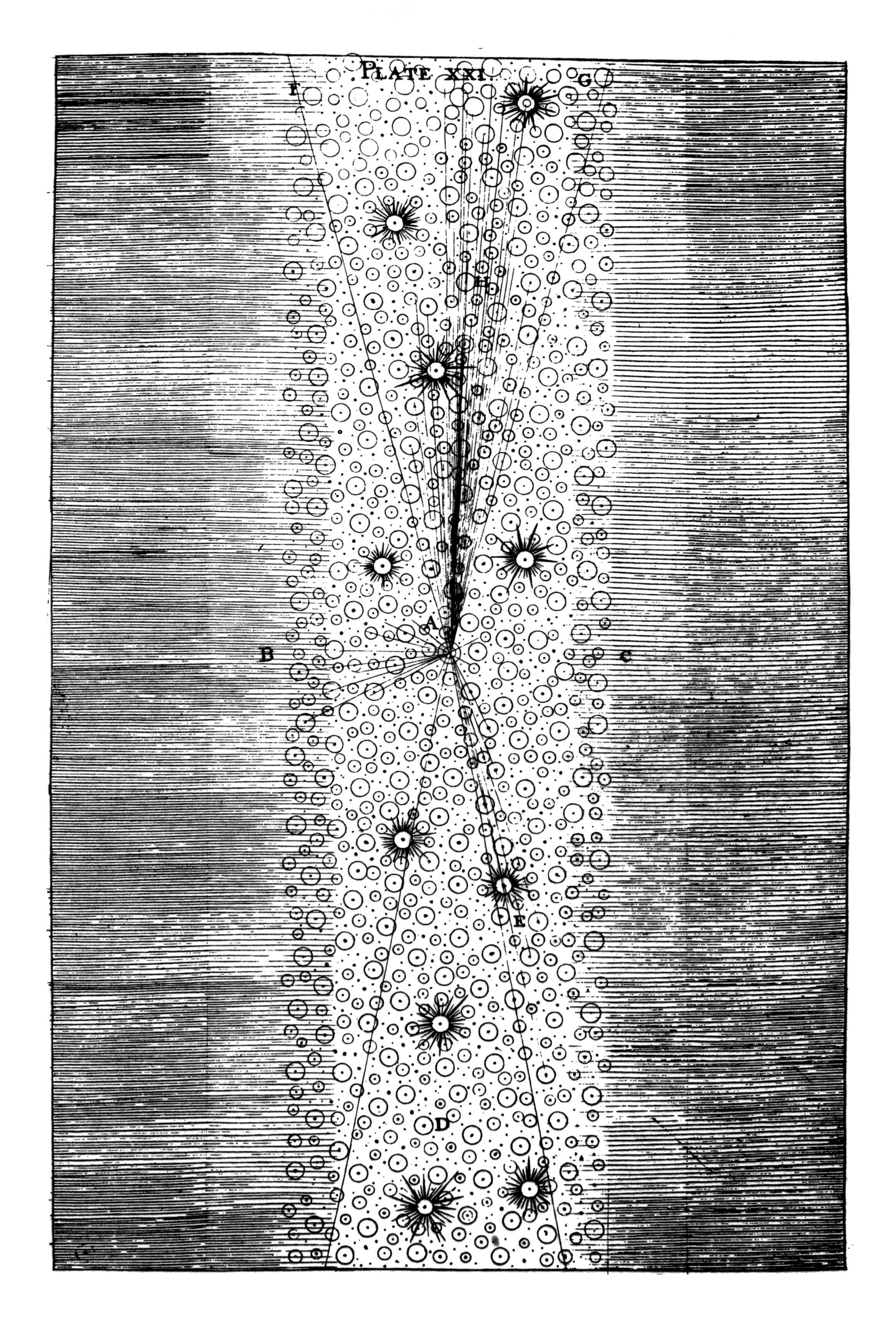
PLATE XXI
F
G
H
A
B
C
E
D

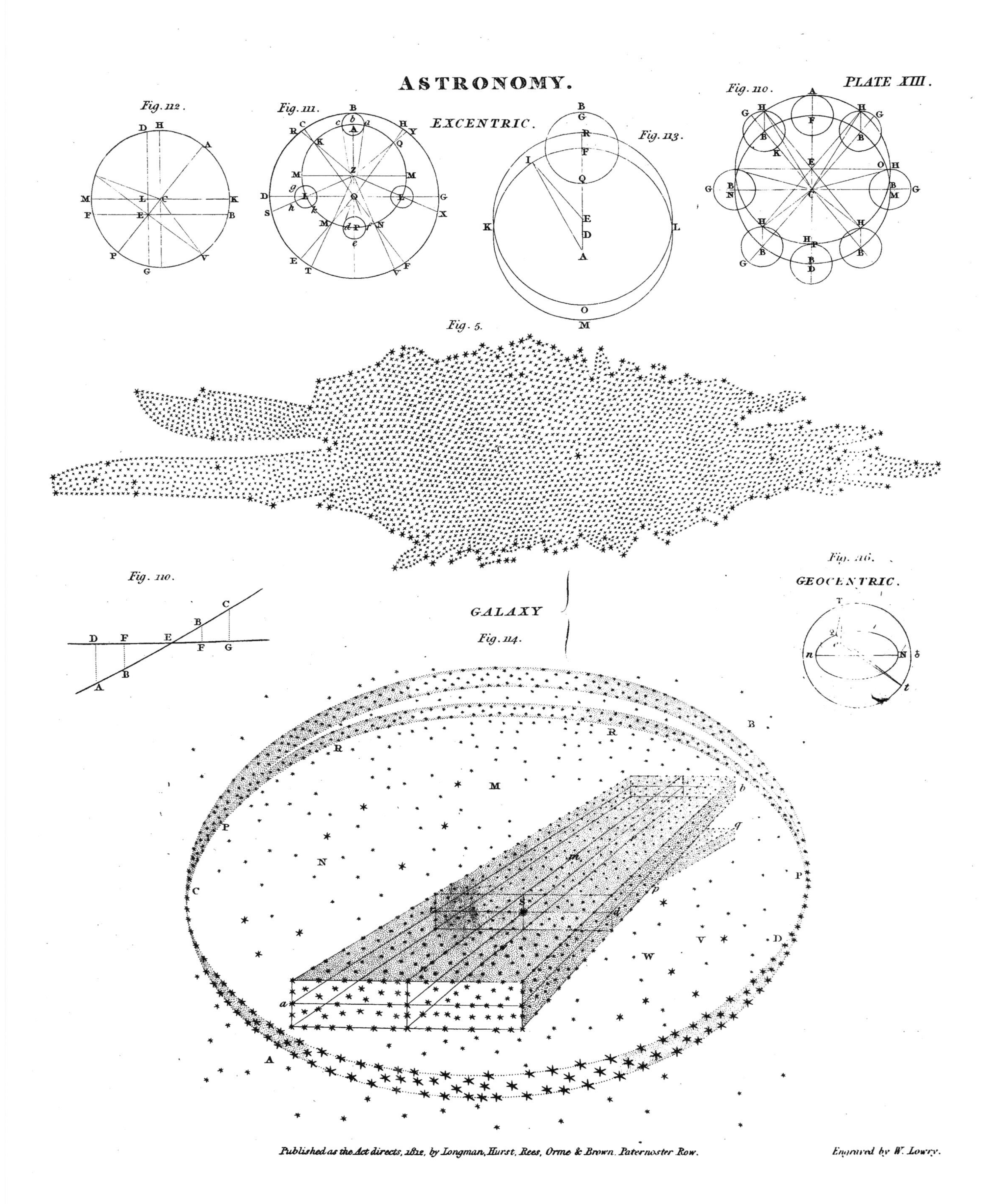

• 1785:

Thirty years after Thomas Wright, who he may not have been aware of, astronomer William Herschel wrote that he believed the Milky Way to be a "detached nebula" similar to the hazy blobs of light he was seeing through his telescopes. He also stated that he thought that some of those nebulae were larger than the Milky Way. But when the time came to depict our galaxy using data he had amassed by counting stars, his map fell far short of deducing the Milky Way's true flattened disc shape. This was in part because his method assumed that the stars are distributed uniformly, something that is not the case, and also that he was able to see all of the stars it is possible to see with a given power of telescopes, which is also not the case, due to the obscuring gas and dust of the galaxy. The irregular form in the middle is the Milky Way that Herschel charted using this "star gauging" technique; the larger star near the center is supposed to be the solar system. (The lower figure is a cross section, also with the sun's putative location.) Herschel's map, which is frequently presented as the first attempt to visualize the shape of the Milky Way, is far more famous than any of Thomas Wright's illustrations.

DECORATION ZU DER OPER: DIE ZAUBERFLÖTE ACT I SCENE VI.
Berlin bei L.W.Wittich

• 1847–49:

The idea of the stars forming a dome above our heads has rarely been more effectively literalized than in this set, for Act 1, Scene 6 of Mozart's *The Magic Flute*, by Prussian architect, painter, and designer Karl Friedrich Schinkel. Titled *The Hall of Stars in the Palace of the Queen of the Night*, it reveals a sense of celestial order probably not surprising from someone who designed the famous Iron Cross military decoration for the Kingdom of Prussia.

LA VOIE LACTÉE

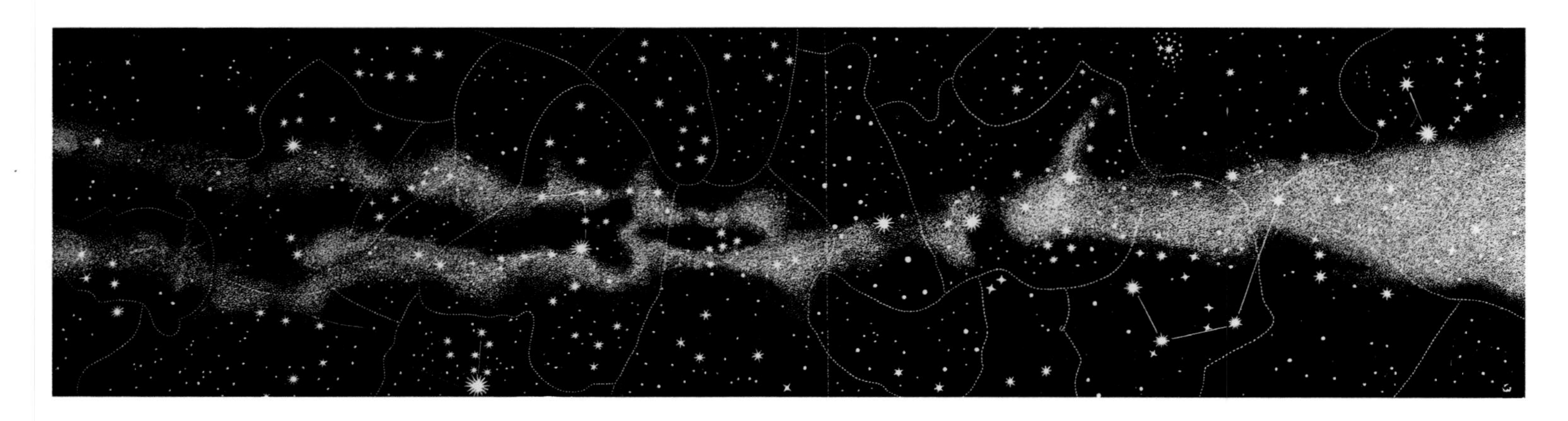

Partie comprise dans l'hémisphère Nord.

Partie comprise dans l'hémisphère Sud. (P. 18).

• 1866:

These two views of the Milky Way as seen from the northern and southern hemispheres come from *L'Espace selestial* (Celestial Space) by French astronomer Emmanuel Liais. Having served as director of the National Observatory in Rio de Janeiro from 1874 to 1881, Liais had plenty of opportunities to observe the southern skies. The Milky Way, as Thomas Wright had correctly deduced, appears as a thin band across all 360 degrees of sky visible from Earth because the solar system is embedded in its flattened disc.

• 1874–76:

Facing page: This print by French artist-astronomer Étienne Trouvelot is about as nuanced and convincing a study of the Milky Way as could be produced without contemporary photographic techniques. It comes from an 1881 Charles Scribner's Sons collection of chromolithographs, but is based on pastel studies made in the mid-1870s. Note the subtle reflection of the galaxy on the waves, and the distant clipper ship under full sail, with stars shining through its rigging. Trouvelot was an absolute master, within a very specific genre that he himself had created. For more by Trouvelot, see pages 98–99, 127, 128–29, 158, 188–91, 270–71, 295, and 297–99.

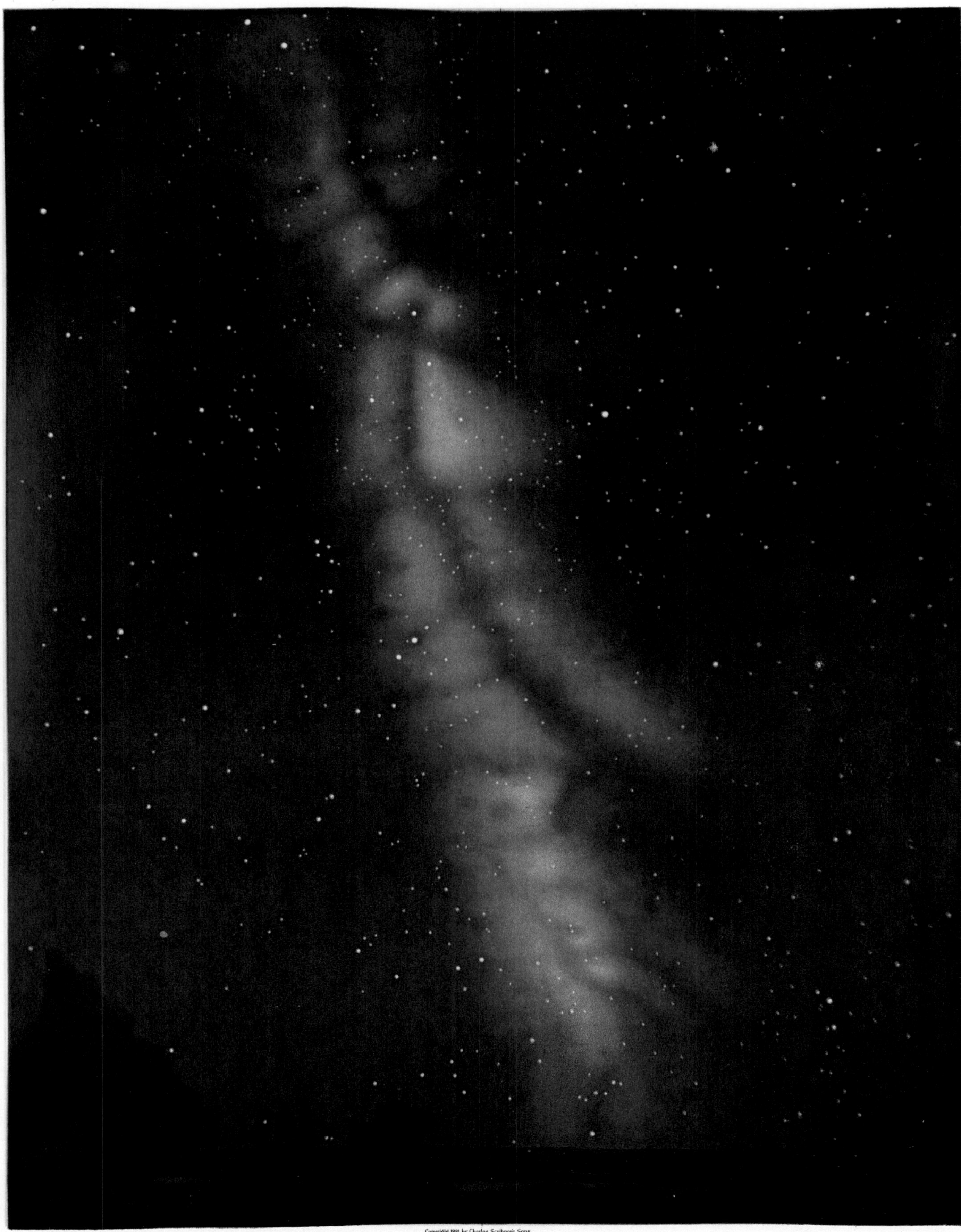

PLATE XIII.

E. L. Trouvelot

PART of the MILKY WAY.

From a Study made during the years 1874, 1875 and 1876.

THE MIDNIGHT SKY AT LONDON, LOOKING SOUTH, JUNE 15.

THE MIDNIGHT SKY AT LONDON, LOOKING SOUTH, FEBRUARY 15.

LOOKING SOUTH, NOVEMBER 15. *(Buenos Ayres.)*

LOOKING SOUTH, AUGUST 15. *(Table Mountain, under the South-east Cloud.)*

• 1869:

Throughout the 1860s, Edwin Dunkin, an English astronomer and staff member at the Royal Observatory Greenwich, meticulously calculated the inclination of the Milky Way and exactly what stars would be above the horizon on specific nights from both London and southern hemisphere destinations. At the end of the decade he published *The Midnight Sky*, with thirty-two plates combining foreground views and stellar cartography. **Top left and right:** London on June 15 and February 15. **Bottom left:** Buenos Aires on November 15. **Bottom right:** Table Mountain, Cape Town, August 15. All views are looking south. Note the presence of the Magellanic Clouds in the two lower prints: The nearest companion galaxies to the Milky Way are visible from the southern hemisphere.

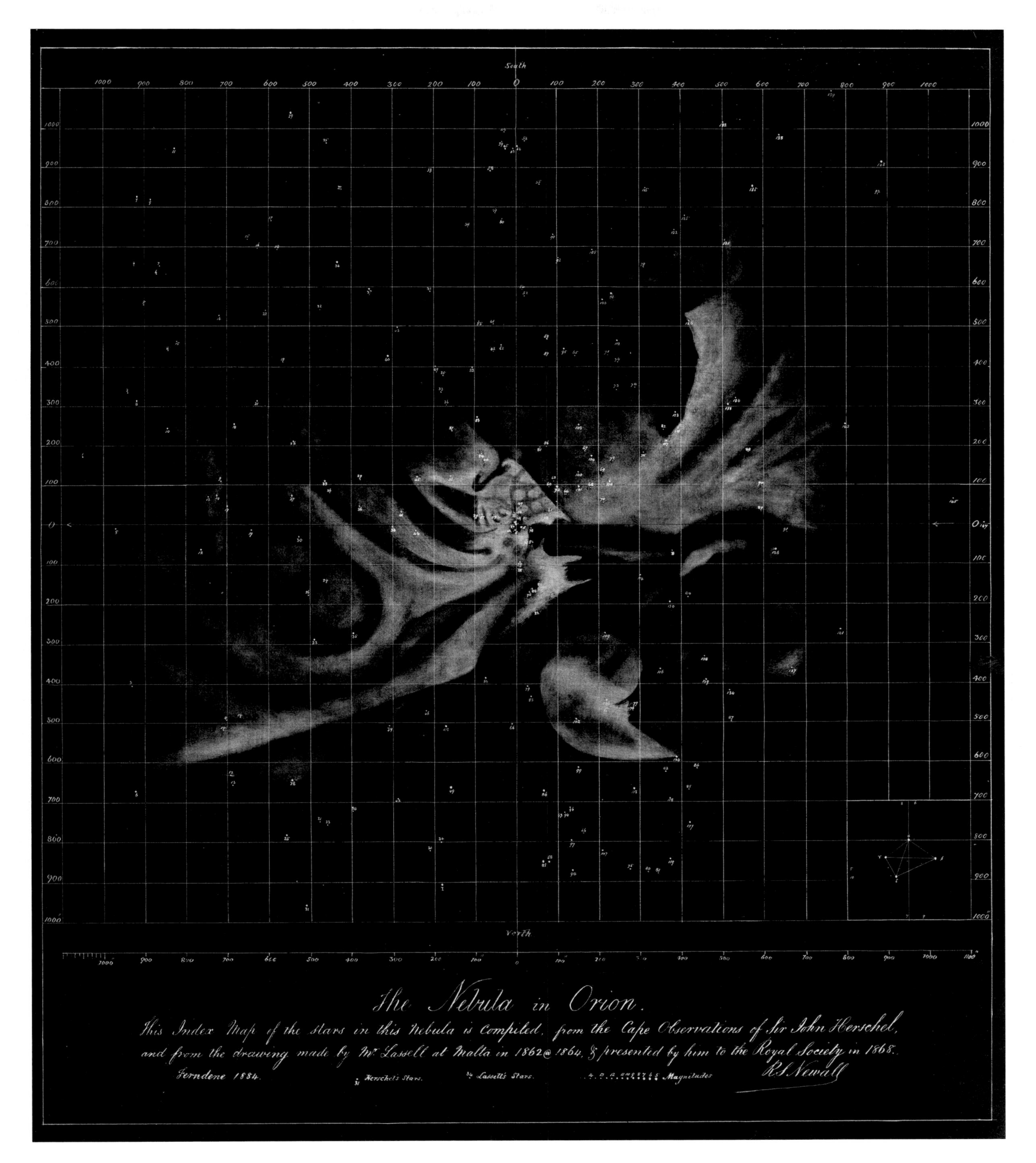

• 1884:

This detailed map indexing stars within the Orion Nebula is by Scottish engineer and astronomer Robert Sterling Newall. Famous for improving and laying submarine telegraph cables, in some ways the very first strands of the Internet, Newall was also an astronomer, and used wealth acquired through his engineering achievements to commission one of the largest refracting telescopes in the world. He based this map not on his own observations, however, but on those made by John Herschel in Cape Town in the early 1830s.

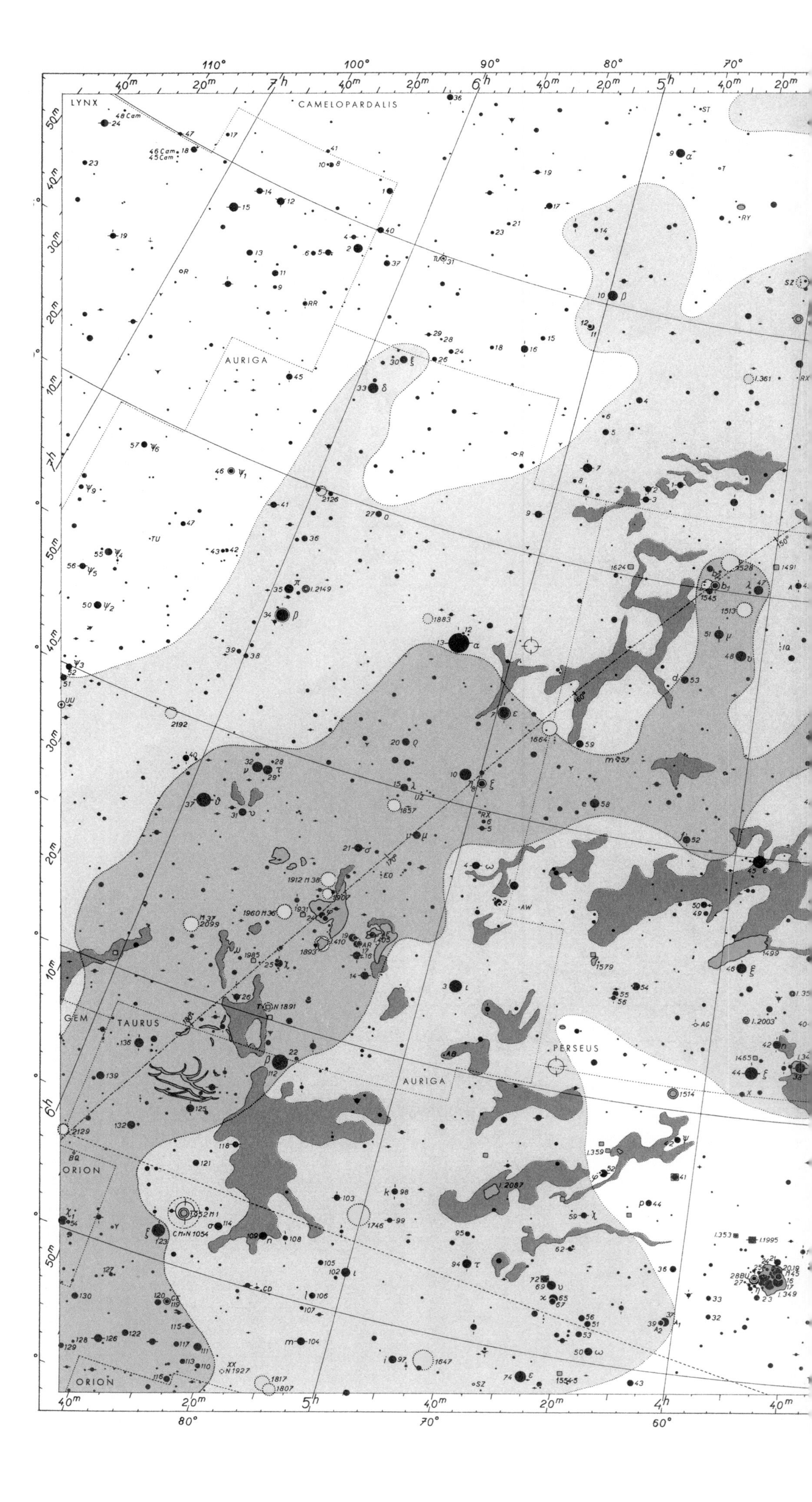

• 1948:

Throughout the 1940s, Czech astronomer Antonín Bečvář worked at the Skalnaté Pleso observatory in the Tatra Mountains, Slovakia, compiling a vast index of deep-sky objects, including stars, galaxies, nebulae, and interstellar dust clouds. In 1948, the Czech Astronomical Society published the first edition of his *Skalnaté Pleso Atlas of the Heavens*, which contained sixteen hand-drawn charts by Bečvář. The atlas was a major step forward in stellar cartography, and when introduced to the international market the same year, it became an instant success, with most observatories and many thousands of amateur astronomers acquiring copies. This chart centers on Perseus, a constellation in a region of the Milky Way characterized by particularly dense molecular clouds, seen here in different hues of blue. The annual Perseids meteor shower originates in this section of the night sky. To the right on this map, a neighboring constellation, Andromeda, features the distant Andromeda Galaxy, visible as a red oval. Below it and to the left, another galaxy, Triangulum, can be seen. Both are members of our Local Group of galaxies.

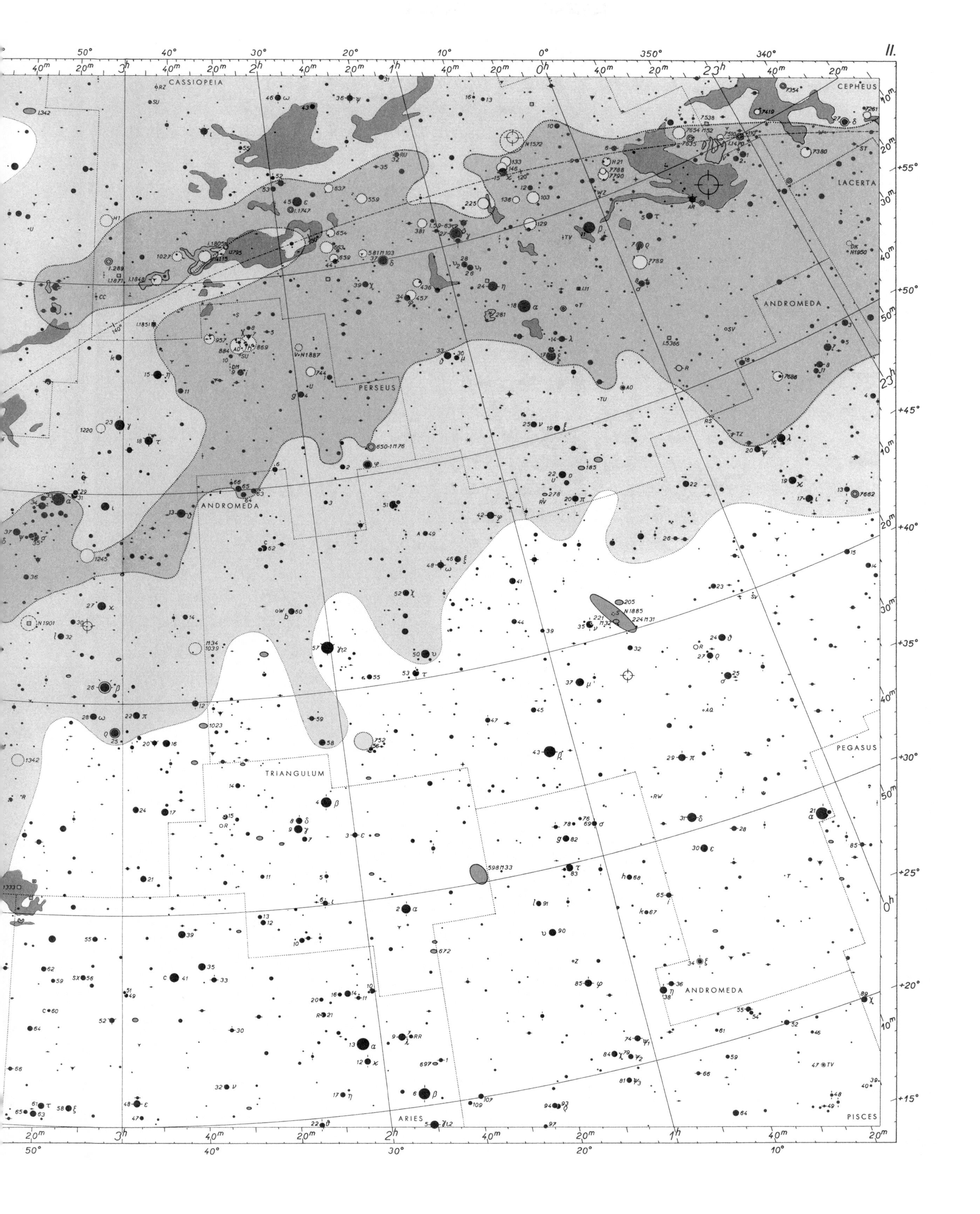
II.
CASSIOPEIA
CEPHEUS
LACERTA
ANDROMEDA
PERSEUS
TRIANGULUM
PEGASUS
ARIES
PISCES

• 2007:
Although extensive stellar mapping took place throughout the twentieth century, attempts to depict the structure of our own galaxy have been surprisingly rare. In part this is because so much of it is obscured by dense clouds of dust and molecular gas, making such rendering speculative. While the basic design of the galaxy is known—we're living in a spiral galaxy about 100 to 120,000 light-years across, and are about a quarter of the way from the nucleus—a lot of the details are subject to debate. This *Galactic Orientation Map* by software engineer and celestial cartographer Winchell D. Chung Jr. positions the solar system within the Orion Spur, and surrounds it with sixteen prominent nebulae and other features within our section of the Milky Way.

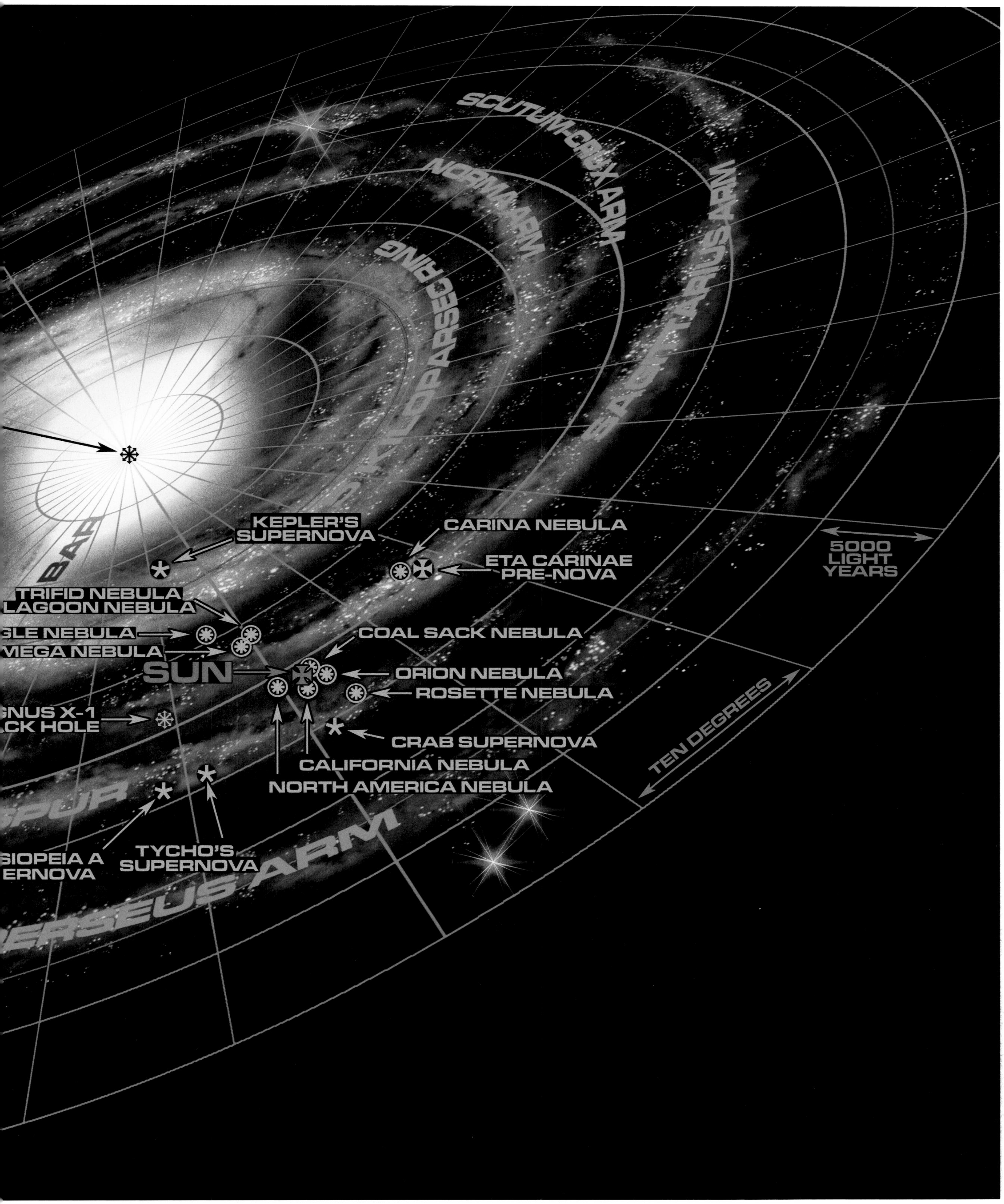

SCUTUM-CRUX ARM
NORMA ARM
SAGITTARIUS ARM
3-KILOPARSEC RING
BAR
KEPLER'S SUPERNOVA
CARINA NEBULA
ETA CARINAE PRE-NOVA
5000 LIGHT YEARS
TRIFID NEBULA
LAGOON NEBULA
GLE NEBULA
MEGA NEBULA
SUN
COAL SACK NEBULA
ORION NEBULA
ROSETTE NEBULA
GNUS X-1
CK HOLE
CRAB SUPERNOVA
TEN DEGREES
CALIFORNIA NEBULA
NORTH AMERICA NEBULA
SPUR
SIOPEIA A
ERNOVA
TYCHO'S SUPERNOVA
ERSEUS ARM

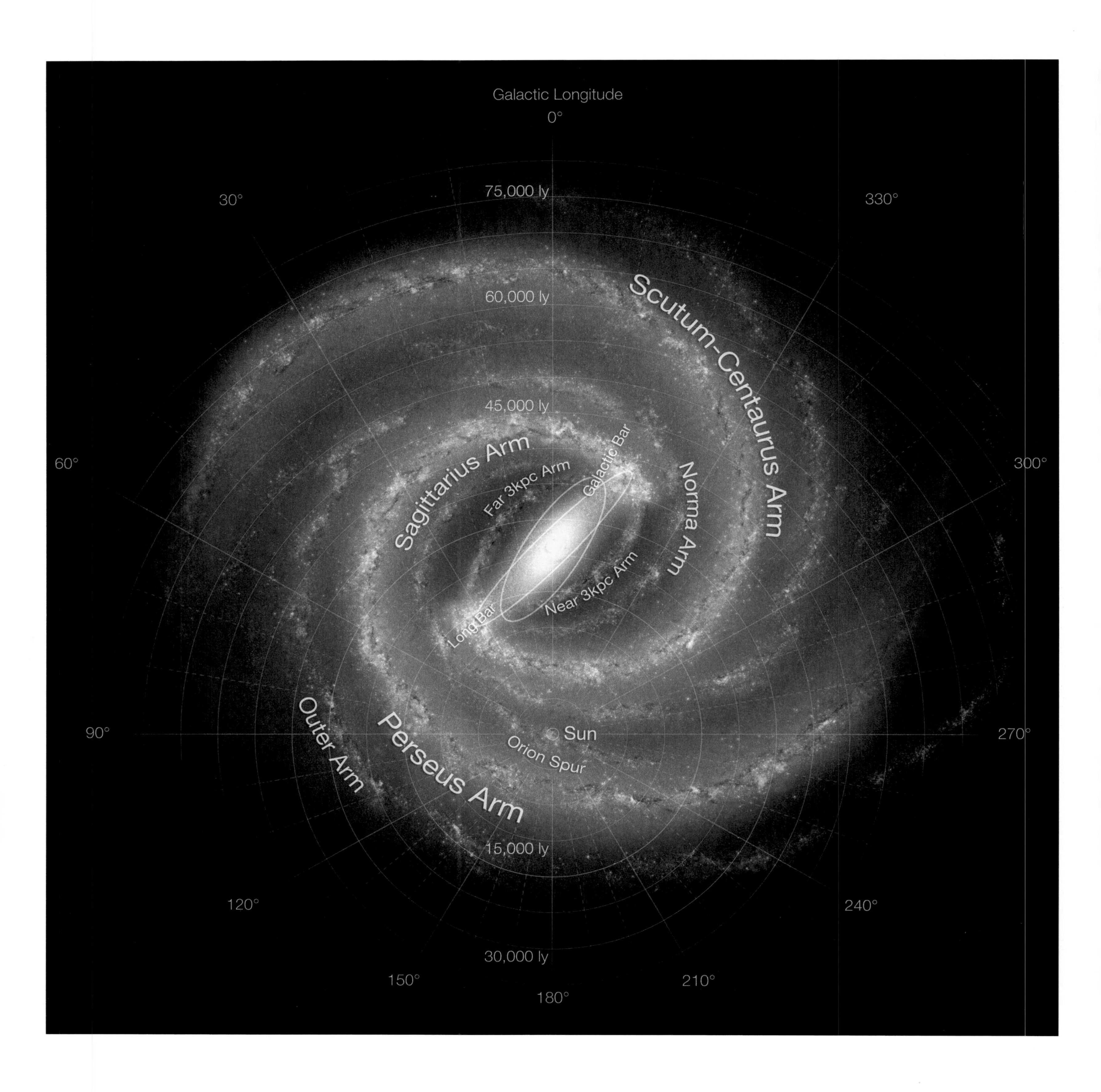

• 2005–8:

This map by Spitzer Space Telescope visualizations scientist Robert Hunt is by far the best-known contemporary rendering of the Milky Way. Because Spitzer is an infrared telescope, it can probe regions of the galaxy so thick with dust clouds that visible wavelengths of light can't penetrate them. Shortly after Spitzer's launch in 2003, its images of previously obscured regions of the galaxy began to reinforce a previously argued conclusion: that the Milky Way has only two major spiral arms, Perseus and Scutum-Centaurus. In this map, the Sagittarius and Norma constellations have been reduced to the status of tributaries. In keeping with its ongoing demotion through the centuries, Earth, until recently thought to orbit the sun within a substantive secondary arm, is here relocated to the Orion "spur." While it has been supposed by many astronomers since the 1990s that the Milky Way exhibits a bar shape across its nucleus, Spitzer data indicated that the bar was more extensive than previously imagined. Although backed by the research of a major NASA space telescope, many aspects of Hunt's map remain controversial; contrary evidence suggests a return to a four-armed galaxy and even one without a bar.

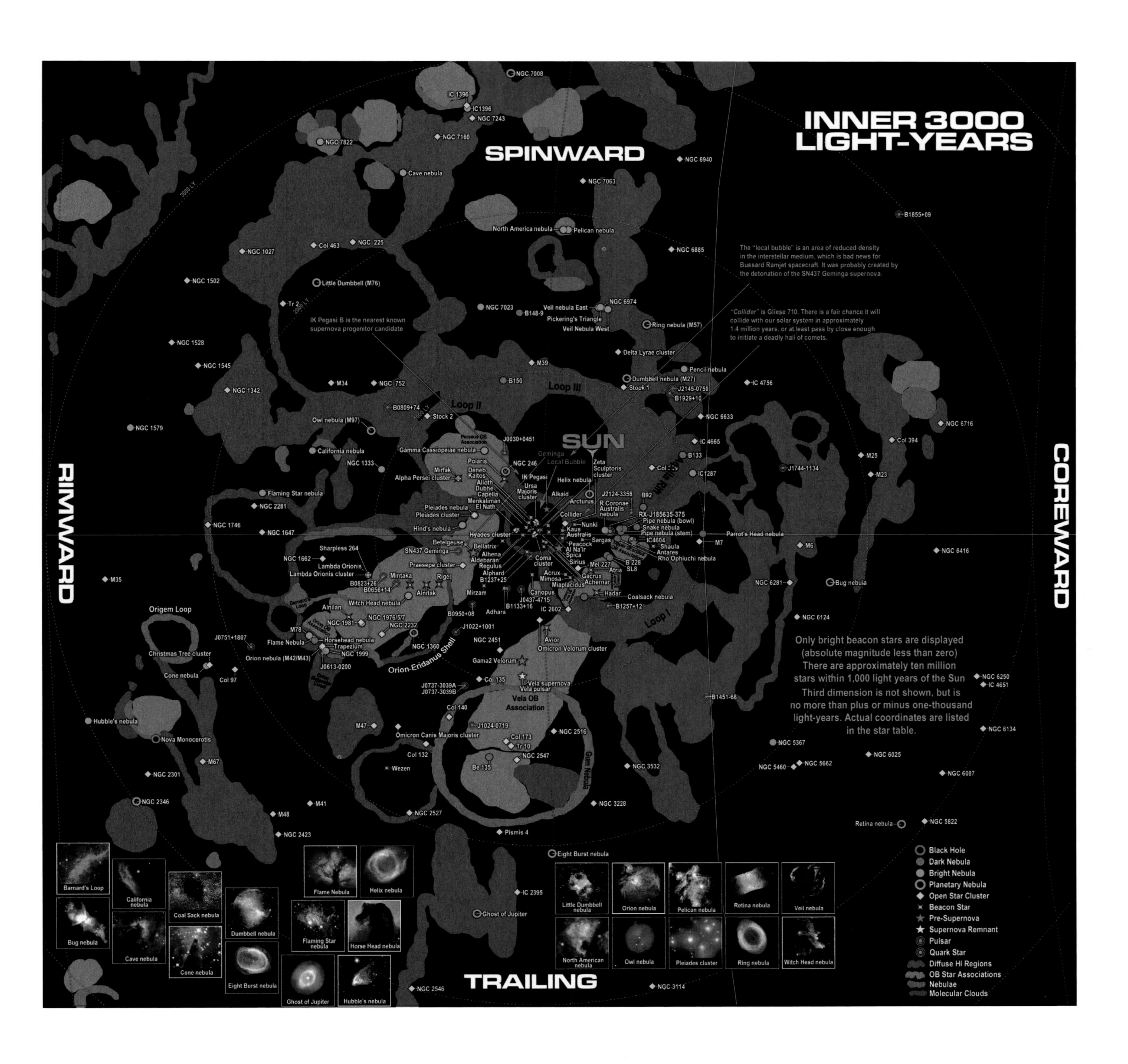

INNER 3000 LIGHT-YEARS
SPINWARD
TRAILING
RIMWARD
COREWARD
SUN
Local Bubble
Loop I
Loop II
Loop III
Orion-Eridanus Shell
Gum Nebula
Origem Loop
The "local bubble" is an area of reduced density in the interstellar medium, which is bad news for Bussard Ramjet spacecraft. It was probably created by the detonation of the SN437 Geminga supernova.
"Collider" is Gliese 710. There is a fair chance it will collide with our solar system in approximately 1.4 million years, or at least pass by close enough to initiate a deadly hail of comets.
IK Pegasi B is the nearest known supernova progenitor candidate
Only bright beacon stars are displayed (absolute magnitude less than zero) There are approximately ten million stars within 1,000 light years of the Sun Third dimension is not shown, but is no more than plus or minus one-thousand light-years. Actual coordinates are listed in the star table.
Black Hole
Dark Nebula
Bright Nebula
Planetary Nebula
Open Star Cluster
Beacon Star
Pre-Supernova
Supernova Remnant
Pulsar
Quark Star
Diffuse HI Regions
OB Star Associations
Nebulae
Molecular Clouds
Barnard's Loop
California nebula
Coal Sack nebula
Dumbbell nebula
Flame Nebula
Helix nebula
Bug nebula
Cave nebula
Cone nebula
Eight Burst nebula
Flaming Star nebula
Horse Head nebula
Ghost of Jupiter
Hubble's nebula
Little Dumbbell nebula
Orion nebula
Pelican nebula
Retina nebula
Veil nebula
North American nebula
Owl nebula
Pleiades cluster
Ring nebula
Witch Head nebula

• 2007:

Above: Only the brightest "beacon" stars are shown in Winchell Chung's map of a three-thousand-light-year-diameter section of the Milky Way, centering on the solar system; this is because there are about ten million stars within the first one thousand light-years alone. As the photographs at the base of the map indicate, many of the nebulae best known to astronomers lie within three thousand light-years of Earth. The galactic center is to the right.

• 2008:

Overleaf: Because of the abundance of hydrogen in the Milky Way, one imaging technique in astronomy is particularly effective; it limits the data collected to a narrow wavelength best suited to recording the hydrogen, known as the H-alpha spectral line. This view toward the center of the Milky Way is based on H-alpha observations by astrophysicist Douglas Finkbeiner, as presented in "Milky Way Explorer," an interactive map at cartographer Kevin Jardine's comprehensive site, GalaxyMap.org. (Each ring centers on a catalogued nebula.)

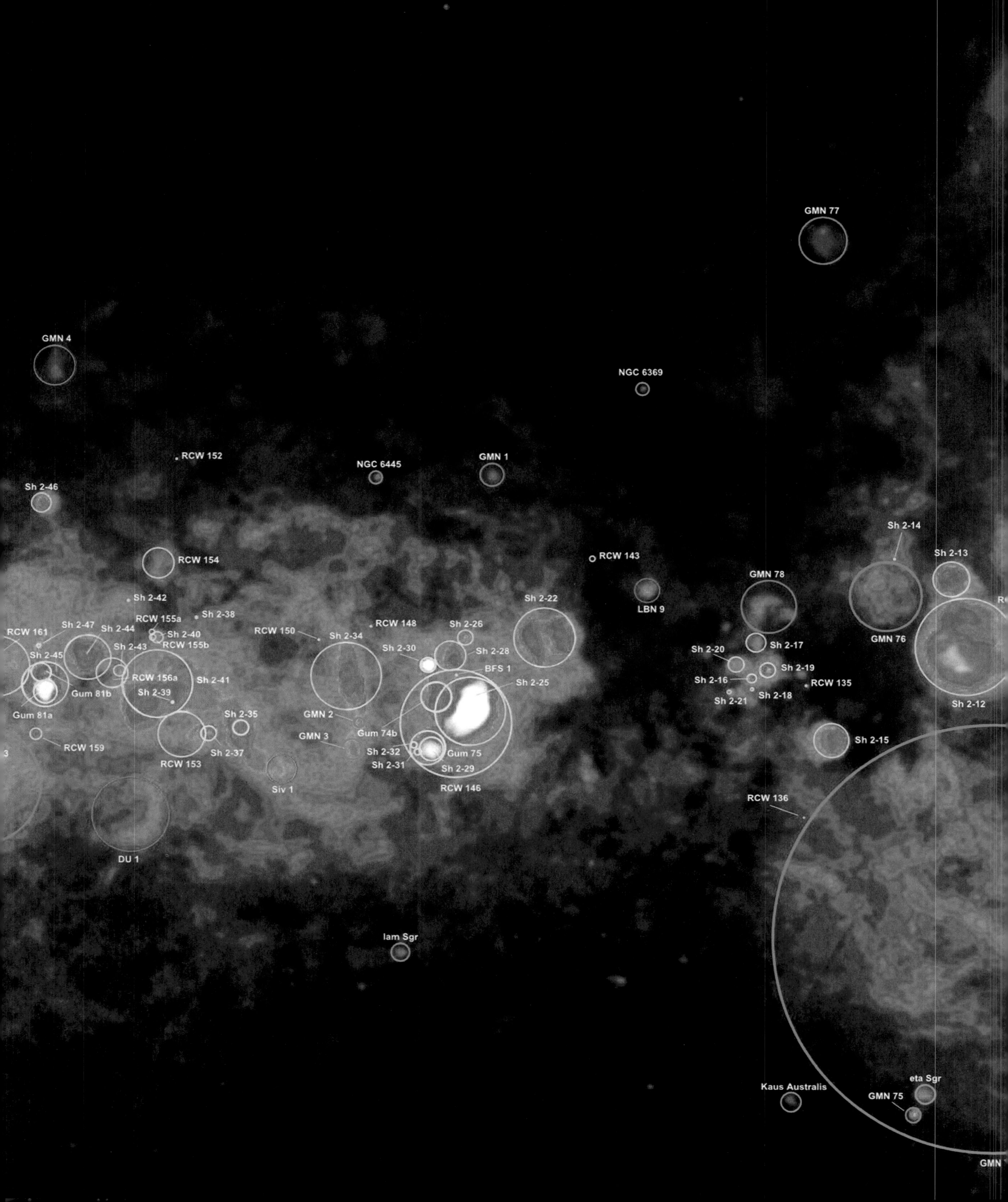

GMN 77
GMN 4
NGC 6369
RCW 152
NGC 6445
GMN 1
Sh 2-46
Sh 2-14
RCW 143
Sh 2-13
RCW 154
GMN 78
LBN 9
Sh 2-42
Sh 2-22
RCW 155a
Sh 2-38
RCW 148
Sh 2-26
RCW 161
Sh 2-47
Sh 2-44
Sh 2-40
RCW 150
RCW 155b
Sh 2-34
Sh 2-17
GMN 76
Sh 2-30
Sh 2-28
Sh 2-43
Sh 2-20
Sh 2-19
Sh 2-45
BFS 1
RCW 156a
Sh 2-41
Sh 2-25
Sh 2-16
RCW 135
Gum 81b
Sh 2-39
Sh 2-18
Sh 2-21
Sh 2-12
Sh 2-35
Gum 81a
GMN 2
Gum 74b
GMN 3
Sh 2-15
RCW 159
Sh 2-37
Sh 2-32
Gum 75
RCW 153
Sh 2-31
Sh 2-29
Siv 1
RCW 146
RCW 136
DU 1
lam Sgr
eta Sgr
Kaus Australis
GMN 75

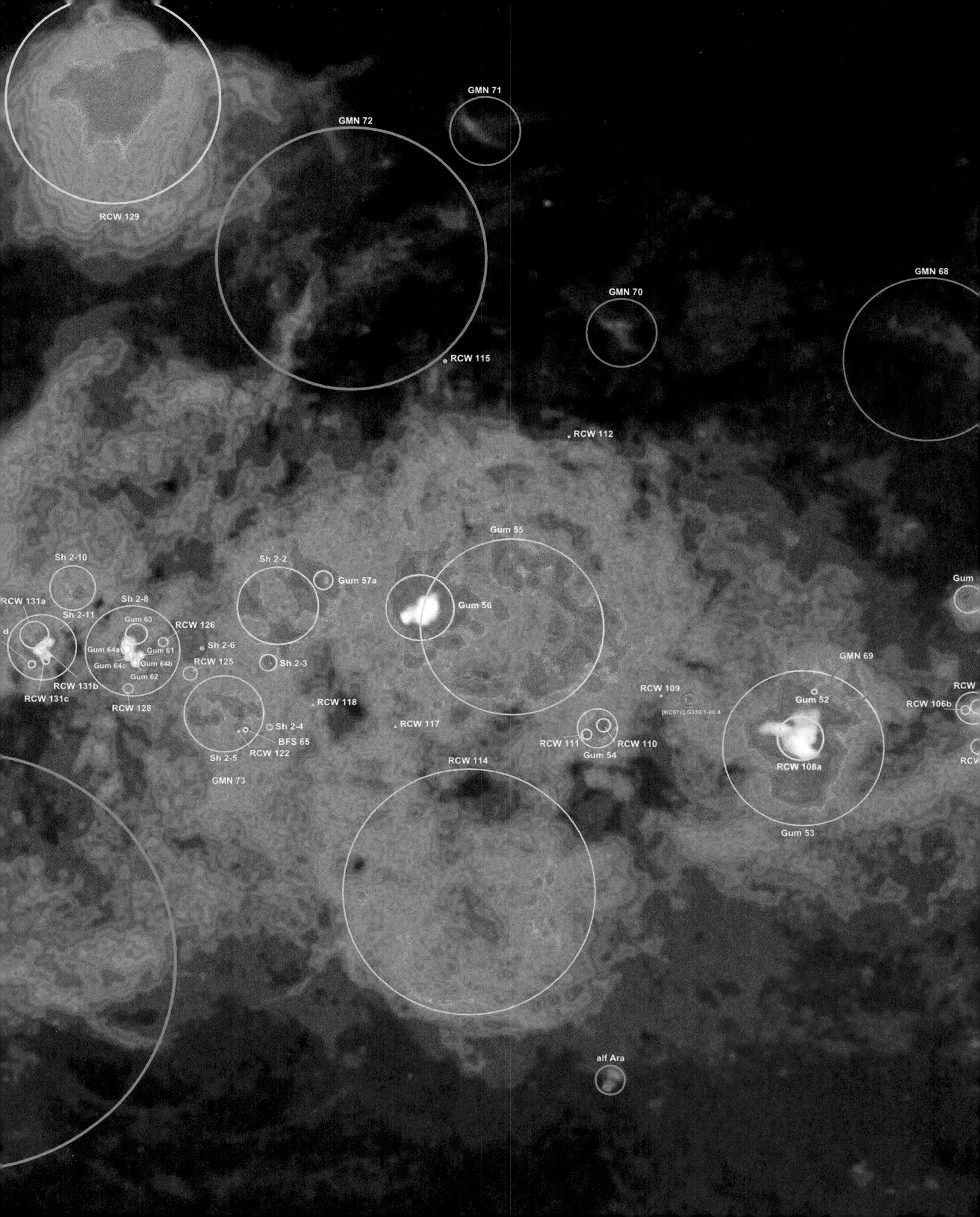

RCW 129
GMN 71
GMN 72
GMN 70
GMN 68
RCW 115
RCW 112
Gum 55
Sh 2-10
Sh 2-2
Gum 57a
Gum 56
RCW 131a
Sh 2-8
Sh 2-11
Gum 63
RCW 126
Gum 64a
Gum 61
Gum 64c
Gum 64b
Gum 62
Sh 2-6
RCW 125
Sh 2-3
RCW 131b
RCW 131c
RCW 128
RCW 118
RCW 117
Sh 2-4
BFS 65
RCW 122
Sh 2-5
GMN 73
RCW 114
RCW 109
[KC97c] G339.1-00.4
RCW 111
RCW 110
Gum 54
GMN 69
Gum 52
RCW 106b
RCW 108a
Gum 53
alf Ara

CHAPTER 8

Eclipses and Transits

The graves stood tenantless and the sheeted dead
Did squeak and gibber in the Roman streets:
As stars with trains of fire and dews of blood,
Disasters in the sun; and the moist star
Upon whose influence Neptune's empire stands
Was sick almost to doomsday with eclipse

—SHAKESPEARE, *HAMLET*

ECLIPSES HAVE BEEN CONSIDERED VERY BAD NEWS throughout history. The intimations of doom associated with the eclipse lurk in the etymology of the word itself, which derives from the Ancient Greek *ékleipsis*, meaning "the abandonment," "the downfall," or "the darkening." The Indians, the Chinese, and the Turks imagined that a snake was consuming the sun during a solar eclipse; when one occurred, they made as much noise as possible, even shooting arrows into the air, to drive it away. Lawrence of Arabia took advantage of this phenomenon, storming an Ottoman cliff-top fortress with his Arab forces during a lunar eclipse on the night of July 4, 1917, while the Turkish troops inside were banging pots and firing rifles "to rescue their threatened satellite."

Both solar and lunar eclipses, clearly, are sensational events. Among the earliest recorded eclipse descriptions are verses attributed to the mythological Hindu bard Atri in an ancient Vedic Sanskrit hymn known as a Rigveda. It's thought to date back more than one thousand years B.C. A clay tablet that was found in Syria in 1948 dates to the New Kingdom period of ancient Egypt, and describes a total solar eclipse that occurred on March 5, 1223 B.C. Eclipse observations from the Zhou Dynasty in China starting in 720 B.C. have been dated reliably, and the Old Testament Book of Amos contains an explicit solar eclipse description that has been corroborated by a similar reference in an Assyrian historical chronicle, the *Eponym Canon*, dating to June 15, 763 B.C.

During the same era, the Babylonians used centuries of amassed astronomical records to discover the eighteen-year saros cycle, the period it takes for the sun, the moon, and the Earth to return to the same approximate configuration, allowing for virtually identical eclipses to recur with a kind of elongated metronomic regularity in those intervals.

Solar eclipses occur when the moon passes in front of the sun, blocking its rays either partially or totally, depending on the position of the observer and how close the moon is to perigee or apogee. If the moon is near apogee, meaning its maximum distance from the Earth, it fails to block the sun entirely and even the best-positioned terrestrial observers still see a bright ring around the moon. This is called an annular eclipse. Alternatively, if it's near perigee, it blocks the sun entirely and a total eclipse can be seen within the path of the moon's umbra, or area of total shadow, as it traverses the Earth. Partial eclipses are seen on either side of that band, within the so-called penumbra.

Lunar eclipses occur when the Earth's shadow crosses the face of the moon. Because the Earth's umbra is more than 5,500 miles wide—the width of three moons—lunar eclipses can take hours to unfold and are visible from the half of Earth oriented toward its satellite during that period, which is always the night side. By contrast, the moon's shadow on Earth is only a narrow band, and so a total solar eclipse only lasts for a few minutes at any one place within the path of totality.

Solar eclipses are among the most dramatic manifestations of cosmic grandeur visible from Earth. At the moment of totality, night falls in the middle of the day, stars can be seen, and the sun's outer atmosphere or corona, which is otherwise invisible due to the glare of the photosphere, visibly streams into space. Giant solar prominences can usually be discerned as well, extending up from the sun's surface, as in the depiction by artist-astronomer Étienne Trouvelot on page 271. If a solar eclipse is observed from a high point such as a mountain summit, sometimes the border of the moon's umbral shadow can be observed rushing across the surrounding landscape at an extraordinary speed. Their ephemerality and surreal power make it unsurprising that solar eclipses have impacted human cultures for as long as we have records.

Lunar eclipses are less transformative affairs. If they're total rather than partial, however, during the period of totality the moon glows a striking red-orange in the sky—a color caused by indirect illumination from all the sunsets and sunrises of the Earth simultaneously. (A lunar eclipse as seen from Earth would appear as a solar eclipse from the surface of the moon.)

Because both kinds of eclipses were understood to reveal something about the orientation of the three most important celestial objects, and because those geometries were known to repeat across spans of many years, they've played a significant role in advancing our understanding of celestial mechanics. The location of both the sun and the moon are known to a high degree of accuracy during eclipses, with both occupying the same half-degree of sky during a solar eclipse, and with the two being exactly 180 degrees apart on either side of the Earth during a lunar one.

Like eclipses, transits of the sun by Venus are among the most complex astronomical phenomena to predict. They, too, take place in cycles, only these repeat every 243 years, with gaps of 121.5 and 105.5 years between paired transits 8 years apart. Because they're hard to observe without a telescope, it's unlikely that ancient astronomers were aware of them, though some may have suspected their occurrence. Mercury, being far closer to the sun than Venus, transits much more frequently, with thirteen or fourteen occurring every century.

In one of the more extraordinary stories in the history of astronomy, the first person to predict a transit of Venus accurately, and then observe as it unfolded, was a twenty-year-old English prodigy named Jeremiah Horrocks. Although he had studied the works of Copernicus, Kepler, and Brahe at the University of Cambridge between 1632 and 1635—or starting at the age of thirteen or fourteen—Horrocks left without graduating, probably for financial reasons. He remained committed to astronomy, however, and was a firm believer in Copernican heliocentrism and in Johannes Kepler's theory of elliptical planetary orbits. Horrocks corresponded with another young astronomer, William Crabtree, as they both laboriously corrected a set of tables of planetary and stellar motion that Kepler had published in 1627 called the *Rudolphine Tables*, which were based in part on observations by Tycho Brahe.

In 1629, the year before his death, Kepler published a pamphlet alerting astronomers to an upcoming transit of Mercury, which did in fact occur in 1631 and was observed by Pierre Gassendi. (The French astronomer was surprised by the planet's tiny size and perfectly round shape. This was a period when planets were still almost entirely mysterious; Tycho Brahe even suspected they generated their own light.) In the pamphlet, Kepler also predicted a transit of Venus in 1631, but warned that it would likely not be visible from Europe. (In fact, it would have been seen from the eastern Mediterranean, but there is no record of its being observed.) He concluded by stating that Venus would just miss the sun in 1639, and that the next transit would therefore not occur until more than a century later, in 1761.

Although Horrocks venerated Kepler, he wasn't convinced that the great astronomer was correct. Horrocks had been monitoring the planet's position for some time from Liverpool and the tiny Lancashire hamlet of Hoole, and he believed he'd found an error in Venus' latitude in the *Rudolphine Tables*. According to his calculations, transits of Venus should always occur in pairs, with eight years in between; if he was right, 1639 would in fact be the second and last transit of the century. He wrote Crabtree, advising him to observe the sun on November 24, 1639 (actually December 4; England was then still using the Julian calendar). And he converted his telescope into a helioscope by rigging it to project the sun's disc on a sheet of paper. Horrocks predicted the transit would start at about three.

Sunday, November 24, turned out to be disconcertingly cloudy. But at 3:15, "the clouds, as if by divine interposition, were entirely dispersed," and Horrocks saw "a most agreeable spectacle"—the black silhouette of Venus against the disc of the sun. The transit was already underway when the clouds parted, meaning his prediction had been entirely accurate, and he watched it unfold until sunset, only half an hour later. Horrocks carefully made three measurements on the six-inch graduated circle he had inscribed on the paper. It was the second transit by Venus of the sun since the invention of the telescope, and the first that had actually been observed.

For an astronomer of twenty to disagree with the great Johannes Kepler, and to be proven correct, was clearly a phenomenal achievement. As for Crabtree, he, too, witnessed the transit through a break in the clouds. He and Horrocks were probably the only two people on Earth to see it. Later, Crabtree wrote that he had been so moved that he'd succumbed to a "womanly display of emotion."

By the time Venus was again ready to surf the straight shot of energy that cascades ceaselessly between the sun and Earth—the 1761 transit that Kepler *had* predicted—an international posse of astronomers were positioned across the globe to observe it. Astronomers from England, France, and Austria traveled to some of the most remote locations on Earth, the goal being to use their widely spaced observations to measure the solar parallax, thus definitively establishing the distance between the sun and Earth.

Although complicated by the Seven Years' War between England and France, the Venus transit effort of 1761 was the first real international scientific collaboration in history. Apart from the logistics of sending the world's best astronomers to the Indian Ocean, Siberia, South Africa, the South Pacific, and the Americas, among other places, a good deal of equipment was required. Measurements of the exact longitude of the observers' locations were critical and required extensive preparatory observations.

On the day of the transit, some unexpected visual phenomena complicated the effort. As Venus neared the sun's edge, a ring of light formed around the planet—a refraction of sunlight through the planet's previously unknown atmosphere. This was followed by the "black drop effect"—a tenuous, wavering teardrop that seemed to connect the planet to the sun's limb, probably due to distortions caused by our own atmosphere, as well as optical imperfections in the telescopes themselves.

While both of these aberrations made it difficult to calculate exact timings as the planet crossed the sun's limbs at each end of the transit—with differences between those timings being the basis of the parallax measurement—a reasonably good estimate was in fact ultimately made of the sun's distance from Earth using observations of the 1761 transit. When that data was augmented by timings from Venus' next transit, in 1769, an estimate within 1 percent of the modern value was achieved. A key contemporary component of celestial measurement, the AU—or astronomical unit—had been established.

As for Jeremiah Horrocks, although he wrote an elegant treatise titled *Venus in sole visa* (Venus Seen on the Sun), it remained unpublished at his sudden death, from unknown causes, at the age of only twenty-two. It was finally published by Polish astronomer Johannes Hevelius twenty-one years later, and subsequently in England as well. Crabtree died at the age of thirty-four, only three years after Horrocks, during the First English Civil War.

Apart from being unprecedented, their Venus transit observations were an important validation of Copernican ideas, and are considered the beginning of British astrophysics. What other contributions Horrocks might have made if he had lived a longer life will of course never be known. In the late nineteenth century, a marble plaque was unveiled in Westminster Abbey. It read in part:

> *In memory of Jeremiah Horrocks . . . who died on the 3rd of January 1641, in or near his 22nd year; having in so short a life detected the long inequality in the mean motion of Jupiter and Saturn; discovered the orbit of the Moon to be an ellipse; determined the motion of the lunar apse; suggested the physical cause of its revolution; and predicted from his own observations the transit of Venus, which was seen by himself and his friend, William Crabtree, on Sunday the 24th of November 1639; this tablet, facing the monument of Newton, was raised after the lapse of more than two centuries, December 9th 1874.*

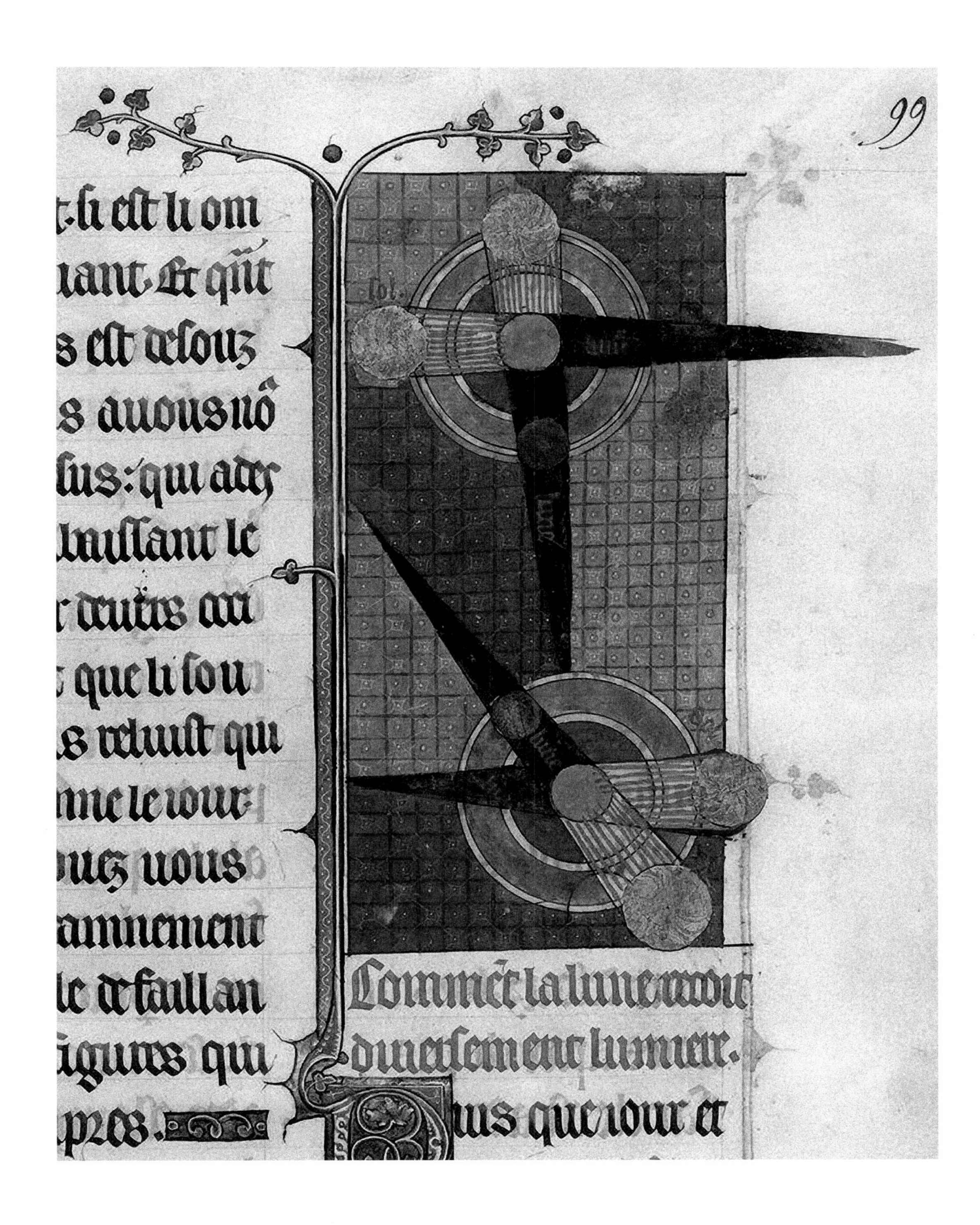

• 1320–25:

While this illumination from French poet-priest Gossuin de Metz's *L'Image du monde* (Image of the World) dates back to the early fourteenth century, one would be hard-pressed to find a more beautiful representation of lunar eclipses from any more of a contemporary source. They were painted by one of a group of artists known collectively as the Master of the Roman de Fauvel. De Metz's encyclopedia, which dates to 1245, combined many sources of information, including Ptolemy's *Almagest*, and contains such startling medieval insights as "Seen from the sky, the Earth would be in size like the smallest of stars." (This 750 years before the Voyager spacecraft's famous "pale blue dot" picture proved de Metz's point.) In 1480, a translation of L'*Image du monde* became the first illustrated book to be printed in English. This copy belonged at one time to the Duke of Berry.

• 1444–50:

Above: A canto in the *Paradiso* section of Dante's *The Divine Comedy* contains an explicit description of a solar eclipse, here illuminated by Giovanni di Paolo. On arriving at the moon, Dante (dressed in blue) asks his guide to explain its dark markings, which he believes must be due to dense and "rare" (thinner) areas. There follows a dialogue on the phenomenon in which Beatrice points out that Dante's theories on the subject must be wrong, because if the dark patches on the moon were due to a lack of density, you would be able to see right through the moon in those patches—whereas during a solar eclipse, she says, the sun doesn't shine through. Explicitly referring to the use of science to confirm a theory, she scolds Dante: "Your thought is submerged in error"—a passage written three hundred years before a telescope was first turned toward the moon. For other works by di Paolo, see pages 33, 75, 116, 144, and 178–79.

• 1499:

Right: 1499: This hand-colored woodcut print from an early printed edition of Johannes de Sacrobosco's *Tractatus de sphaera* (Concerning the Sphere) depicts a lunar eclipse. The book presented a coherent account of the complex planetary motions of the Ptolemaic universe to more than two centuries of readers; the first edition was in 1230. (For another illustration from this Venetian edition, see page 41.)

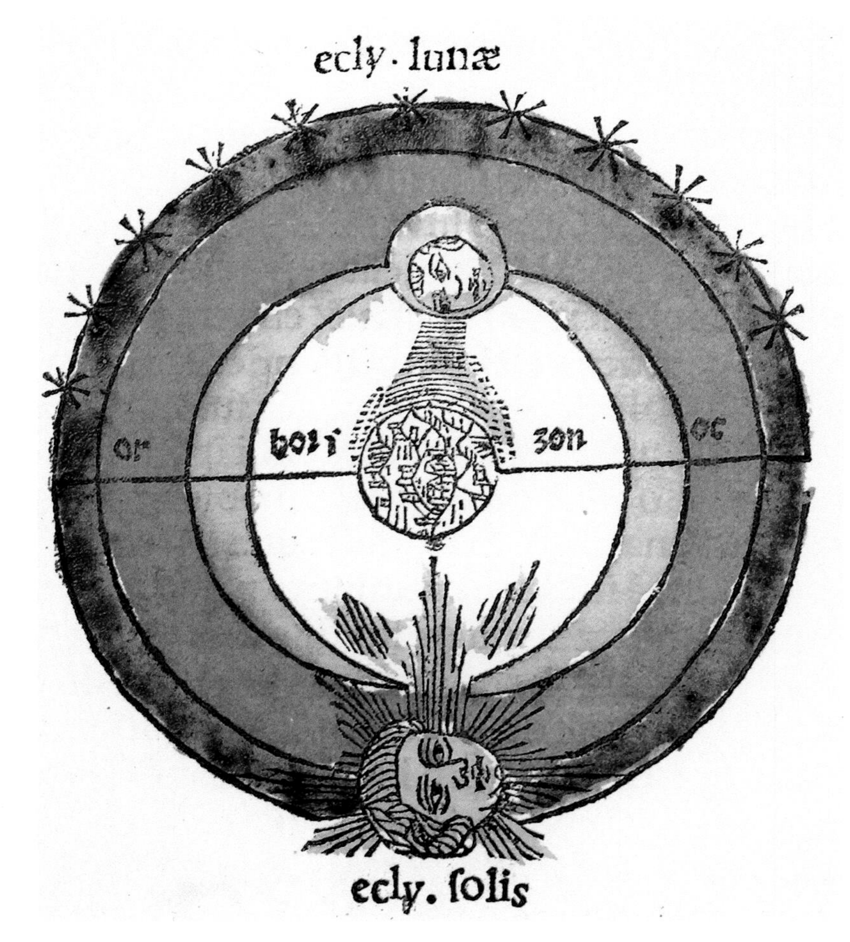

• 1478:

Facing page: Illuminated eclipse prediction tables by German miniaturist Joachinus de Gigantibus, from the scientific treatise *Astronomia* by Tuscan-Neopolitan humanist Christianus Prolianus. On the left margin in vertical Latin the years of the upcoming eclipses are specified, with the predicted months, dates, and eclipse lengths in horizontal script on either side of the projected visual appearance at or near totality. Gold leaf is only used for the solar eclipses. Another illumination from this book can be seen on page 117.

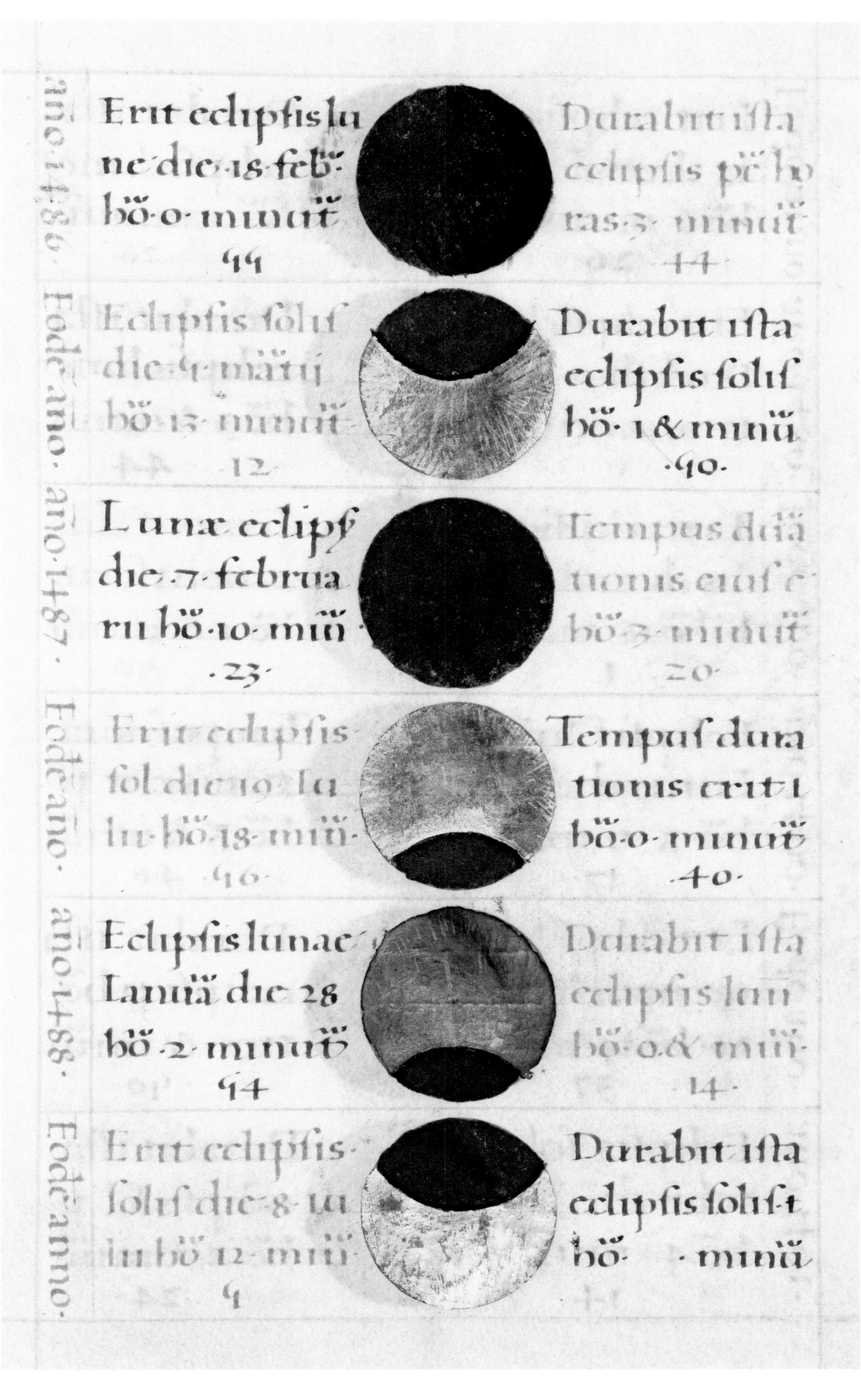
āno 1486
Erit eclipsis lu
ne die 18 febr
hō 0 minut
44
Durabit ista
eclipsis per ho
ras 3 minut
44
Eodē āno
Eclipsis solis
die 4 martij
hō 13 minut
12
Durabit ista
eclipsis solis
hō 1 & minu
40
āno 1487
Lunæ eclips
die 7 februa
rii hō 10 min
23
Tempus dura
tionis eius e
hō 3 minut
20
Eodē āno
Erit eclipsis
sol die 19 Iu
lii hō 18 min
46
Tempus dura
tionis erit 1
hō 0 minut
40
āno 1488
Eclipsis lunae
Ianuā die 28
hō 2 minut
44
Durabit ista
eclipsis lun
hō 0 & min
14
Eodē anno
Erit eclipsis
solis die 8 Iu
lii hō 12 min
4
Durabit ista
eclipsis solis 1
hō minu

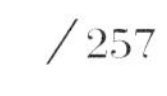

• 1547–52:

In July 2008, London Old Masters dealer James Faber bought a remarkable book at auction in Munich. The bound manuscript was packed with 167 watercolor and gouache paintings, each illustrating a miraculous event. Faber noted that the last miracle described was in 1552, and commissioned a thorough analysis of its paper and materials. The results confirmed his hunch that what became known as the *Augsburger Wunderzeichenbuch* (Augsburg Miracles Book) in fact dated to the mid-sixteenth century, and was almost certainly made in Augsburg at the height of the Reformation. Approximately sixty of the paintings depict astronomical subjects, including eclipses, comets, and other celestial phenomena, all presented in a drily reportorial Old German. **Above:** The text reads: "In 1483, the locusts flew through the welsch land [Southern Europe], devastated the countryside around Brixen, and if Margrave Louis of Mantua had not prevented it, they would have destroyed the whole seed in Lombardy. He had them killed, burned and chased away. Afterwards, an eclipse of the sun was seen and then came a great dying, so that more than twenty thousand people died in Brixen and around thirty thousand died in Venice."

Above: The text of this *Augsburger Wunderzeichenbuch* illustration reads: "In A.D. 1362, at the time of Otto, the emperor from Saxony, a stone—wondrous and big—fell from the sky in heavy wind and rain. And on many people, little blood-red crosses appeared and a great eclipse of the sun appeared."

• 1554:

Overleaf: Bohemian astronomer Cyprián Lvovický earned his reputation correcting tables of celestial motion by prior astronomers Regiomontanus and von Peuerbach—earning the respect and friendship of Tycho Brahe in the process. At a time when astronomers frequently doubled as astrologers, Lvovický interpreted major historical events via astrological omens, paying particular attention to eclipses. These illuminations from his *Eclipses luminarium* (Eclipses of the Luminaries) are part of a manuscript containing predictions of eclipses from 1554 to 1600. Many of Lvovický's eclipses included landscapes within *trompe-l'oeil* frames, as here. **Above:** Eclipse of the moon on the night of March 2, 1569. Note Jupiter in the sky on the upper left above the torchlit group, and the eclipsed moon on the far right, above a man on horseback. **Facing page:** Partial solar eclipse on April 8, 1567. The brave soul in a tree on the left isn't a spectator trying to get a closer look; he's using an ax.

Figura Eclipsis Lunæ, Sole constituto in longitudine media ecentrici sui.

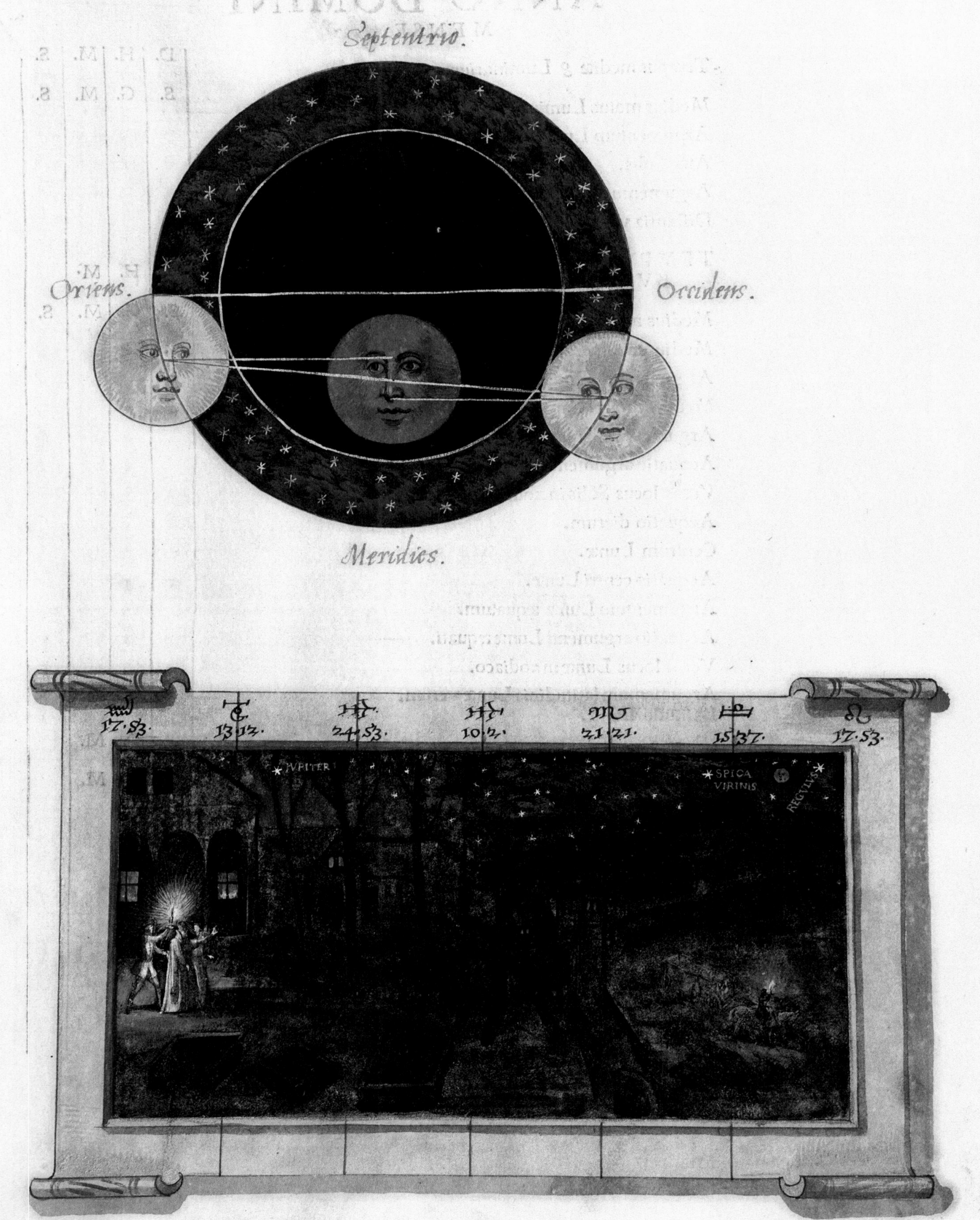

Figura Eclipsis Solis constituti in longi-tudine media ecentrici.

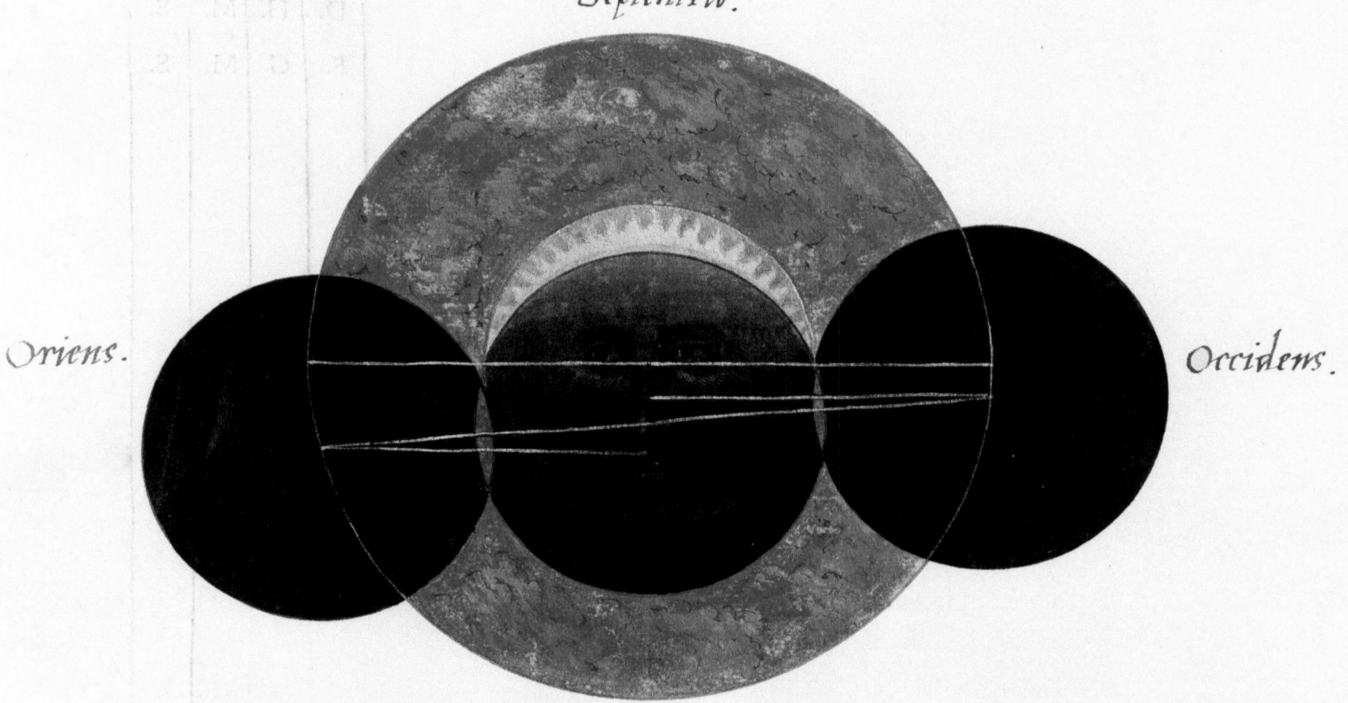

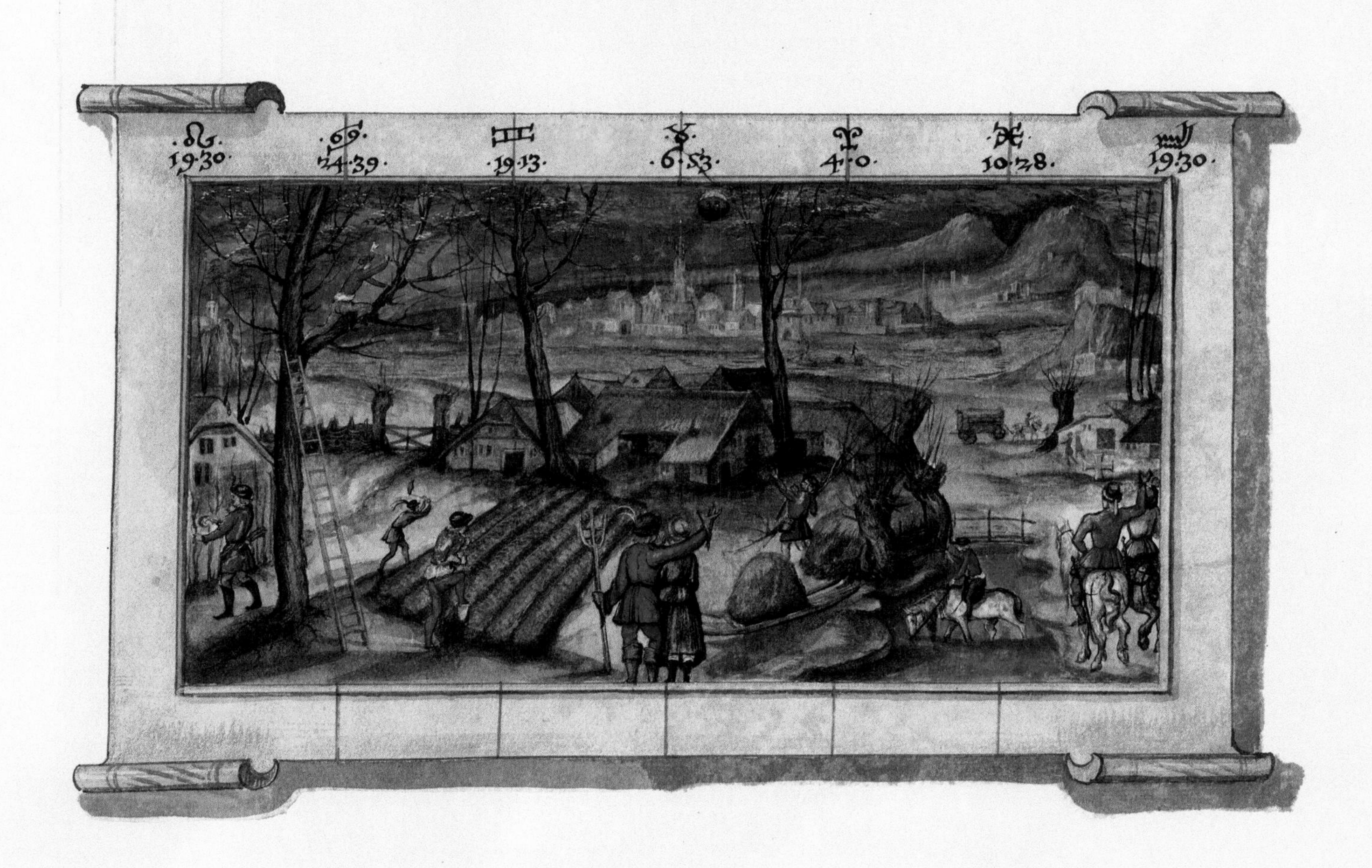

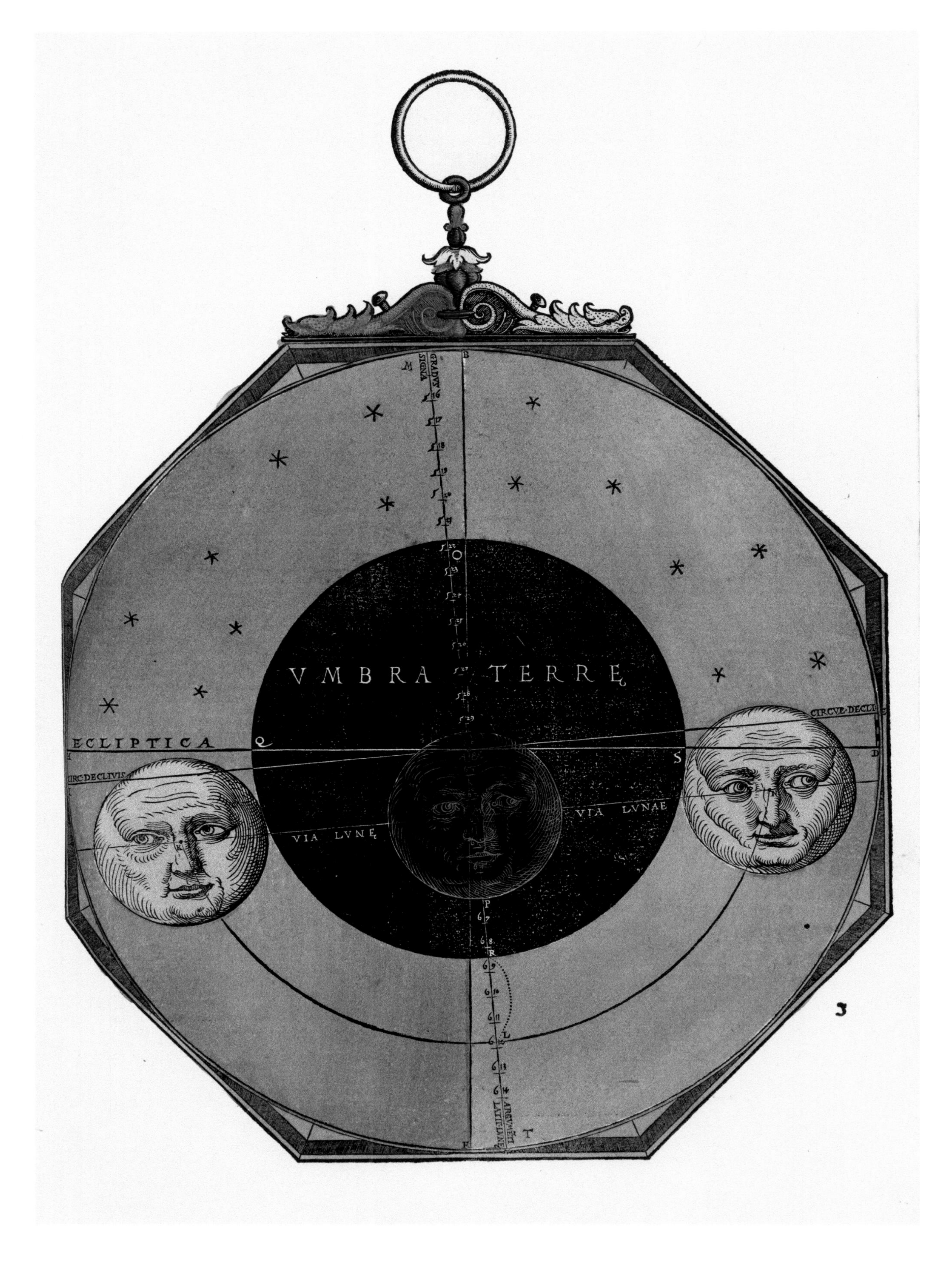

• 1540:

In this lunar eclipse representation from German printer and cosmographer Peter Apian's *Astronomicum Caesareum* (Caesar's Astronomy), the black circle represents the shadow of Earth, and the ecliptic line, or path, the sun appears to follow in the sky, cuts across the shadow horizontally. By contrast, the path of the moon through the Earth's shadow is at an angle to the ecliptic. Earth's shadow follows the ecliptic line, and is always oriented toward the antisolar point. Unlike similar prints from Apian's book, this isn't an actual volvelle, meaning it has no moving parts. It commemorates the eclipse of the moon that took place on October 6, 1530—the date the book's dedicatee, Charles V, was crowned Holy Roman Emperor. For more from the book, see pages 43, 77, 119, 180, and 227.

• 1570s:

In this painting attributed to French Renaissance court painter Antoine Caron, a solar eclipse unfolds above an idealized Paris. One title of the work, *Astronomers Studying an Eclipse*, was given much later, by Anthony Blunt of the Courtauld Institute of Art in London, who acquired the painting in 1947. It's now—probably erroneously—titled *Dionysius the Areopagite Converting the Pagan Philosophers*, and on view at the Getty Center in Los Angeles; the bearded figure in Greek garb in the foreground is most likely meant to be Greek-Alexandrine astronomer Claudius Ptolemy. No total solar eclipses occurred in Paris during Caron's lifetime, but in January of 1544 a partial eclipse covered 96 percent of the sun, and recollections of that event may have served as the basis of this painting. At least five of the figures are equipped with tools suitable to astronomers of the day, from armillary spheres to compasses and dividers. (Note the *putto* on the steps, taking notes and surrounded by his own compass, square, and straightedge.)

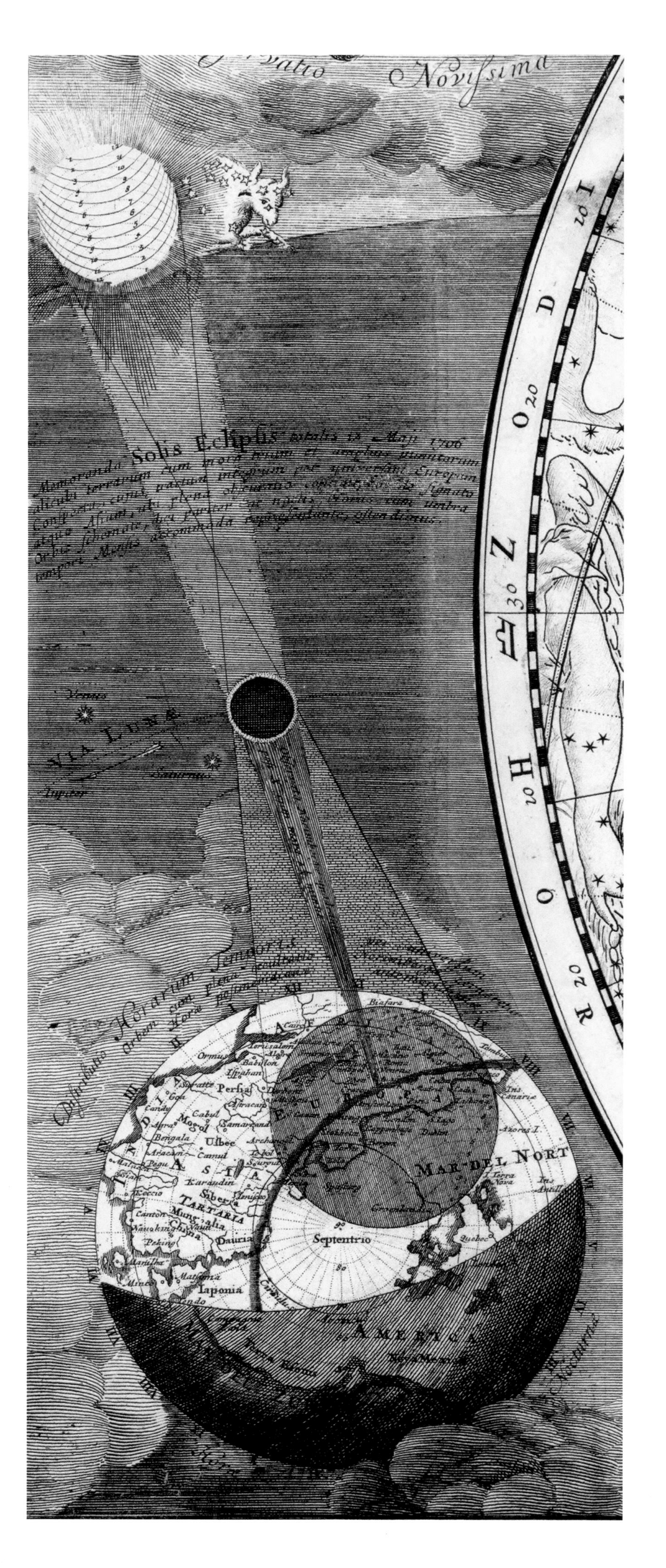

• 1700:

On May 12, 1706, a dramatic total solar eclipse plunged a swath of Europe into darkness. The path of totality can be seen here within the wider penumbral shadow. This depiction of the event, a detail view of a polar projection of the southern celestial hemisphere made for Amsterdam cartographer Carel Allard, adopts an outside view—the viewer is implicitly positioned tens of thousands of miles above Earth's North Pole. Images like this gradually set the stage for the idea that travel to such distances might one day be possible.

• 1715:

Facing page: This is one of the earliest maps to show in detail the path of a total solar eclipse, as calculations had only just become accurate enough. For the eclipse of April 22, 1715, English astronomer Edmond Halley, best known for computing the orbit of the eponymous comet, put out a single-page broadsheet map similar to this one. It was relatively accurate but by no means perfect. With the approach of another total solar eclipse on May 11, 1724, Halley issued a corrected map for the 1715 eclipse that also contained the predicted path of the 1724 one, as seen here. The 1724 eclipse duly crossed the English Channel, heading southward toward France.

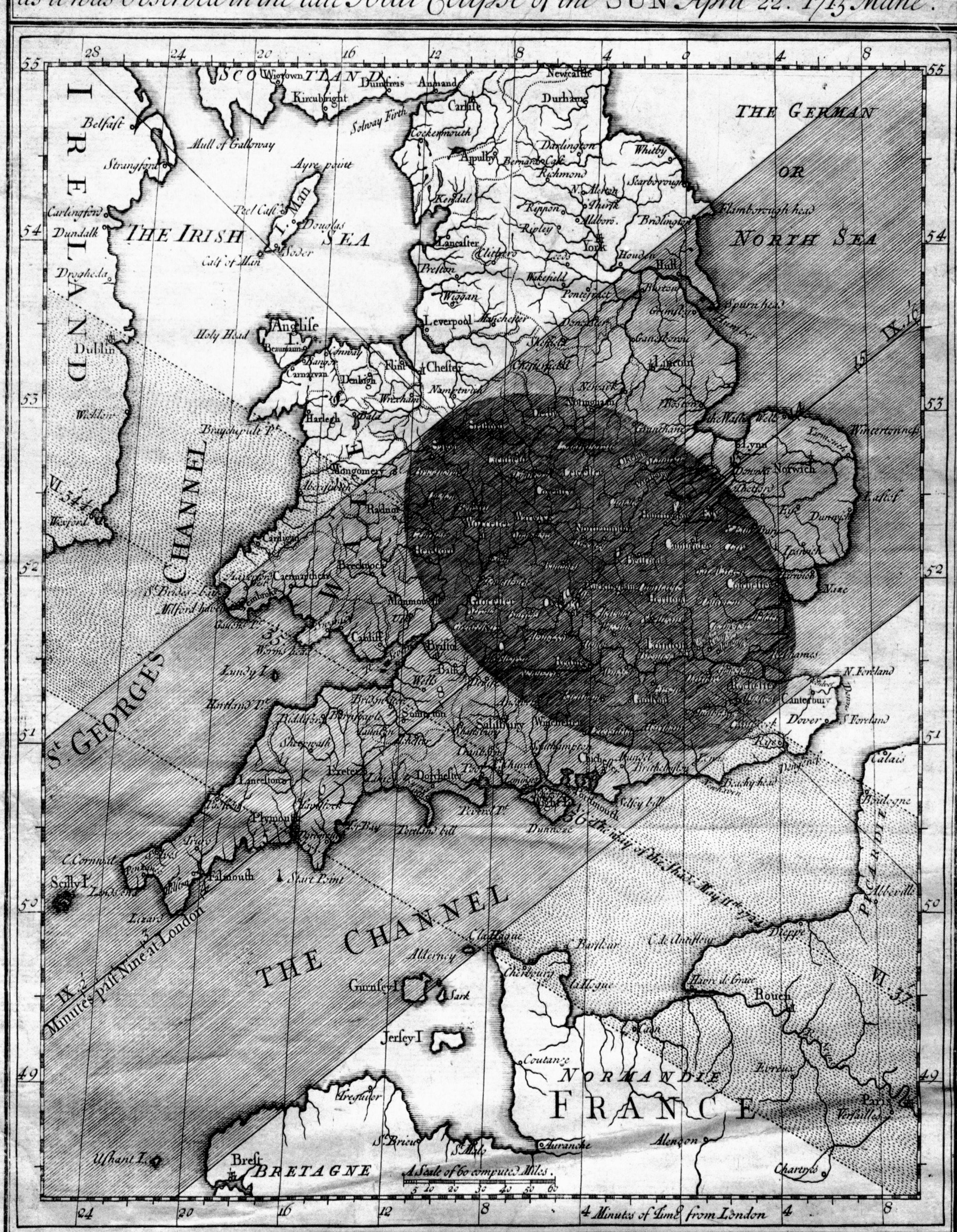
A Description of the Passage of the Shadow of the Moon over England
as it was Observed in the late Total Eclipse of the SUN April 22d. 1715 Manè.
SCOTLAND
Wigtown
Dumfreis
Annand
Newcastle
Kircubright
Solway Firth
Carlisle
Durham
IRELAND
Belfast
Mull of Galloway
Cockermouth
Darlington
Whitby
Strangford
Ayre point
Apulby
Richmond
Scarborough
THE GERMAN OR NORTH SEA
Carlingford
Peel Castle
I. Man
Douglas
Kendal
Rippon
Thirsk
Flamborough head
Dundalk
THE IRISH SEA
Ripley
York
Bridlington
Calf of Man
Lancaster
Clithero
Leeds
Houden
Hull
Drogheda
Preston
Wakefield
Wigan
Pontefract
Spurn head
Holy Head
Anglise I.
Leverpool
Manchester
Dublin
Beaumaris
Conway
Flint
Chester
Lincoln
Bangor
Carnarvan
Denbigh
Namptwich
Wrexham
Newark
Nottingham
Derby
Wicklow
Harlech
Bala
Boston
Brachypult Pt.
Grantham
the Wash
Wells
Lynn
Wintertonness
Norwich
Montgomery
Shrewsbury
Litchfield
Leicester
Wexford
Radnor
Worcester
Warwick
Northampton
Cardigan
Hereford
Huntington
Cambridge
Dunwich
Brecknock
Bedford
Ipswich
Caermarthen
Harwich
St. Brides Bay
Milford Haven
Monmouth
Glocester
Oxford
Colchester
Cardiff
Bristol
London
Worms head
Bath
Lundy I.
Wells
Thames
N. Foreland
Reading
Canterbury
Hartland Pt.
Dover
S. Foreland
ST. GEORGES CHANNEL
Taunton
Salisbury
Winchester
Southampton
Chichester
Brighthelmstone
Rye
Dungeness
Calais
Exeter
Launceston
Dorchester
Beachy head
Boulogne
Plymouth
Dartmouth
Tor Bay
Portland bill
Peverel Pt.
Dunnose
Selsey bill
PICARDIE
C. Cornwall
Falmouth
Start Point
Scilly I.
Lizard
Abbeville
The way of the Shadow May 11th 1724
Minutes past Nine at London
THE CHANNEL
Alderney
C. la Hague
Barfleur
C. de Antifer
Dieppe
Cherbourg
La Hague
Havre de Grace
Rouen
Gurnsey I.
Sark
Jersey I.
Coutance
Evreux
NORMANDIE
FRANCE
Paris
Versailles
Treguier
St. Brieu
St. Malo
Avranche
Alençon
Ushant I.
Brest
BRETAGNE
A Scale of 60 computed Miles
Chartres
Minutes of Time from London

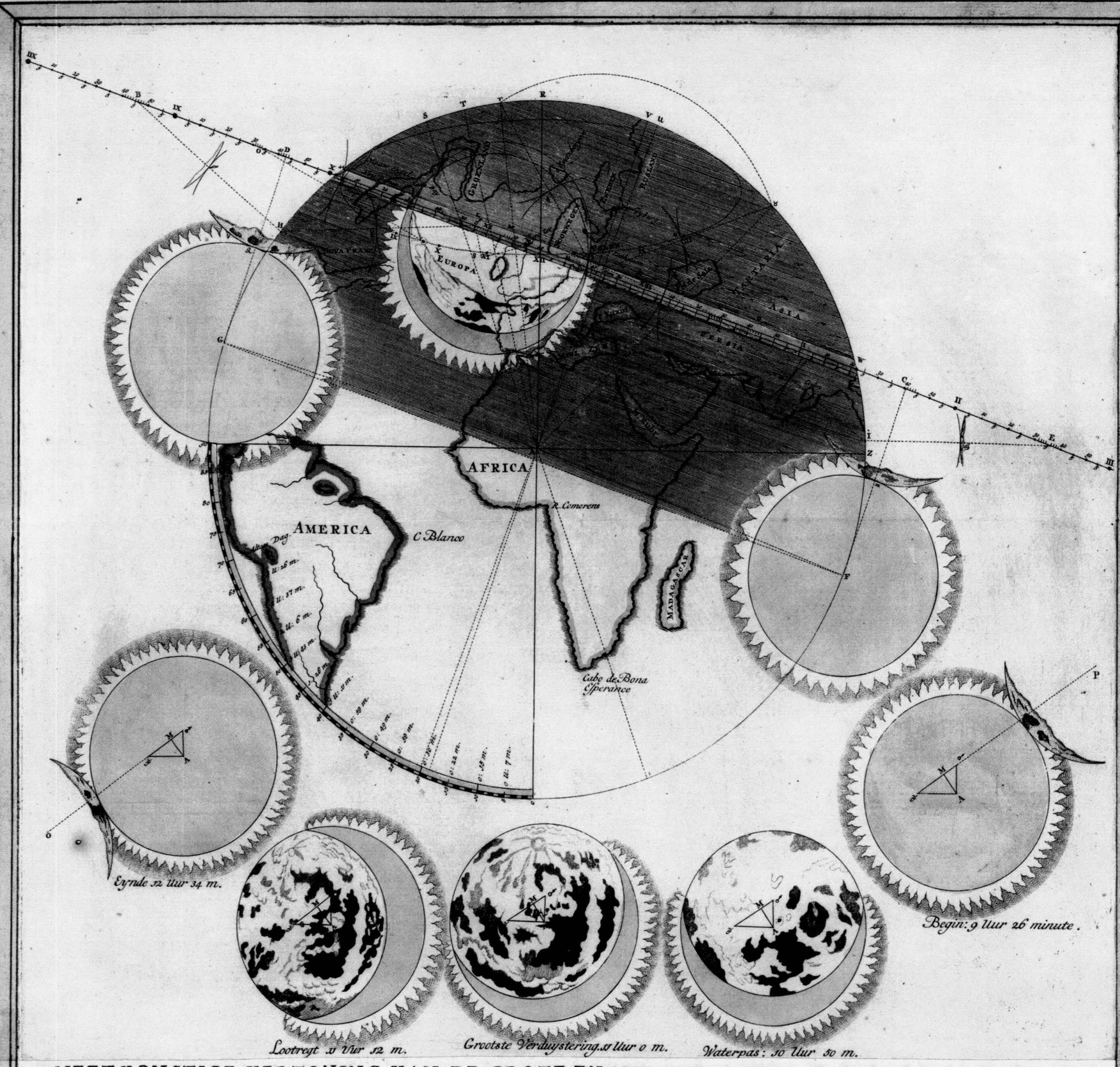

MEET-KONSTIGE VERTONING VAN DE GROTE EN MERK-WAARDIGE ZONS-VERDUISTERING.

Die wezen zal den 25. Julius 1748. Hoe dezelve zig boven de Stad AMSTERDAM en andere omleggende STEDEN zal vertoonen. Als meede een nette Aanwyzinge, waar de Middelpunts-Schaduwe zal passeeren; en hoe verre deeze Eclips ten Zuiden en Noorden zal konnen gezien worden. Alles door de Meetkonst betoont, door SIMON PANSER,

Stads Mathematicus, Leermeester der Wis- Sterre- en Zeevaartkunde tot EMBDEN.

VERKLARINGE over deeze bovenstaande Afbeeldingen.

Daar is in de geheele Astronomia, of Sterrekunde, niets, dat de wonderlyke Sneedigheid van het menschelyke vernuft, en dezelfs schrandere doorzichtigheid, meer aantoont; als de klare uitlegginge van de verduisteringen van Zon en Maan, en de Naauwkeurige voorzegginge derzelver, zoo als die by de Sterrekundige word opgemaakt.

Ik heb in den Jare 1738. een Astronomische Hemel-Spiegel in 't Ligt gegeven, waarop deeze Zon-Eclips vertoonde: van gedagten zynde, om over dit Verschynsel niet meer te reppen: Maar met het uitkomen van den Almanach, zoo zag wel haast, dat het de pyne waart was, om deeze seldzame Verduysteringe der Zon eens van nieuws wederom te hervatten; om dat het met de Tyd in den Almanach wel 1 Uur 28 Minuten verschilde: Ja hier komt nog by, dat eenen *Joh. van der Boot*, in zyn Tractaat: *De Eeuwigduurende, en Onveranderlyke Zon- en Maans Tafelen.* Het begin stelt 10 Uur 30 Min. verschillende wederom 1 Uur 4 Min. En hy zeid, dat dezelve byna geheel zal verduisteren, of ten uitersten zeer weinig ligt over de Noordzyde zal behouden. Daar dog dezelve nog omtrent 1⅓ Duim ligt: niet over de Noordzyde, maar over de Zuidzyde behoud.

Dit is de reden, waarde Leezer, dat my bewogen heeft om 't werk zeer naauwkeurig nog eens naar te gaan: En ik twyffele gantschelyk niet of deeze Uitpassinge zal met de Observatien zeer na overeen komen. Aangaande de Nieuwe Maans Tyd heb ik naar de Tafels van *De la Hire* bereekent. En dit is 't geene ik vooraf te zeggen hadde.

In de Generale Figuure verbeeld het Rond de Schyve des Aardkloots, zoo als dezelve zig zoude vertoonen, als wy op de Maan geplaatst waren: waar van A de Zon is, staande in Top. R de Noord-As. RS. RU. de Zons- of ☾-Declinatie. B.M. W.E. De Maans Weg, lopende regthoekig over de Maans Asboog A. V. Het Ovaal is de Weg van Amsterdam, dewelke zig uit het Oogpunt de ☾. dusdanig zou swajen, welkers weg: Als meede die van de ☾ in Uuren en Minuten verdeelt zyn: Welkers evene Tusschentyd, genomen met de ½ Midlynen van Zon en Maan de begeerde Tyden kan bepalen. O ☉ is het Begin (hier is telkens de Equatie des Tyds 6 Minuten afgetrokken) te 9 Uuren 26 Minuten. De Horens Waterpas in 10 uuren 50 minuten in ♂. Het midden in M ten 11 uuren 0 minuten. De Horens der Zon lootregt in ♃ ten 11 uuren 12 minuten. Het Einde in P ten 12 uuren 34 minuten. En zal op zyn grootst 10⅓ Duim over de Noordzyde verduisteren. NB. Dit ziet op de Generale Figuur.

Nu staat ons aan te toonen, waar de Schaduwes-middelpunt over de Oppervlakte des Aardkloots zal passeeren, en verbeeld de Breede Riem van X tot W. De Schaduwes-middelpunt beslaat in de Breete omtrent 40 Mylen, welkers onderleggende Plaatsen rontsom de Maan, de glinsterende Zons-Cirkel, onder de gedaante van een gulden Ring, zien vertoonen, om reden, dewyl den Maans-middellyn zig naar ons gezigt minder vertoont als die van de Zon: of 't welk even eens is, om dat de schaduwachtige Kegel van de Maan, zyn uiterste Spits nog ruim 2 7/10 Halve Aardkloots-middellyn van de Aarde blyft.

De Ware Tyd der Samenstand is gevonden Voormiddag te 11 uren 43 min. tot Amsterdam. Op dezelve Tyd komt de Maan in zyn Weg in B. en treed in ♎ voor de Zonne: alsdan valt de Byschaduw in H op den Aardbol. En neemt met de Zons opgang haar aanvang 's morgens ten 5 uuren 1 min. op deeze Tyd is de ☾ Schaduws-middelpunt in B. Te Amsterdam 8 uuren 46 min. dat is 3 uuren 45 min. vroeger, of 56 graden 15 min. ten Westen, dat is, op 325 graden 45 Lengte. Om nu de Pools Hoogte te vinden, zoo haalt van 't Punt H regthoekig op de N. As AR. van daar regthoekig door AS. of AU tot op de Rand van de Aardkloot: zoo zal van dat Punt, tot aan U of S 't Compl. van de Pools Hoogte zyn, gelyk men kan afmeeten op het Quadrant: waar op van 5 tot 5 graden de Zons opgang, van ieder Pools hoogte ook genoteert staat: men vind dan naar onderregtinge dat de Plaats H leid op 35 gr. 30 min. Noorder Breedte.

Van daar rukt de Maan op de Middelpunts-weg tot in ⇸ alwaar men de Zon 's morgens ten 4 uuren 31 min. als een gouden Ring zal zien opkomen, alsdan is het te Amsterdam 10 uuren 2 min. zynde 5 uuren 31 min. vroeger, dat is de Plaats ⇸ op 299 graden 15 min. lengte. En men vind 46 graden Noorder Aspunts Hoogte.

XII. Is de Plaats, alwaar de Zonne annulaar zal verduisteren, en dat op de middag ten 11 uuren 20 min. dat is ten Oosten 10 graden op 32 graden Lengte, en 53 graden Noorder Breedte.

Het Punt in de ☾ Asboog A V. is de Plaats, alwaar men de Zon 1 uur 7 min. naar de middag ringswyze zal zien verduisteren, leggende op 44 graden Lengte, en 50 graden Noorder Breedte.

W. Is de Plaatze, alwaar men de Zonne-Centraal 's avonds ten 6 uuren 17 min. zal zien ondergaan, alsdan is de ☾ in zyn Weg volgens de Tyd te Amsterdam 1 uur 22 min. Dan leid de Plaats W op 96 graden Lengte, en 12 graden 30 minuten Noorder Breedte.

I. Is de Plaats, alwaar de Schaduwe van de Aarde scheid, met de Zons Ondergang ten 6 uuren 1 minut, zynde te Amsterdam 2 uuren 37 min. dat is op 73 graden Lengte en 1 graad Noorder Breedte.

Indien men nu de Plaatzen naarzoekt, zo vind men, dat deeze Middelpunts Duistering zyn begin neemt in de Westindische Zee, tusschen La Barmuda en de Canarische Eilanden, lopende tusschen Nieuw Engeland en Nova Francia door, over Caap Breton, bezuiden Hudzons Bay, Groenland over Duitsland, Polen, Bohemen, tusschen Pontus Euxinus en het Persische Meir, over Tartaria, Persia tot in Oostindien, daar de Schaduwe van de Aarde scheid; niet verre van Sumatra. Ten Zuiden loopt de Schaduwe tot aan Florida en Madagascar; en ten Noorden word dezelve met den Sigteinder afgesneeden in 't onbekende Noorder gedeelte van den Aardkloot. Deeze Middelpunts-schaduwe heeft over de Aardkloot geloopen van ⇸ tot W. 2279 mylen, dat is in ieder uur 683 1/10 myl. En alzo make ik met deze Beschryvinge een Einde. Doch men moet betragten, dat alle Plaatzen die benoorden ⇸, W leggen, de Zon over de Zuidzyde; maar bezuiden ⇸. W. over de Noordzyde zullen zien verduisteren. Hier meede afbreekende, wensche den kunstbegeerigen Leezer veel plaisier tot deeze raare Verschyninge. Verblyve

Embden, den 16. May 1748.

U. E. D. W. Dienaar

SIMON PANSER,

Stads Mathematicus &c.

t'Amsterdam, by R. en J. OTTENS, Kaart- en Boekverkopers in de Kalverstraat.

PLATE II.

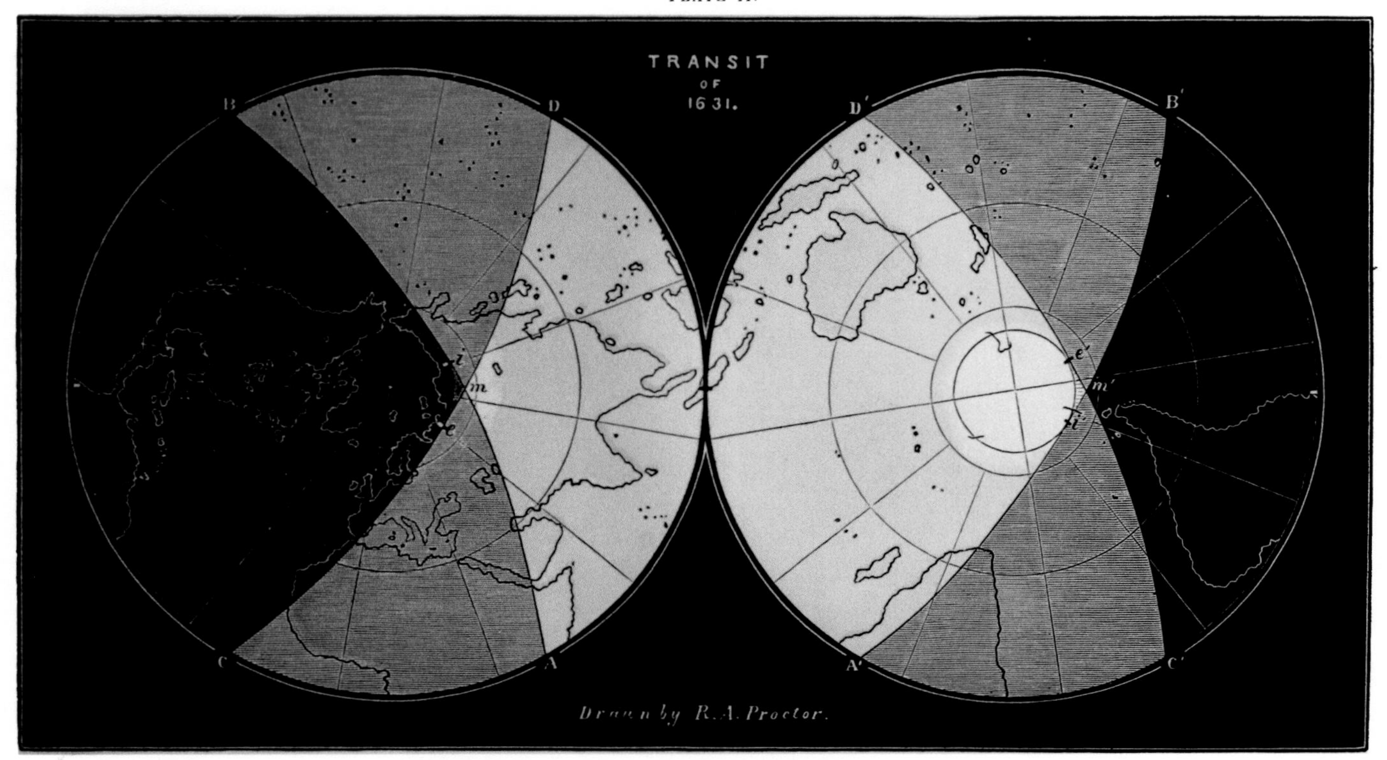

A B and A' B' mark the boundary between the sunlit and dark hemispheres a the beginning of the transit.
C D and C' D' mark the boundary between the sunlit and dark hemispheres at the end of the transit.

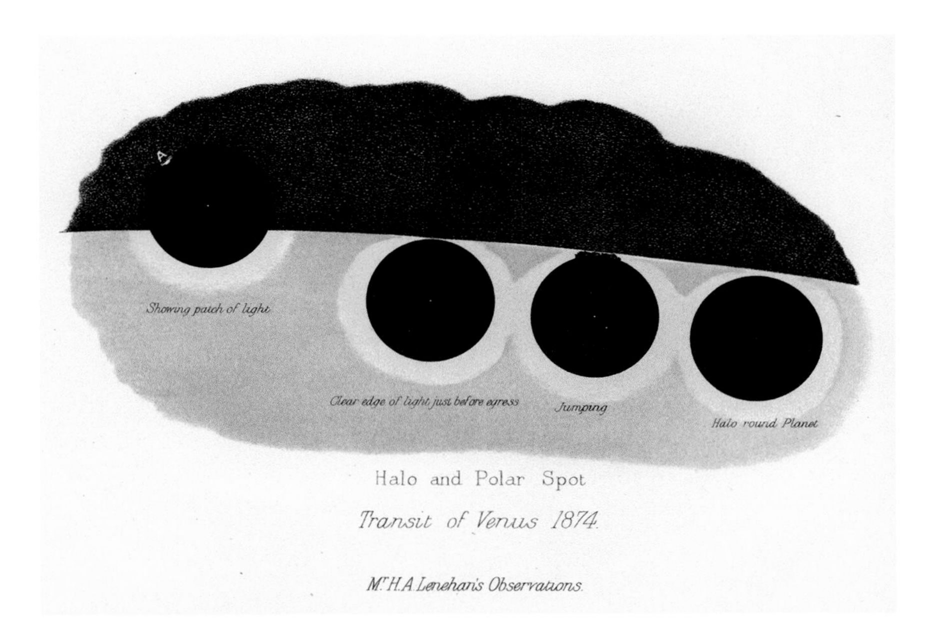

• 1748:

Facing page: Dutch broadsheet by mathematician Simon Panser depicting the annular solar eclipse of July 25, 1748. During an annular eclipse, the moon's apparent size is slightly smaller than the sun, which forms a bright ring around the silhouetted moon. Panser based his estimation of the start and end time of the eclipse on astronomical tables by French astronomer Philippe de La Hire. Here, details on the moon are depicted as being visible as it passes across the sun's disc, although they would have been entirely obscured by the sun's glare (and the fact that the night side of the moon faces Earth during a solar eclipse).

• 1875:

Above top: Transits of Venus across the face of the sun are far more rare than solar eclipses. They occur in pairs eight years apart, and are always separated by more than a century. The transits themselves can last six hours or more. In these polar projections of Earth from English astronomer Richard Proctor's book *Transits of Venus*, the hemisphere from which the transit of 1631 was visible is shown, with the shifting line between the day and night sides of our turning planet during Venus' lengthy transit indicated by the two shaded areas.

• 1892:

Above: The 1874 transit of Venus was the first of a pair to occur in the nineteenth century. This illustration from *Observations of the Transit of Venus* by New South Wales Government Astronomer Henry Russell depicts the transit as described by Sydney Observatory astronomer Henry Lenehan, who reported that he "distinctly saw a clear band of light" as Venus completed its initial passage across the sun's limb. Russell's book constituted the Sydney component of a worldwide effort within the British Empire to record the transit from multiple locations.

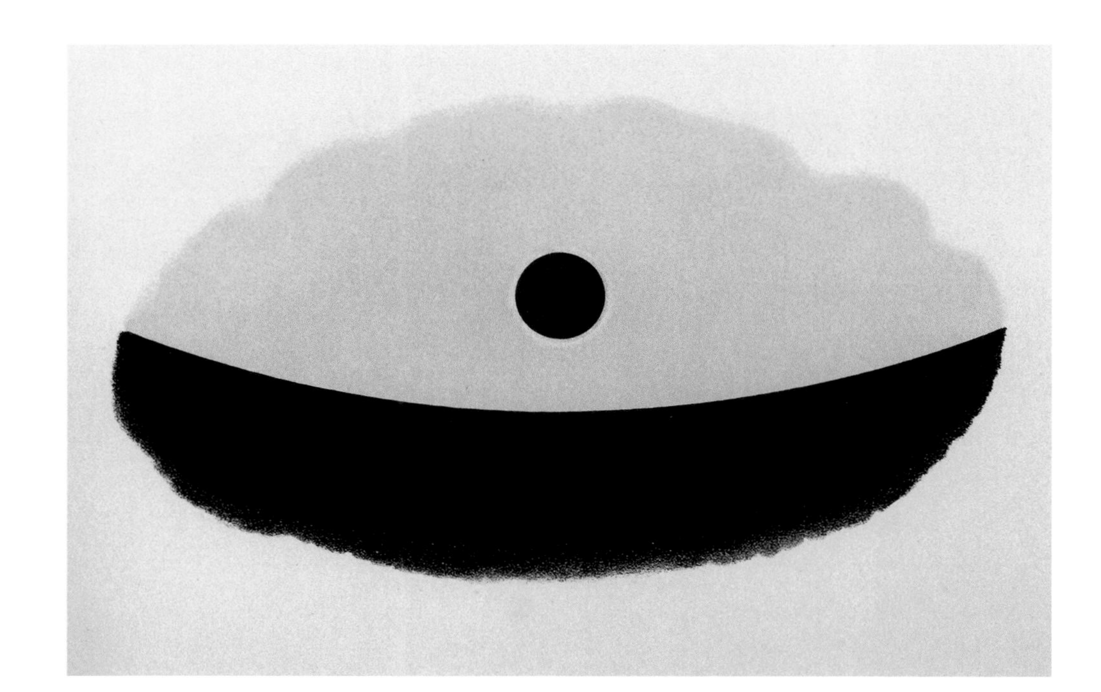

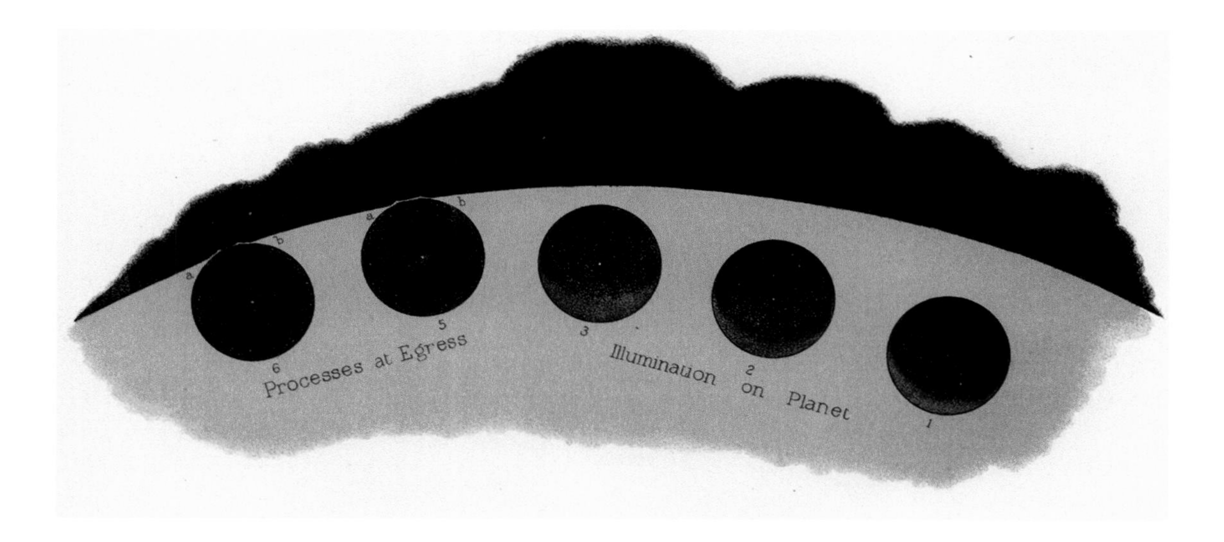

• 1892:

Three more illustrations from Henry Russell's *Observations of the Transit of Venus*, reflecting observations reported by his team of seven observers. **Top:** Amateur astronomer Geoffrey Hirst reported seeing a "narrow fringe of dull red light" all around Venus. **Middle:** Like Henry Lenehan, geologist Archibald Liversidge also reported seeing a "hazy grey filament like a streak of smoke" between the planet and sun at the moment of full ingress to the sun (mistakenly called "egress"). He also reported that Venus "appeared illuminated on the inner side." **Right:** A. W. Belfield and Archibald Park saw a "gorgeous deep blue towards the circumference" of Venus during the transit.

Facing page: For his part, Sydney amateur astronomer Alfred Fairfax "distinctly saw . . . a most brilliant line, very narrow" filled with "forms and colors" around Venus as it transited the sun's limb. Still, reacting to this plate, he wrote, "I would not for one moment say that the drawing is correct to scale; in fact the halo was so narrow I could not make a drawing like it." From Henry Russell's *Observations of the Transit of Venus*.

Transit of Venus 1874.

Sydney, N. S. W.

Mr. A. Fairfax's Observations.

270 /

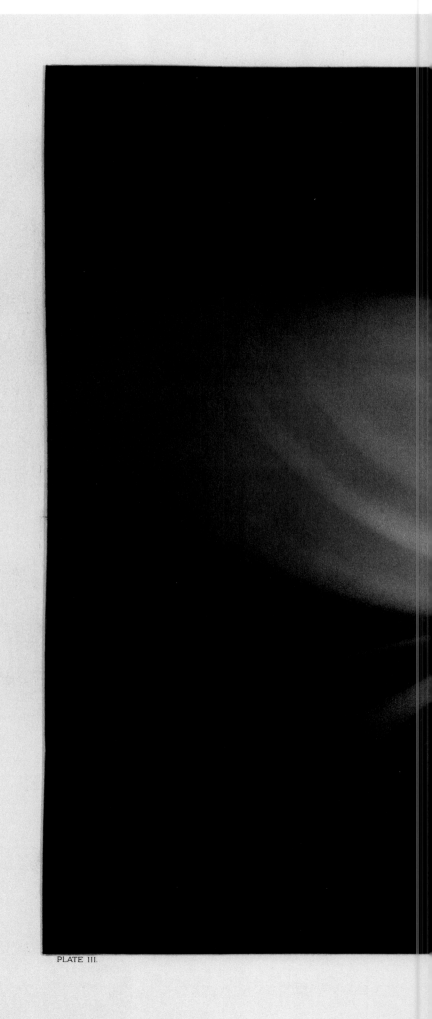

• 1881:

On July 29, 1878, artist-astronomer Étienne Trouvelot traveled to Creston, in the Territory of Wyoming, to observe a total solar eclipse. At the moment of totality, the sun's corona could be observed, as could solar prominences, as seen here. Trouvelot captures the faint red tint of the solar chromosphere, as well as the diaphanous outer atmosphere of the corona streaming away on all sides. From the Charles Scribner's Sons limited-edition collection of Trouvelot chromolithographs. For more by Trouvelot, see pages 98–99, 127, 128–29, 158, 188–91, 241, 295, and 297–99.

E. L. Trouvelot

TOTAL ECLIPSE of the SUN.

Observed July 29, 1878, at Creston, Wyoming Territory.

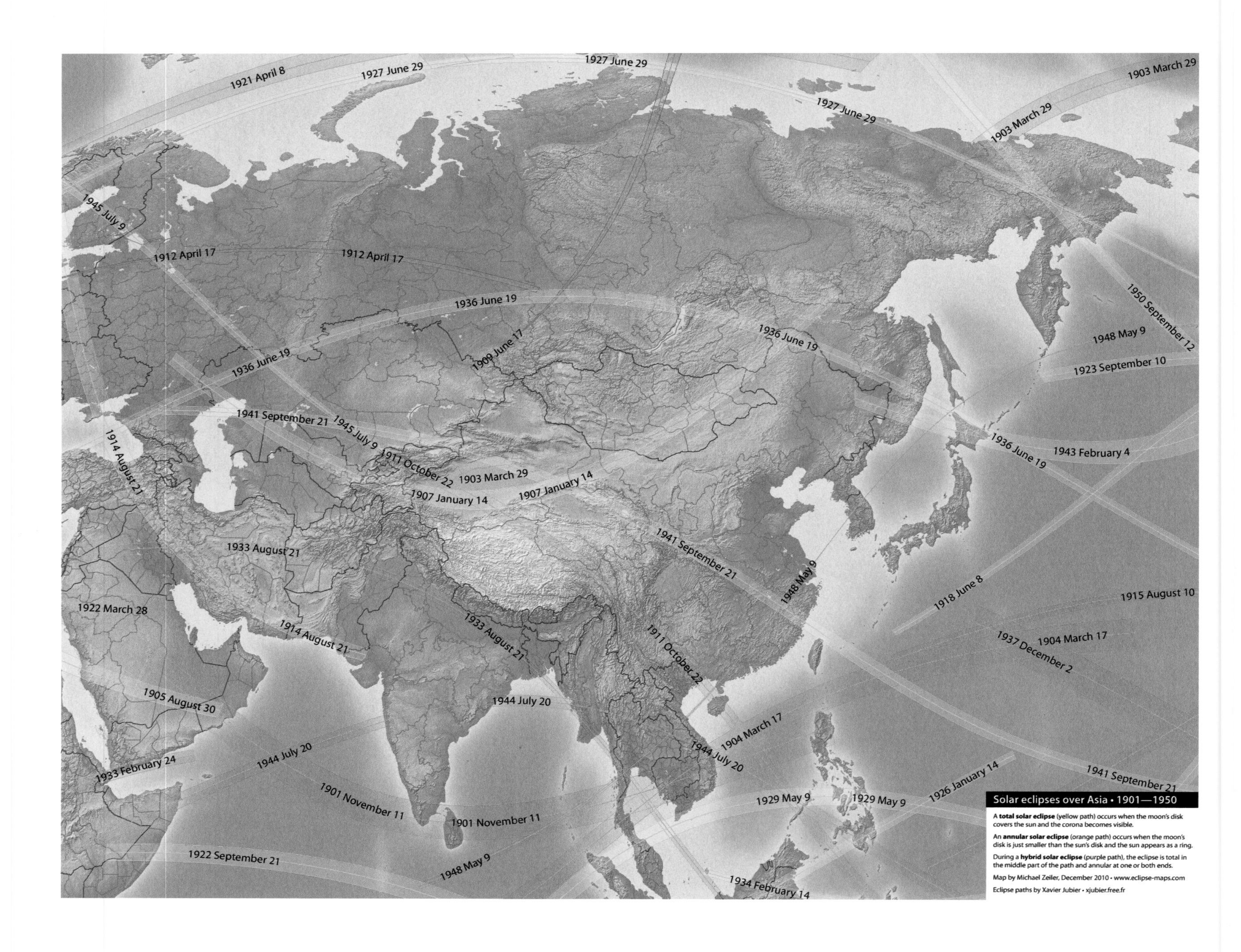

• 2010:

Calling a total eclipse "the most amazing spectacle in nature," Santa Fe-based geographic information systems specialist Michael Zeiler has produced an extensive series of maps covering both past and future eclipses and transits. Based on data supplied by Paris engineer Xavier Jubier, Zeiler's eclipse cartography lives at eclipse-maps.com. The above map records the path of all the solar eclipses that occurred in Asia from 1901 to 1950. The yellow paths are for total eclipses, the orange for annular eclipses (when the sun appears as a ring around the moon), and the purple for hybrid eclipses (when the eclipse is only total in the middle parts of the eclipse path).

Annular Solar Eclipse of 2012 May 20

Magnitude of eclipse

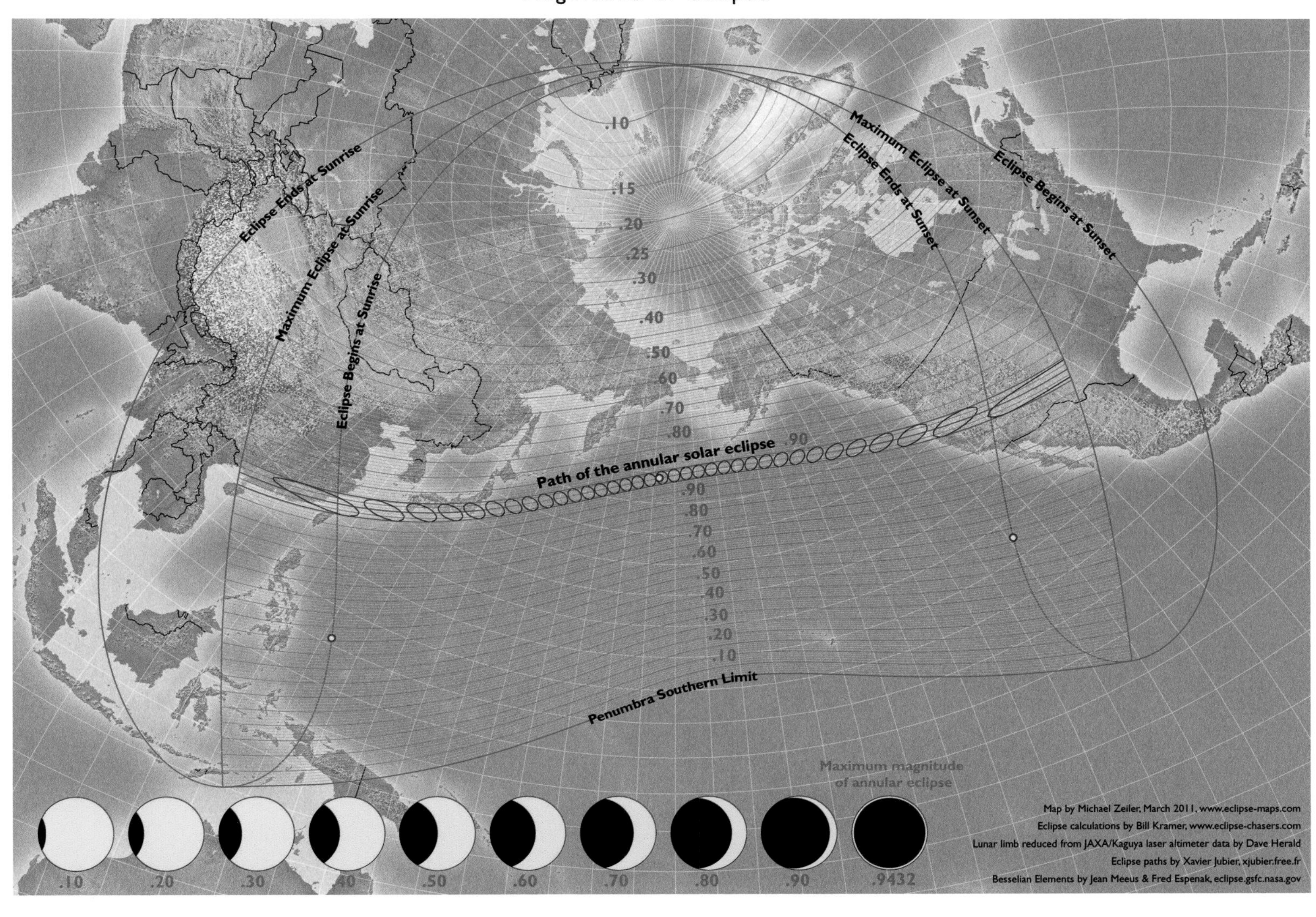

• 2011:

This complex annular solar eclipse map depicts ascending and descending gradations of magnitude depending on distances from the central eclipse track. The chain of ovals within the main body of the track depicts the shape of the shadow itself, with its distortion a product of the Earth's curvature and the angle of the shadow. The table at the bottom depicts the sun as it might appear at each of the roughly latitudinal lines above and below the path of near-totality. Eclipse cartographer Michael Zeiler manages to convey a lot of information with elegant simplicity.

Transit

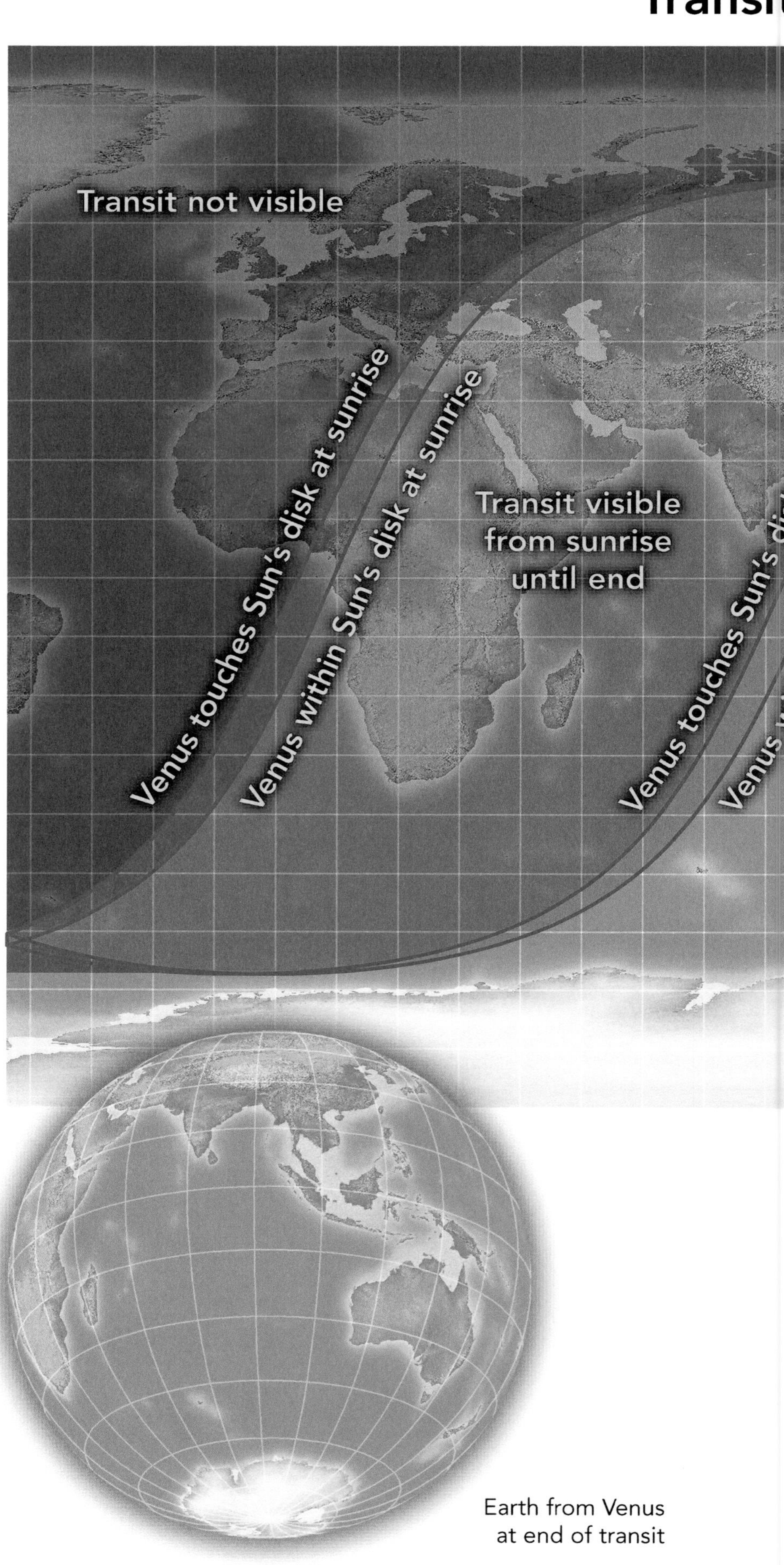

• 2012:
As seen earlier in this chapter, it has been possible to calculate future eclipses and transits with accuracy for many centuries. With this map, Michael Zeiler presents areas on Earth from which the next transit of Venus will be visible, on December 10–11, 2117. It's a map almost identical to the one for the transit of 1874 (not presented here) that was described by Australian observers on pages 267–69.

# of Venus, 2117 December 10-11

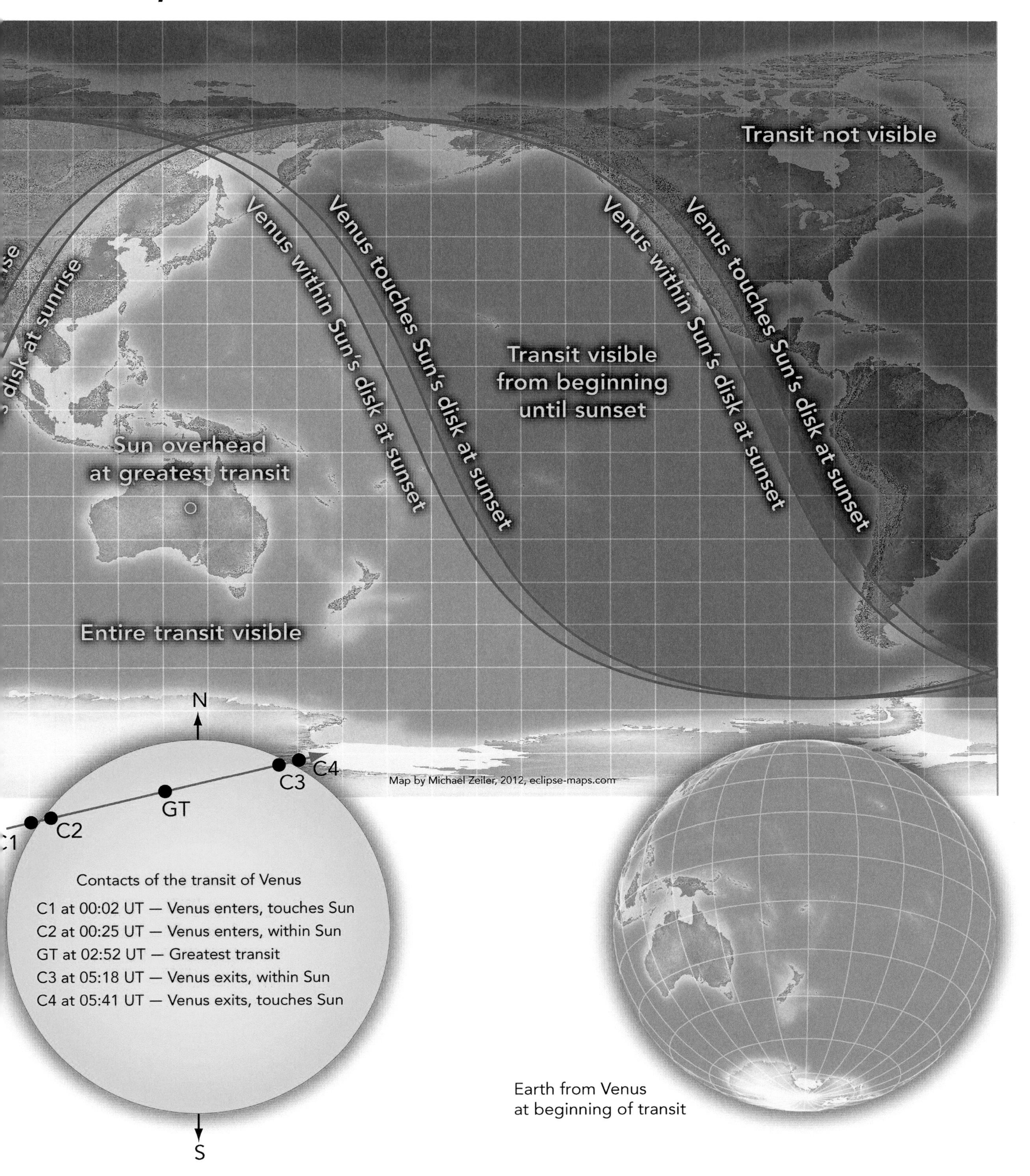

Earth from Venus at beginning of transit

CHAPTER 9

Comets and Meteors

come back, and we'll rebuild our home
or leave forever, like a comet
sparkling and cold like frost
discarding the dark, and sinking back into darkness again

—Bei Dao, *Comet*

Like eclipses, comets were once thought to portend doom, and were associated with civilizational disasters, be they pestilence, plague, or invasion. But unlike eclipses, most visible comets have such extended, loopy orbits that they weren't generally understood as cyclical, and so couldn't be predicted. (The latter remains the case, with some exceptions: Most comets still appear unexpectedly.) Comets are essentially big dirty snowballs from the far reaches of the solar system that have been nudged, via obscure gravitational disturbance, onto trajectories toward the inner planets. As they near the sun, their frozen gases and water begin to vaporize, sending one or more long plumes extending across space. These tails can unfurl up to 200 million miles long, with sensational results.

The length of a comet's tail depends on such factors as the size of the cometary nucleus, the percentage of its frozen gases and water to solid material such as rock and dust, and how close its trajectory brings it to the sun. Some sun grazers come so close they're vaporized entirely; others return to the outer solar system in a severely diminished state.

Particularly noticeable comets have been called "great," and can have a substantial cultural impact. The Great Comet of 1811, for example, which was visible for an unprecedented ten months, was one of the most spectacular in history. This extremely prominent comet had a particularly large coma, or cloud of bright gas and dust, around its nucleus, subsequently turning up as a backdrop for William Blake's painting *The Ghost of a Flea* and as a gloomy reference in Tolstoy's *War and Peace*, where it was said to herald the end of the world. In fact the comet was widely rumored to have foreshadowed both Napoleon's invasion of Russia and the War of 1812. It seems we always suffer enough disasters to justify any cometary omen.

Meteors are of course a very different phenomenon, and are caused when fragments of material enter the Earth's atmosphere at speeds as high as 160,000 mph. Depending on their size and composition, they either burn up, as usually happens, or survive their meteoric plunge and impact the ground. Those that do so leave behind one or more meteorites—scorched pieces of rock, or sometimes chunks of iron and nickel. Occasionally meteors such as the Great Daylight Fireball of 1972 burn with such velocity and at such an angle through the atmosphere that they exit on the other side, continuing on their merry way after their fleeting hypersonic visit to Earth (albeit after scorching off much of their mass).

Aristotle thought that both comets and meteors were phenomena of the atmosphere, in part because neither restrict their activities to the zodiac, which is where the planets can always be seen. In fact, only meteors are the result of high-speed entry into the atmosphere, though there are two ways to look at that. The Earth travels around the sun at a velocity of 66,700 mph, and so even if an innocuous piece of space debris were to be minding its own business in an utterly static position within our orbital path, we would in effect hit *it*, turning it into a kinetic fireball from our perspective. Some have gone so far as to compare meteors to bugs that go splat on the windshield, a simile accurate enough from an orbital dynamics perspective, but somehow entirely wrong when we consider their spectacularly kinetic incandescent passage in the night sky.

Although Aristotle was wrong about comets being part of the atmosphere, in fact those meteors that arrive in annual showers, such as the Perseids and Leonids meteor showers that peak every August 12 and November 17, are thought to be debris left behind after the passage of a comet. Cometary tails give banner-size evidence of materials dispersal, and the more solid debris in that trail remains within the comet's trajectory, with the Earth swinging right through it once a year if the orbits coincide.

While some in our scientized global civilization might enjoy chuckling patronizingly at the primitive superstitions of those believing that a comet's arrival foreshadows doomsday, we should consider the sobering example of Comet Shoemaker-Levy 9. In 1992, it broke into twenty-one large fragments, which, across six days in July 1994, slammed into Jupiter at such high velocities that fireballs an estimated 2,000 miles high were observed, and bruised areas in the planet's gaseous atmosphere were seen ranging in size from 3,700 to 7,500 miles across—meaning from approximately the diameter of the Earth to over twice that size. The largest Shoemaker-Levy impact released an energy estimated at 6 million megatons, or six hundred times the total power of all the nuclear weapons on Earth combined.

It might be better, in other words, to laugh at those banging drums and shooting at the sky during an eclipse. The by now widely accepted theory concerning the extinction of the dinosaurs during the Cretaceous–Paleogene extinction event 66 million years ago holds that the impact of a massive comet or asteroid was the cause. Meanwhile, astronomers are arriving at a realization that distinctions between asteroids and comets are not so clear; some asteroids contain a lot of frozen volatile gases and show signs of water vapor, and some comets stick to the main asteroid belt between Mars and Jupiter, only exhibiting cometary comas during part of their orbits. So-called extinct comets—objects that have exhausted all their volatiles during repeated passes of the sun—have for all extents and purposes become asteroids.

Comets and meteors are thought to have brought vast quantities of water to Earth over the aeons, and may even be responsible for most of terrestrial water. It's also considered possible that the organic molecules that are a necessary prerequisite to life were deposited here from space across millions of years. So there's something dialectical about comets, those enigmatic bringers of life and death, and that's also evident in their visual depictions over the years.

Although they have appeared in art as early as the Bayeux tapestry, circa 1070, in which an oddly kite-shaped comet is celebrated as an omen heralding the successful invasion of England by the Normans, the first relatively faithful rendering of one is in Giotto di Bondone's 1305 *Adoration of the Magi* in Padua (see page 279). Here a comet plays the role of the star of Bethlehem—clearly a harbinger of good tidings and life. (Although he worked at the end of the medieval period, Giotto is generally considered the first Italian Renaissance artist, in part because of his accurate representations of nature.) But most of the more than thirty comets whipping through the painted pages of the sixteenth-century *Augsburger Wunderzeichenbuch* (Augsburg Miracles Book) are described as heralding some variety of disaster; the same is true for the thirteen watercolors of the Flemish manuscript *Kometenbuch* (Comet Book). A sampling from both can be seen in this chapter.

Aristotle's idea that comets and meteors are atmospheric phenomena fit in well with a cosmological model in which the planets were all embedded in revolving spheres, and the only change possible was supposed to take place below the orbit of the moon. It wasn't until Danish astronomer Tycho Brahe measured the parallax of the Great Comet of 1577 that it was proved that comets are well outside the atmosphere— Brahe estimated at least three times as far as the moon. This undermined Aristotle's dictum, of course, because comets are clearly highly changeable phenomena, and their trajectories ride right through those supposedly impenetrable spheres.

Although he was the founder of the field of celestial mechanics, Brahe's onetime employee Johannes Kepler failed to grasp that most comets ride highly elongated curving trajectories, contending instead that they always follow a straight path. (He believed the apparent curve in their trajectories was an illusion caused by the Earth's motion; Kepler was a confirmed Copernican.) But when considering the findings of German mathematician Peter Apian, who in 1531 observed that cometary tails always point away from the sun, Kepler grasped an essential fact about the nature of comets. "The head is like a conglobulate nebula and somewhat transparent," Kepler wrote in 1625. "The train or beard is an effluvium from the head, expelled through the rays of the sun into the opposed zone and in its continued effusion the head is finally exhausted and consumed so that the tail represents the death of the head."

In 1687, Isaac Newton wrote in *Principia* that comets follow parabolic orbits around the sun, applying his inverse square law of universal gravitation to the sungrazing Great Comet of 1680, but comets were largely peripheral to the book. Newton's friend Edmond Halley, however, who both helped instigate and paid to publish *Principia*, used the Newtonian principles to parse through historical records of dozens of comets, calculating the gravitational effects of the outer planets on their trajectories. Among those he'd observed was one in 1682. In studying the records of both Apian and Kepler, Halley began to discern similarities between its trajectory and the orbital parameters they had described for comets they'd observed in 1531 and 1607, respectively.

Halley concluded that Apian's and Kepler's comets were likely to be one and the same as the one that appeared in 1682, and he calculated a periodicity of seventy-six years, meaning it would return again in 1758. His published prediction was met with hoots of derision on the part of many commentators, who noted that he'd timed its return safely past his own lifetime, and so would escape the humiliation of being proved wrong. Halley died in 1742 at the age of eighty-six, and the comet he'd identified duly returned, with only days to spare, on December 25, 1758. It was the first time an object other than the planets had been proved to orbit the sun.

Halley's Comet is the only one visible to the naked eye with an orbit of less than two hundred years, and also the only one that might appear more than once in a human lifetime. The relative precision with which we know its orbital period has allowed for earlier appearances to be located in the historical records: twenty-nine have been identified, reaching back to 240 B.C. It's even considered possible that its periodicity was discerned in ancient times: A passage in the Talmud describes "a star which appears once in seventy years that makes ship captains err."

As it happens, Halley's Comet appeared in the skies over Europe just prior to William the Conquerer's successful invasion of England in 1066: His captains didn't err. And Giotto saw it three orbits later, in 1301. This makes the cometary kite woven into the Bayeux tapestry and the celestial object caught in frescoed mid-trajectory above the Nativity in Padua's Scrovegni Chapel one and the same vaporous interplanetary snowball. (It can also be seen in its 1835 incarnation as observed by John Herschel from South Africa on page 292, and is probably also behind the strangely bifurcated comet depiction on page 282.)

Halley's Comet returns to the inner solar system in 2061. It will slip between the orbits of Venus and Mercury, making its closest approach to the sun on July 28.

• 1305:

This depiction of a comet as the star of Bethlehem in Giotto di Bondone's *Adoration of the Magi* is probably the first relatively faithful rendering of a comet in Western art. Although he lived in the late Middle Ages, Giotto is generally considered the first Italian Renaissance artist because of the interrogatory accuracy of his representations of nature. He saw Halley's Comet in the skies over Europe in 1301. From Giotto's famous fresco cycle in the Scrovegni Chapel in Padua.

• 1547–52:

Like eclipses, comets were seen as harbingers of disaster throughout European history. Most of the more than thirty comets whipping through the pages of the *Augsburger Wunderzeichenbuch* (Augsburg Miracles Book) are described as heralding some variety of plague, pestilence, war, or other natural or man-made disaster. **Above top:** The text reads: "In A.D. 1184, a comet appeared for three months. That was followed by such heavy rain, storm, wind and peal of thunder as had never been heard before. And the element acted as if it wanted to destroy the city of Rome, and a lot of livestock died in a gruesome way. And the people died from the lightning in the sky which they saw." **Above:** The text reads: "In A.D. 1401, a big comet with a tail appeared in the sky in Germany. That was followed by a great, terrible plague in Swabia."

Above top: The text reads: "In A.D. 1007, a wondrous comet appeared. It gave off fire and flames in every direction. It was seen in Germany and welsch land [Southern Europe] that it fell onto the earth." **Above:** The text reads: "In A.D. 1300, a terrible comet appeared in the sky. And in this year, on St. Andrew's Day, the soil was shaken by an earthquake so that many buildings collapsed. At this time, the first jubilee year was established by Pope Boniface VIII."

• 1587:

Similar to the *Augsburger Wunderzeichenbuch*, though much shorter and with a specific focus, the late-sixteenth-century manuscript *Kometenbuch* (Comet Book) presents thirteen watercolor paintings of comets and meteors. The book originated in Flanders or northeastern France, and is based on an anonymous Spanish treatise titled *Liber de significatione cometarium* (Book About the Significance of Comets), from about 1238. Written in French longhand, *Kometenbuch* states that the comets in most of its paintings derive from descriptions in Ptolemy's *De centum verbis* (One Hundred Terms)—though in fact no comets are mentioned in that text. (However, a manuscript misattributed to Ptolemy does describe eleven comets, and is likely one source for this book.) "This comet is called Pertica," says *Kometenbuch* about this strange pair of forms, "which has a ray large and obscure, and Aliquind says that when she is in the west she appears in the form of a column made of parts of sun, and when she rises in the east she's like a hot star which has its rays divided into two parts . . ." Pertica is said to have appeared in 1531; if the date is correct, that would probably make it Halley's Comet.

The comets in *Kometenbuch* are described as having "pleated" tails. This one, according to its accompanying text, is either named Miles, or Cheval, or Omnes crines, or Ladescodo! Whatever its name, the naive scene unfolding below presents a kind of rude analog to Pieter Bruegel's *Landscape with the Fall of Icarus*, with a rowboat instead of a ship, and the ploughman taking his undoubtedly well-deserved evening relief under a wide-eyed owl. The text reads: "When this comet appeared, she demonstrated terminal power in rays and strength so that it scared and amazed everybody. The men of the world . . . shed the ancient laws and made new ones. They got naked of their status and costumes."

This eerie painting from *Kometenbuch* almost certainly represents a meteor shower, in which individual meteoroids all seem to radiate from one part of the sky—though its flying saucer shape and spraying rays provide some obvious material for UFO theorists. According to the accompanying text, "This comet is named aurora, otherwise known as Matuta."

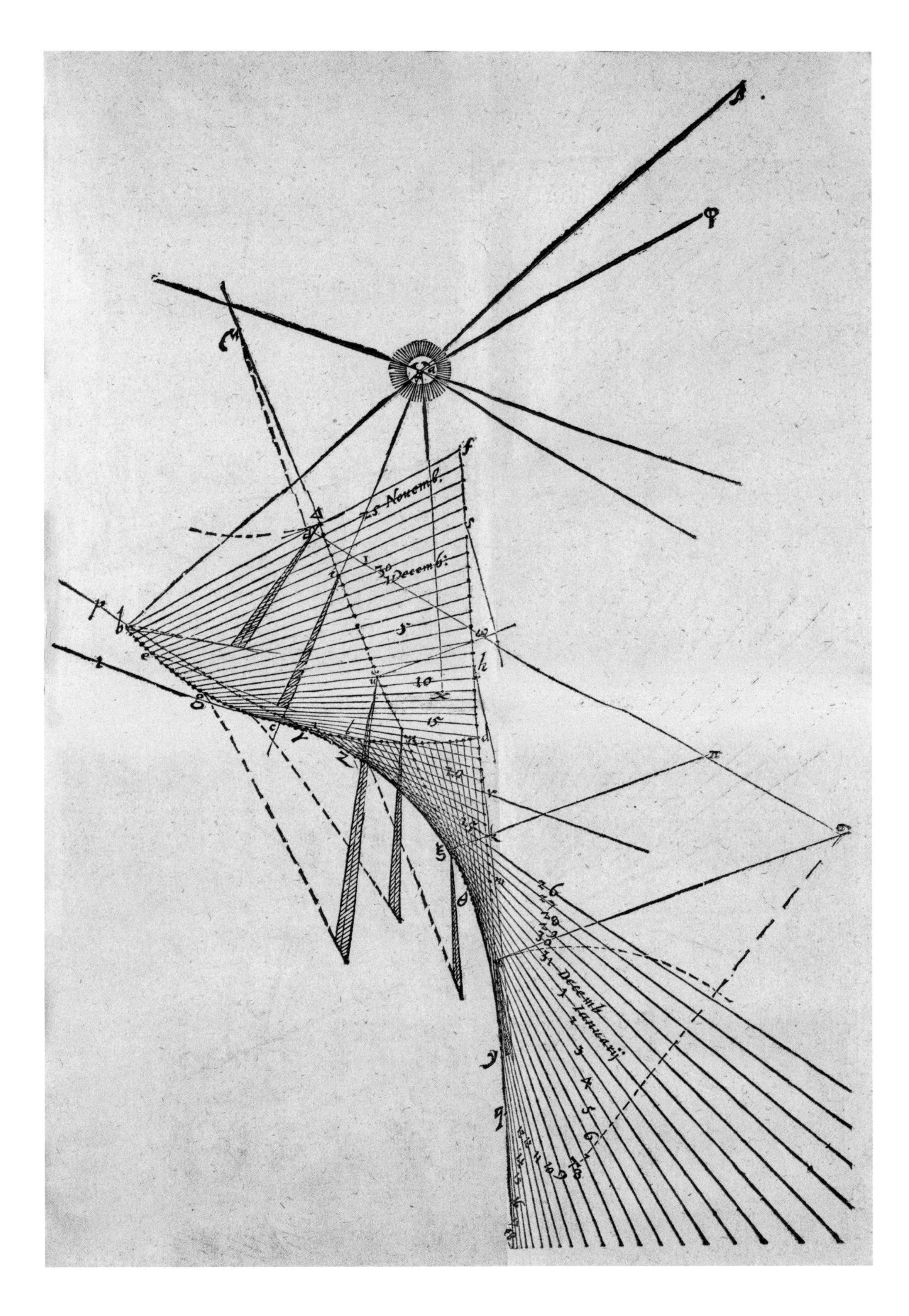

• 1619:

Johannes Kepler's major legacy is his laws of planetary motion. The German mathematician and astronomer was the first to realize that planets move in elliptical orbits—a revolutionary finding that suddenly clarified previously confounding inconsistencies in their motions. Kepler's laws were the most significant breakthrough since Copernicus, and were the starting point for Newton's law of universal gravitation. But when Kepler's focus turned to comets, which exhibit orbits so elongated they're parabolic, he resolutely held to the line that they unwaveringly hold to a line. The astonishing curvilinear waveform in this print from Kepler's treatise *De cometis libelli tres* (The Comets in Three Books) seems to predict twentieth- and twenty-first-century technology, architecture, and design. It's computer graphics centuries before mechanical computation. But if you look closely, you will see that the comet moving toward the sun at an angle from the lower right to the upper left is in fact traveling in a straight line. It had been understood since the mid-sixteenth century that cometary tails always point away from the sun. The curving waveform here is associated with the changing angles of the comet's tail, and specifically a succession of lines at right angles to the tail. This comet was likely the last and most spectacular of the three that appeared in rapid succession in the fall of 1618.

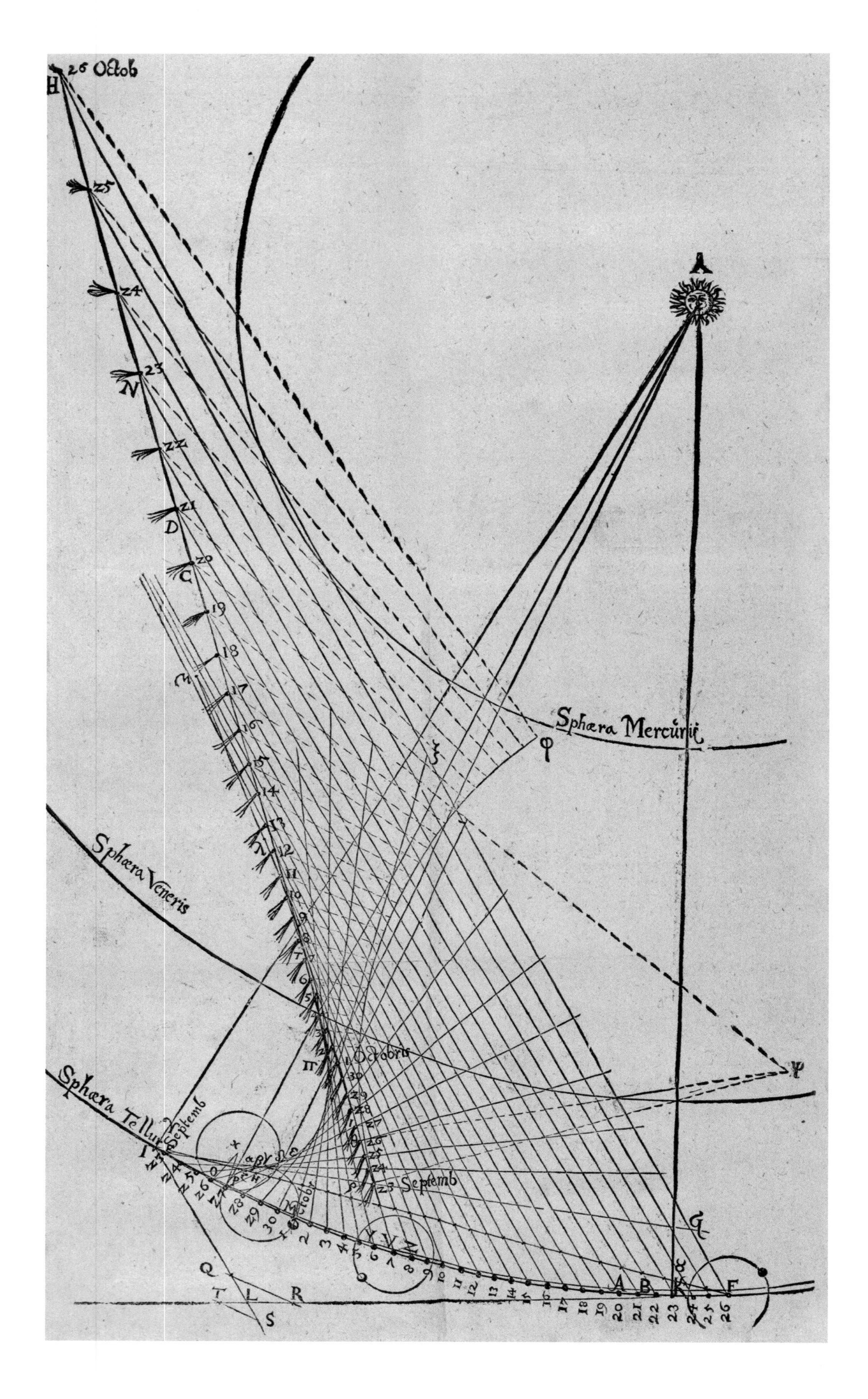

Another view of a comet approaching the sun from Kepler's *De cometis libelli tres*. Kepler's friend and correspondent, astronomer Johannes Remus Quietanus, argued that comets exhibit curving trajectories. Kepler maintained that this was an illusion created by the movement of the Earth—one similar to the illusion that the planets sometimes travel in retrograde, which in fact is due to our planet's orbit. This diagram attempts to illustrate the mechanism of the illusion, with the curving waveform due to changing sight lines from Earth (seen moving along its orbit at the bottom). In depicting the orbits of Earth, Venus, and Mars, Kepler holds to the use of the term *sphaera*, or "sphere," even though comets were a leading piece of evidence that the planetary spheres of Aristotle and Ptolemy didn't exist.

• 1668:

Decades after Kepler's treatise, Polish astronomer Stanislaw Lubieniecki took up the subject of comets, publishing his densely illustrated book *Theatrum cometicum* (Theater of Comets) in Amsterdam. One of two important seventeenth-century works on comets, with the other being Johannes Hevelius's *Cometographia*, it records more than four hundred, from the Old Testament to the late seventeenth century. Here, the trajectory of the Comet of 1664–65 takes it right through the open jaws of the sea monster Cetus (representing the constellation of the same name) and across the ecliptic. That comet, the brightest in more than forty years, renewed the focus of European astronomers on the phenomenon, reopening questions concerning their orbits, origins, and nature. Although Danish astronomer Tycho Brahe had measured the parallax of the Great Comet of 1577, concluding it was located well outside the orbit of the moon, a century later some astronomers still stuck to the Aristotelian position that comets were terrestrial atmospheric phenomena.

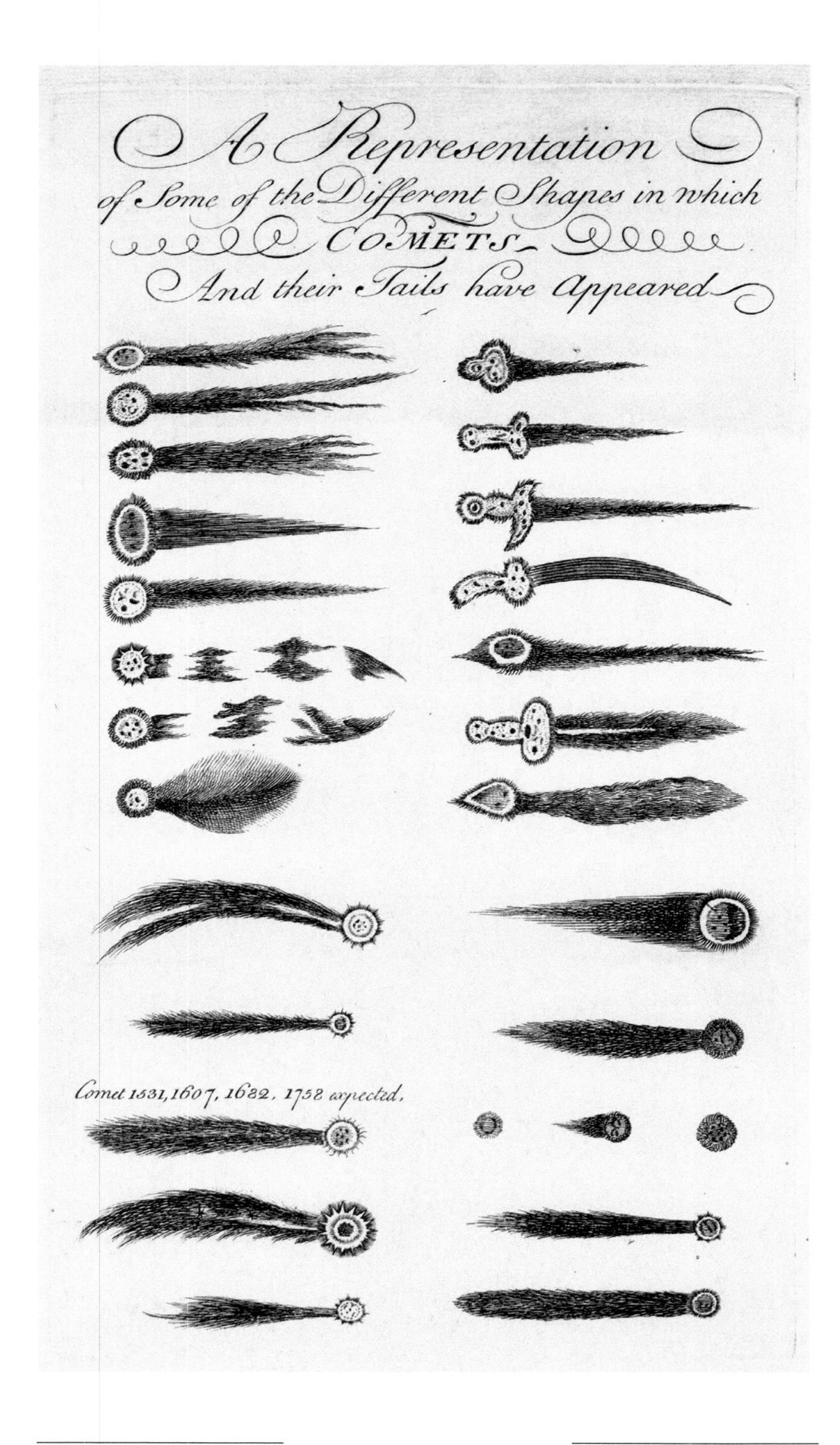

A Representation
of Some of the Different Shapes in which
COMETS
And their Tails have Appeared
Comet 1531, 1607, 1682, 1758 expected.

• 1757:

Throughout history, astronomers and astrologers have seen terrestrial forms in the heavens, be they oxen, sea monsters, or scientific instruments. Comets were no exception, with many descriptions comparing them to swords or brooms. This print, adapted from Polish-Lithuanian astronomer Johannes Hevelius's 1668 book *Cometographia*, was included in a posthumous compilation of writings by English astronomer Edmond Halley. *A Compendious View of the Astronomy of Comets* arranges them in two neat rows for contemplation, with five swords suitable for dueling among them. Halley continues to be best known for his theory that the comets observed in 1458, 1531, 1607, and 1682 were one and the same, and that it would return in 1758. This it duly did, becoming Halley's Comet. The astronomer, who died in 1742, wasn't around to see it.

• 1742:

This beautiful hand-colored engraving by German mapmaker Georg Matthäus Seutter shows the path of the Comet of 1742 as it descends toward the sun. The comet crosses the ecliptic at the skirts of an androgynous Cepheus, arriving at the giraffe-like neck of the obscure constellation Camelopardalis. On the right is a depiction of the comet moving past a very traditional presentation of Earth centering on an armillary sphere. While Seutter followed Kepler in depicting the comet as following a straight line, there the resemblance ceases.

CI 1742. APPARUIT, EX OBSERVATIONIBUS, A DIE 13 MARTY USQUE AD 15. APRILIS, PLUSQUAM.
UIT FIERI, DILIGENTIA NOTATUS, ET SECUNDUM SUAS PROPRIETATES PLERASQ: AB ASTRONOMIA CONSIDERARI SOLITAS, UTPOTE FORMAM, VI-
VELOCITATEM, DESCENSUM AD PERIGÆUM, SEU APPROXIMATIONEM AD TERRAM, ET ELONGATIONEM AB EA; ITEM DISTANTIAM INFRA, VEL
BLIQUUM, QUOTIES, SALTEM INTRA POLUM MUNDI, MERIDIANUM CONSTANTIENSEM, ET LOCORUM SUB ELEVATIONE POLI 47. GRAD.
ERO SECUNDUM ASCENSIONEM RECTAM, ET DECLINATIONEM, LONGITUDINEM, ET LATITUDINEM; ET TANDEM SECUNDUM
NSERTA, CUM IN PLANO TUM IN FIGURA ARMILLARI EXHIBITUS: UBI TUM NUDO INTUITU, TUM OPE REGULÆ, ET CIRCINI OMNIA FACILE
ANTE ASTROPHILO QUODAM, EX ORD: S. BENED: NOMINA QUI CHRISTO GAUDET INESSE SUA. A MATTH: SEUTTERO, S.C.M.G. AUG.V.
Cum Gratia et Privil. S.R. Imp. Vicariat in partib. Rheni Sveviæ et Juris Franconici.
endam Ascensionem rectam
Æquinoctiali.
Zenith
Nadir
POLUS MUNDI ARCTICUS
POLUS MUNDI ANTARCTICUS
HORIZON
COLURUS
ECLIPTICA
ÆQUATOR SEU
TROPICUS
CANCRI
CAPRICORNI
CIRCUL. ANTARCTICUS
Draco
Cepheus
Sagittarius
Herculis
Aquarius
Pisces
Aries
Taurus
Dictus Cometa jam in fine Ianuarÿ fuißet haud dubiè spectabilis, si præprimis ejus via, et descensus ad Terram prævie potuisset sciri; et nisi eo tempore sub radÿs solaribus latiaßet. Cæterum Cursus ejus videtur fuisse è Corona Australi per partes anteriores Sagittarÿ, per pedem Antinoi borealem, et per alam australem, caudamq, Aquilæ. Inde anserem transiens, factoq, inter Lyram et collum Cygni Perigæo, per ejusdem Cygni alam borealem, et secundam Draconis flexuram, Crura Cephei attigit: per hæc quoq, de die in diem lentiùs progrediens, Polò Mundi quinq, circiter gradibus cum dimidio à sinistris relictò, Camelopardali scapulas petÿt, in quibus tandem 28. April. ultimo visus fuit.

• 1840:

The Great Comet of 1811, which was visible for an unprecedented ten months, was one of the most spectacular in history. This extremely prominent comet had a particularly large coma, or cloud of bright gas and dust around its nucleus, and had a substantial cultural impact across Europe, appearing for example in Tolstoy's *War and Peace*. It was likely one model for the superb depiction of a comet as a harbinger of doom in this painting by English Romantic painter John Martin. Building on his reputation for large-scale atmospheric paintings of Old Testament themes, Martin followed up an earlier print on the same subject with *The Eve of the Deluge*. According to art historian Roberta Olson and astronomer Jay Pasachoff, the figure gesturing toward the sky to the left of a seated Noah is Methuselah, and the scroll he's holding contains astrological portents written by his father, Enoch. Striped with clouds directly above Noah's head, the setting sun is one of six or more accurately rendered celestial objects visible in the painting. Note the uneasy dog, looking at the sky.

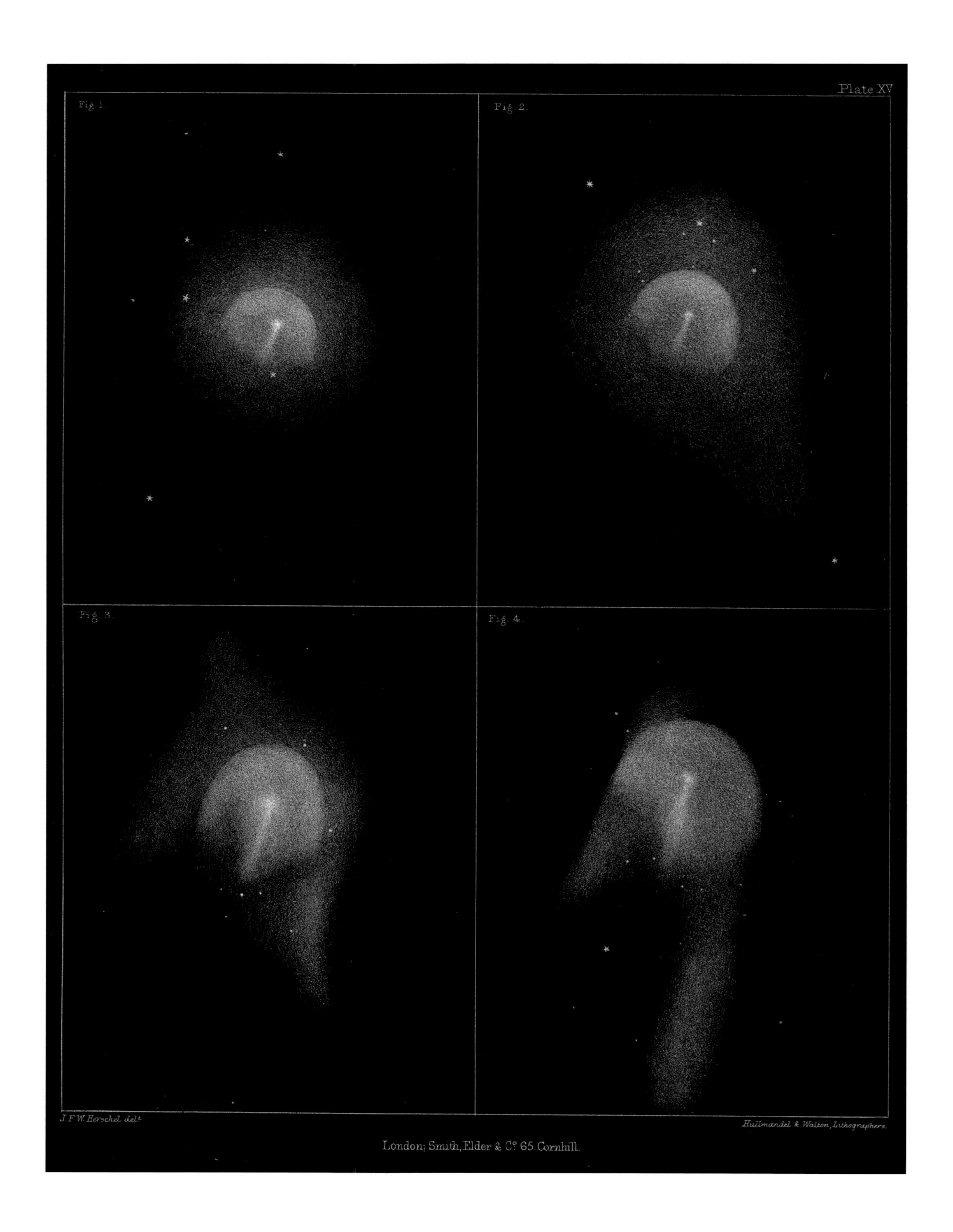

• 1847:

In 1833, English astronomer John Herschel traveled to Cape Town, South Africa, in order to catalog the southern sky. He stayed for five years in the English colony, among other things observing the 1835 return of Halley's Comet from an observatory he set up for his giant twenty-one-foot telescope on the southeastern side of Table Mountain. Herschel observed the comet for seven months, from October 28 until May 5 of 1836, noticing physical changes in its coma and nucleus, as visible here. Herschel developed a controversial theory involving magnetism and electricity to account for the comet's oblong shape. He later published thirteen engravings of Halley's Comet in his *Results of Astronomical Observations at the Cape of Good Hope*; these are four of them. Their almost photographic accuracy is heightened here by rendering them as negatives of the original black-ink-on-white-paper prints. (Herschel, also a chemist and photography pioneer, coined the terms *positive* and *negative* in the context of photography.) For an actual photograph by Herschel, see page 89.

• 1858:

The second sensational nineteenth-century cometary appearance following the Comet of 1811 was 1858's Donati's Comet, named after its discoverer, Italian astronomer Giovanni Battista Donati. Harvard College astronomer George Phillips Bond photographed it through the college's fifteen-inch Great Refractor—the first such attempt—but the results were not outstanding. He also made detailed drawings, however, one of which was turned into this print by engraver James W. Watts. Bond, who became the first American to win the gold medal of the Royal Astronomical Society in part for this work, later wrote, "The nucleus . . . was unusually bright, and rounded on the side toward the sun. An increase of brilliancy in the nucleus was afterwards recognized as the precursor of a fresh eruption from its surface. . . . There were three dark openings in the innermost envelope, between which it was intersected with bright rays." He then commented on this print: "In Plate I the engraver has given an eminently successful representation of the comet as it appeared in the field of the great refractor."

• 1871:

Right: Although sometimes confused with comets, meteors are an entirely different phenomenon. Rather then being distant objects on stately, months-long trajectories around the sun, after which they disappear, sometimes to return decades, centuries, or even millennia later, these kinetic fireballs are the result of the high-speed entry into Earth's atmosphere of a fragment of debris. Meteors come individually or in annual showers, in which dozens or even hundreds are visible within an hour or so. The latter phenomenon is the result of the Earth's annual passage through belts of material thought to have been left behind by comets or asteroids. In this print from English meteorologist and aeronaut James Glaisher's book *Travels in the Air* (written with Camille Flammarion and others), a meteor shower is seen as viewed from a balloon.

FALLING STARS AS OBSERVED FROM THE BALLOON

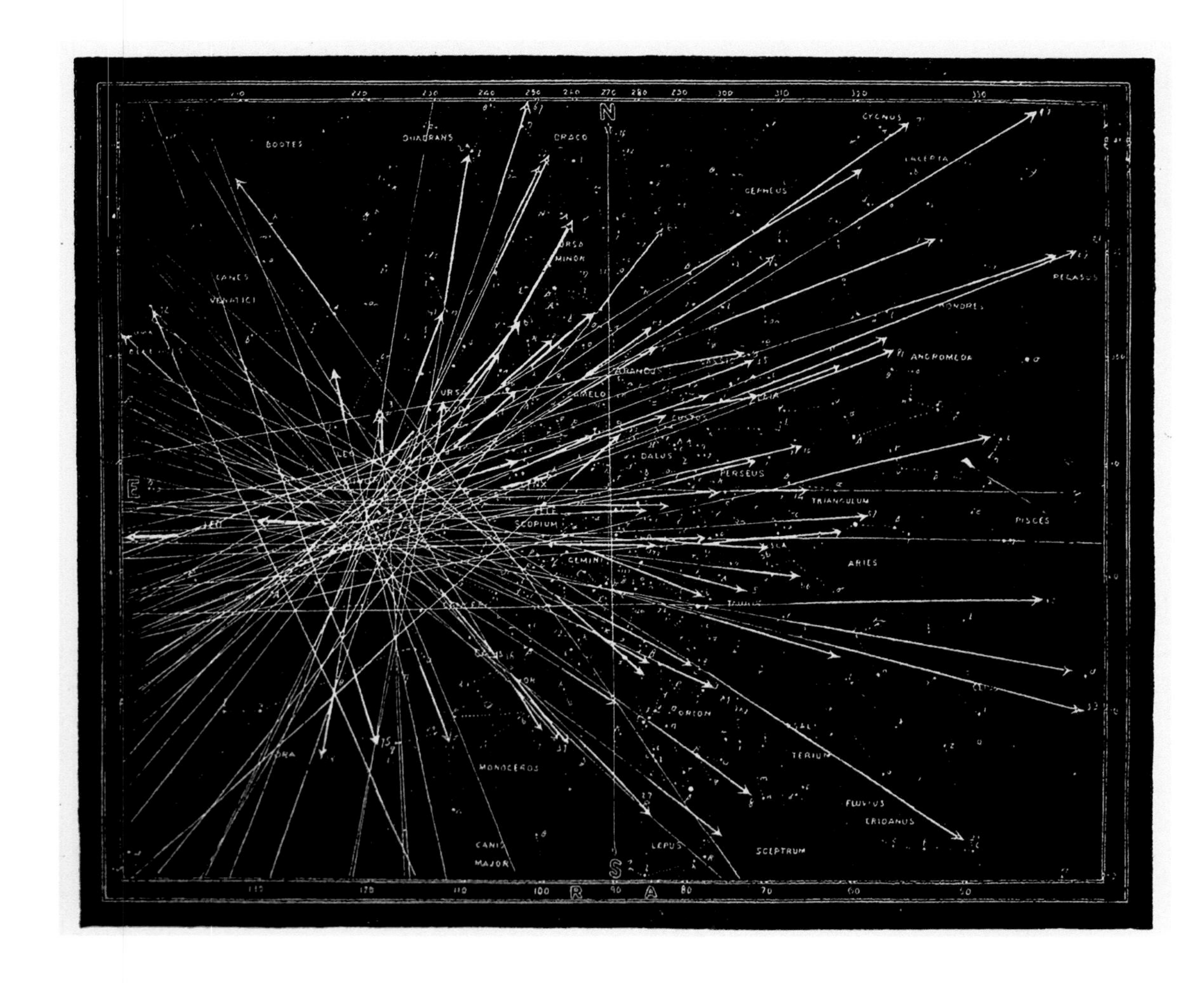

• 1882:

Left: During a meteor shower, individual meteors can all seem to radiate from one part of the sky, as in this print from the popular astronomy book *The Heavens Above* by Joseph Gillet and W. J. Rolfe.

• 1881:

Facing page: A rare example of erroneously presented astronomical phenomena by Étienne Trouvelot. Although artistic license can cover the presentation of dozens of meteors simultaneously—in fact, they are rarely simultaneous, though dozens can shoot across the sky in relatively short periods at the height of a shower—meteors never change their apparent direction, let alone slalom or make U-turns, as portrayed here. "November Meteors" refers to the Leonids, an annual meteor shower between November 15 and 19 every year. They are thought to be the debris of a long-period comet. For more by Trouvelot, see pages 98-99, 127, 128-29, 158, 188-91, 241, 270-71, and 297-99.

PLATE XII.

E. L. Trouvelot

THE NOVEMBER METEORS.

Fig. 104.—An Astrologer's Prediction – the End of the World, caused by the Collision of the Earth with the Stony Nucleus of a Comet.

(After a drawing by Heinrich Harder.)

• 1911:

Above: As seen most recently on pages 292–93, comets and meteors have long been thought harbingers of doom. But rarely has cometary ruination been portrayed more directly than in this print, which could almost be a publicity still from a Hollywood summer blockbuster. Based on a drawing by Heinrich Harder depicting the results of an astrologer's prediction, from early German science journalist Bruno Burgel's book *Astronomy for All*.

• 1881:

Facing page and overleaf: The Great Comet of 1881 was not quite as globally sensational as the two earlier impressive nineteenth-century comets mentioned in this chapter, but as this print by Étienne Trouvelot demonstrates, it was still a sight to behold. Discovered by accomplished Australian amateur astronomer John Tebbutt on May 22, the comet became visible to northern hemisphere astronomers on June 22, with Trouvelot doing the drawings this print is based on only a couple days later, on the night of June 25–26. It was traveling on almost the exact same orbit as the Comet of 1807, and though it was determined not to be the same comet, the shared trajectory indicated an unknown link between the two. As with Donati's Comet, the Great Comet of 1881 exhibited a complex, ever-shifting structure in its nucleus and coma, as can be seen here by its twin-lobed appearance. For more by Trouvelot, see pages 98–99, 127, 128–29, 158, 188–91, 241, 270–71, and 295.

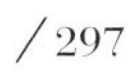

PLATE XI.

E. L. Trouvelot

THE GREAT COMET of 1881.

Observed on the Night of June 25-26. at 1 h. 30 m. A.M.

CHAPTER 10

Auroras and Atmospheric Phenomena

But pleasures are like poppies spread,
You seize the flower, its bloom is shed;
Or like the snow falls in the river,
A moment white—then melts for ever;
Or like the borealis rays,
That flit ere you can point their place

—Robert Burns, *Tam o' Shanter*

MOST OF THE CELESTIAL OBJECTS DEPICTED IN THIS book give little sign of their observational circumstances. The frame isn't supposed to be part of the picture, even if its environmental and cultural influence impinge everywhere on these images in subtle and obvious ways. But virtually every subject presented here was perceived through several lenses: the eyes of the astronomers, or artists; the glass or mirror of the telescopes, depending on the subject; and the domelike filter of the atmosphere, with its sometimes distortion-producing turbulence.

Just outside the boundaries of all these graphics, in other words, is an observatory with a rotunda: Earth. The circumstances of the planet can't really be divorced from them; they're inevitably subjective terrestrial expressions, views of the cosmos from a place fixed neither in space nor time, made by an observer population changing and evolving over the centuries and millennia, but still a place that always carries its terrestrial characteristics—its earthly baggage—along on its cosmic ride.

A spherical body with an equatorial diameter of 7,926 miles, a layer of atmosphere about 60 miles deep, a volume of 259,875,159,500 cubic miles, and a mass in metric tons with so many zeroes on it that there's no sense wasting your time, the Earth's ancient bulk—because it's also 4.54 billion years old, or one-third the age of the universe itself—inevitably produces its own cosmographical effects, some of which are quite sensational.

Churning liquid iron rendered molten by the planet's solid, red-hot core generates a powerful magnetic field, which extends out from the interior and interacts with the solar wind—the steady flow of charged subatomic particles perpetually streaming from the sun. This produces auroral displays: flickering greenish and sometimes reddish sheets of plasma, seen particularly in the Arctic and Antarctic regions, caused by photon emissions in the upper atmosphere.

Lower in the air, and also more commonly seen in colder northern and southern regions, simple hexagonal ice crystals refract sunlight (and also moonlight), causing complex "parhelia" or sun dogs, circumzenithal arcs, twenty-two-degree halos, and a whole range of other atmospheric phenomena described by an arcane terminology: Lowitz and Parry arcs; circumscribed halos; anthelic, tangent, and circumzenithal arcs, and the like.

Well above any terrestrial ice crystals, past the auroras, starting outside the atmosphere but seemingly an extension of it, the diffuse glow of the zodiacal light can sometimes be seen in spring or fall. It's caused by sunlight shining on the vast cloud of interplanetary dust that pullulates in the solar system's plane, always within the flattened disc of the ecliptic—something first explained by French-Italian astronomer Giovanni Cassini in 1683.

All of these phenomena have had their effects, across centuries, on the sometimes credulous citizens of Earth. On the morning of April 12, 1535, for example, the sky above medieval Stockholm filled with a dazzling, complexly interlocking set of arcs, halos, circumzenithal arcs, and even, seemingly, additional suns—or sun dogs—all doubtless caused by an unusually dense layer of hexagonal ice crystals fluttering in the atmosphere. While individual examples of such phenomena had been well known in this latitude for as long as records were kept, this was an unprecedented and even alarming display.

With the city buzzing, a painting of the scene was commissioned by a powerful local clergyman, Olaus Petri, motivated in part by a determination to dispel rumors that the phenomena were divine omens against the Protestant Reformation. But Petri was also deeply troubled by what he viewed as excessive actions by the king against the Swedish Catholic churches, which he believed should be treated with more respect during the transition to Lutheranism, and he wasn't entirely sure how to take April 12's skyful of omens.

That summer, Petri displayed the resulting *Vädersolstavlan*, or sun dog painting, in front of his congregation. (The image on page 303 is a copy from 1636; the original was destroyed.) He warned that the eerie air show over Stockholm could equally well be seen as divine warning against pro-Reformation excesses as an endorsement of the kingdom's theological path. Petri helpfully explained that two types of omen exist: those conjured by God to lead humanity on the straight path, and those produced by the devil to lead us astray. Few clues, he said, are given as to which is which—and he himself didn't know.

The painting he had commissioned, then, was a kind of cipher. It has since become a symbol of Stockholm, and even made it onto the 1,000 kronor banknote.

Although inaccurate in many respects to how we now know the actual display over the city would have appeared, all the phenomena in the *Vädersolstavlan* painting have a natural explanation, and all are caused by very simple kinds of hexagonal ice crystals—not the more complex flakes we're used to picturing when we think of ice crystals fluttering through the air. These produce what is essentially the cold-weather equivalent of rainbows, only, instead of spherical drops falling, angular crystals refract sunlight in complex prismatic ways.

The sun in the upper right of the painting is accompanied by sun dogs on either side, though in reality these are always oriented horizontally and seen at the same distance from the ground as the sun. A 22-degree halo is also depicted in the painting, but it isn't centered on the sun, as it always is in reality. Instead, the sun is positioned along a 120-degree parhelic circle, and the two sun dogs are at the meeting point of that circle and two 90-degree circles. A crescent shape at the center of that halo is called a circumzenithal arc. When this phenomenon appears it's always seen when the sun is lower than 32 degrees, which is not the case here, and is never to its side but rather above; it has been called a "smile in the sky." The three small white spots on the 120-degree circle are anthelia, a variant of sun dogs that could be caused by a different, columnar variant of hexagonal crystal.

Other images reproduced in this chapter represent different variants of these atmospheric phenomena, some with understandable deviations from what trained observers have recorded (and even, more recently, simulated with computers). One of these, by arctic explorer Fridtjof Nansen, depicts a lunar version of parhelia: moondogs. And then there are works here, particularly from the mid-sixteenth-century *Augsburger Wunderzeichenbuch* (Augsburg Miracles Book), that apparently defy rational explanation, as miracles probably should.

Nansen also depicted another phenomenon that has mesmerized Arctic and Antarctic explorers: auroras. These are the visible result of a collision between terrestrial and solar forces. Auroras such as the northern lights or its southern equivalent are giant plasma fields generated when ionized nitrogen and oxygen molecules in the upper atmosphere regain electrons and emit photons. The ionization itself is produced when the stream of charged particles from the sun meets the terrestrial magnetosphere, producing a bow shock wave in which magnetic field lines on the Earth's dayside present an obstacle to the free electrons and positive ions streaming from the star at the solar system's center. In the resulting turbulence, magnetic field lines break and reconnect, permitting the solar plasma to enter and mix with the magnetosphere. On the night side of the planet, the so-called magnetotail—the leading player in polar auroras—extends almost 4 million miles downstream into space, channeling the glowing plasma into northern and southern lobes. (For supercomputer simulations of this complex interaction, see the top and bottom illustrations on page 135.)

The auroras themselves can result from plasma flowing, somewhat counterintuitively, *upstream* from the magnetotail toward and around the Earth, where it extends back into the solar wind on the dayside. When some of that plasma flux diverts down toward the planet's surface along the Earth's magnetic field lines, it sheds energy when it slams into the high atmosphere more than fifty miles up, producing auroral displays. These frequently appear as glowing curtains extending in a roughly east-west orientation; the curtain shapes are caused by the magnetic field lines, which are close to vertical near the poles. When those curtains ride directly down on an observer, a kinetic spray of greenish rays can appear to expand in all directions across the sky in a phenomenon called a corona.

During periods of particularly intense solar activity, auroral displays are stirred into raging storms. Two of the most sensational in history, the great geomagnetic storms of 1859, were produced by coronal mass ejections—giant bursts of magnetized solar wind, frequently accompanied by solar flares. During late August and early September of that year, auroras were visible much farther south than normal, with the northern lights reportedly so bright above Boston on September 2 that newspapers could be read outdoors in the middle of the night. A comparable auroral display today would be accompanied by the sizzling sound of satellite electronics frying and terrestrial power grids failing.

In 1899, the Danish painter Harald Moltke was invited to participate in a winter expedition to Iceland organized by the Danish Meteorological Institute specifically to study auroras. Although the expedition was outfitted with the latest spectrograph and other scientific instrumentation, photographic emulsions weren't nearly good enough to do justice to the northern lights, nor were they yet in color. So Moltke's participation on the expedition to Akureyri, a small town on the northern coast, was considered very much a part of the research enterprise. It was hoped that he would be able to capture some of the essence of what was still a mysterious phenomenon.

Fortuitously, the winter of 1899–1900 proved to be a period of high auroral activity, and on night watches, Moltke set to work sketching the northern lights in real time by pencil on cardboard, while also noting their time and characteristics. In the mornings, working in a small studio, he used this as the basis for his paintings. At first, he tried to use pastels, achieving marginal results. He soon realized that only oil paints could really capture the glowing light and spectral beauty of the phenomena he was observing.

Moltke refined his technique through the long Icelandic winter, and when the steamship *Botania* came up the coast again the following April to collect the expedition, he'd produced nineteen exceptional works. They amounted to a new genre, one combining scientific inquiry and genuine artistic vision, a type of painting where the ground is secondary to a shimmering electric firmament: skyscapes. In one extraordinary example, an undulating sheet of auroral flame joins an abstract planetary vista to unspeakably powerful, wordless cosmic forces. It's picture and frame fused; Earth in space, space brought to Earth. It's also a kind of ongoing, ever-renewable *Fiat lux*.

"Auroras are not like anything else on our globe," Moltke wrote. "They're mysterious. They go beyond human fantasy to a degree where you instinctively resort to expressions such as 'supernatural,' 'divine,' 'miraculous.' Only little by little did I learn to reproduce these hovering, dancing revelations; only little by little did I realize that in all their arbitrariness, there were still laws obeyed even by these wild, intemperate phenomena."

And it was good.

• 1535:

The earliest-known painted representation of Stockholm, the so-called Vädersolstavlan painting, is also thought to be the earliest depiction of a so-called "sun dog" or parhelion event. (The name means "sun dog painting.") On the morning of April 12, 1535, the sky above Stockholm filled with arcs, halos, circumzenithal arcs and even, seemingly, additional suns—all probably caused by an unusually dense layer of hexagonal ice crystals in the atmosphere. With the city buzzing, a painting was commissioned by a local pro-Protestant Reformation clergyman determined in part to dispel rumors that the phenomena were divine omens against the Protestant Reformation in Sweden—but who also saw them as a possible divine warning against pro-Reformation excesses. (This is actually a 1636 copy of the original painting, which was destroyed.)

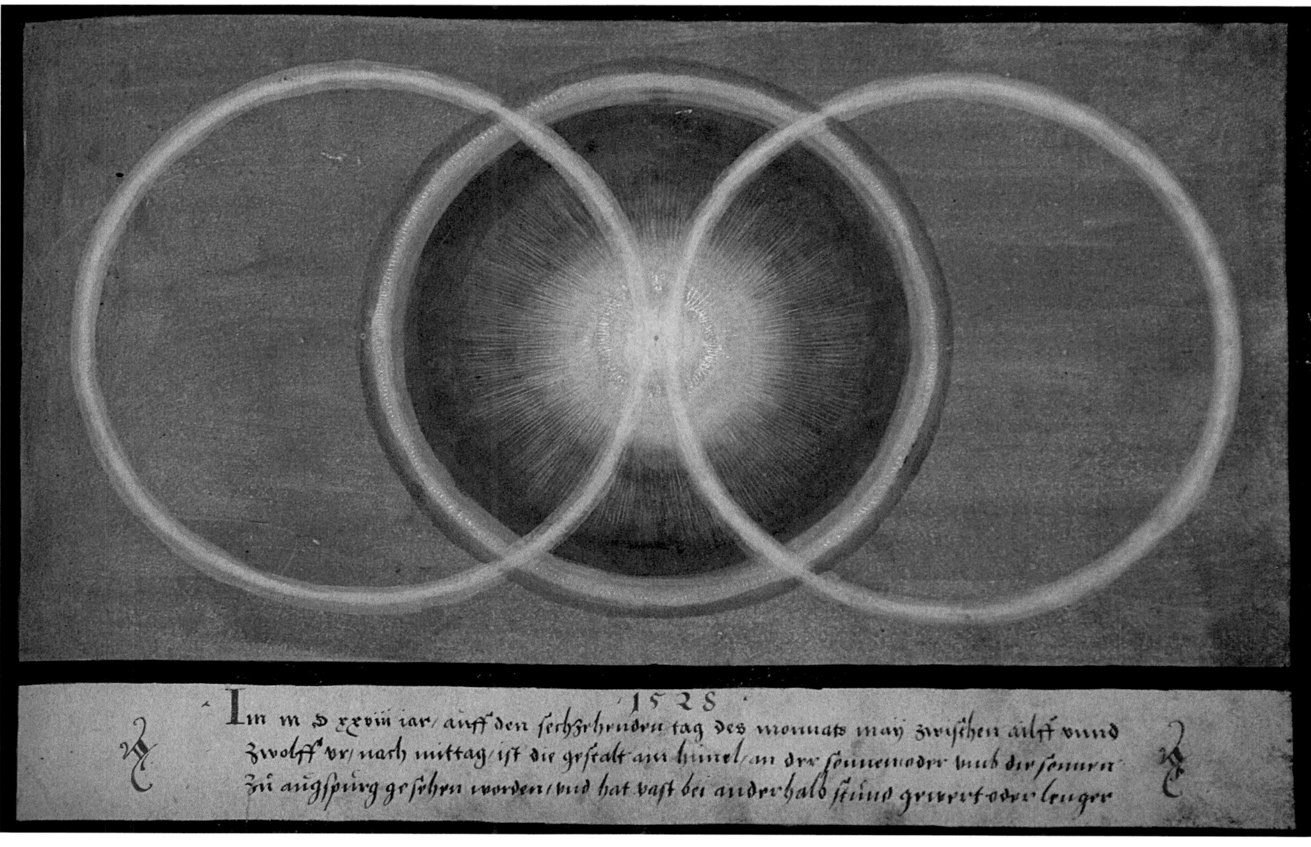

• 1547–52:

Many of the 167 paintings in the *Augsburger Wunderzeichenbuch* (Augsburg Miracles Book) also present odd atmospheric phenomena such as phantom suns or parhelia, halos, zodiacal lights, auras, auroras, and also occurrences far harder to explain, such as apparitions and fires in the sky. **Above top:** This painting may depict the zodiacal light, a faint glow sometimes visible in the sky in spring or fall, although its vertical appearance is mystifying, because the zodiac is always at an angle to the horizon except for near the equator. The phenomenon is due to dust in the zodiacal cloud—a flattened disc of interplanetary dust on the plane of the solar system. The text reads: "In 1515, in the month of May, this phenomenon was seen near Berlin and the Prince-Elector Margrave Joachim of Brandenburg is said to have seen it himself, too." **Above:** This could be a variant on a sun dog centering on a parhelic circle. The text reads: "In 1528, on the sixteenth day of the month of May, between eleventh and twelfth hour after midday, this shape was seen in the sky at the sun or around the sun in Augsburg and stayed for almost one and a half hours or longer."

Above top: This *Augsburger* miracle could conceivably have been due to a rare aurora, usually visible farther north. The text reads: "In 1542, a fire, which burns like a big fire pan in the clouds at the 12th hour in the night, which many trustworthy people saw for a long time, was seen in the sky here in Augsburg."

Above: No scientific explanation suggests itself for this startling image; sometimes a miracle is really a miracle. The text reads: "In 1531, a bloody bust portrait and a sword in its hand were seen at Strasburg and other regions, also a fiery castle and opposite to it an army platoon on horseback, as it is painted here."

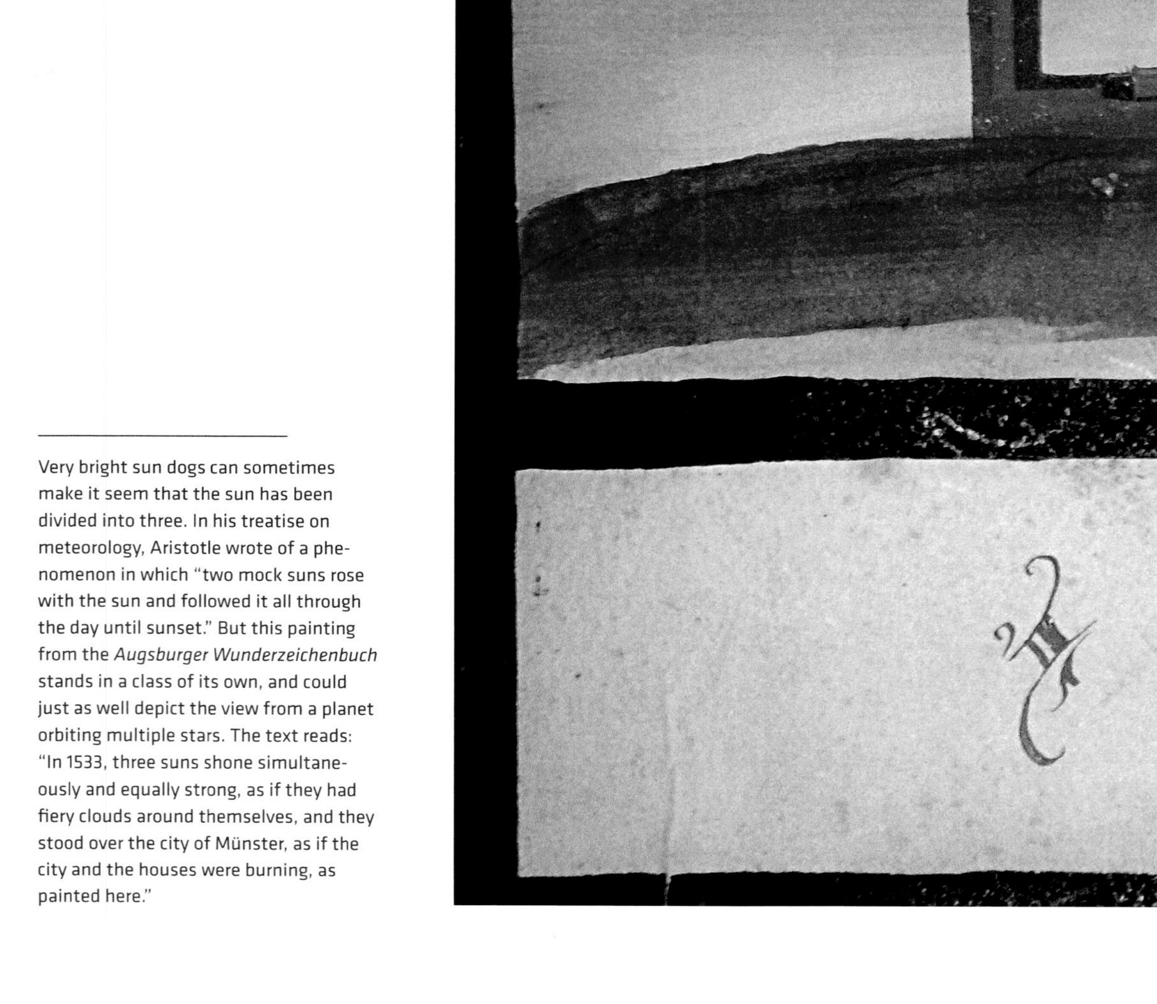

Very bright sun dogs can sometimes make it seem that the sun has been divided into three. In his treatise on meteorology, Aristotle wrote of a phenomenon in which “two mock suns rose with the sun and followed it all through the day until sunset.” But this painting from the *Augsburger Wunderzeichenbuch* stands in a class of its own, and could just as well depict the view from a planet orbiting multiple stars. The text reads: “In 1533, three suns shone simultaneously and equally strong, as if they had fiery clouds around themselves, and they stood over the city of Münster, as if the city and the houses were burning, as painted here.”

1533
ennd drey sonnen in gleichem schein als ob sie
r vmb sich / vnnd stunden vber der stat vnnr
nd heuser brennend / wie hir gemaltt

• 1580:

From the sixteenth century forward, the increasing availability of printing presses across Europe allowed for the publication of sensational single-page broadsheets publicizing various events, be they earthly or celestial. This example from southeastern Germany came with an accompanying headline: "News of the halo phenomena of the sun over Altdorf near Nürnberg on 12 January 1580 from one o'clock in the afternoon until about sunset." Although it predates him by two centuries, the print clearly shows a rare halo now known as the Parry arc, after arctic explorer William Edward Parry, who recorded his observations of one during an expedition in search of the Northwest Passage in 1820. The arc, seen here as an elongated white "U" over the sun, results when a twenty-two-degree halo meets an upper tangent arc. It's thought to be the result of airborne hexagonal column crystals. Sun dogs can be seen on either side, and the rainbow colors of the arcing lines indicate a phenomenon observed frequently when such arcs and halos are seen in the first place—though then the red edge is always closest to the sun, reversing the order seen here.

Aurore boréale irrégulière (p. 153).

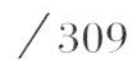

La lumière zodiacale sous le tropique (p. 182).

Couronne boréale.

• 1866:

Four engravings of celestial atmospheric phenomena from *L'Espace selestial* (Celestial Space) by French astronomer and botanist Emmanuel Liais. **Top left**: The shimmering curtains or other forms of aurora borealis are caused by interactions between the Earth's magnetosphere and charged particles in the solar wind. **Top right:** A sun dog, or parhelion effect, due to ice crystals in the air. **Bottom left:** The zodiacal light is the result of illuminated interplanetary dust on the plane of the solar system. **Bottom right:** Crown-shaped aurora called a corona, caused by solar-charged particles slamming into the magnetosphere.

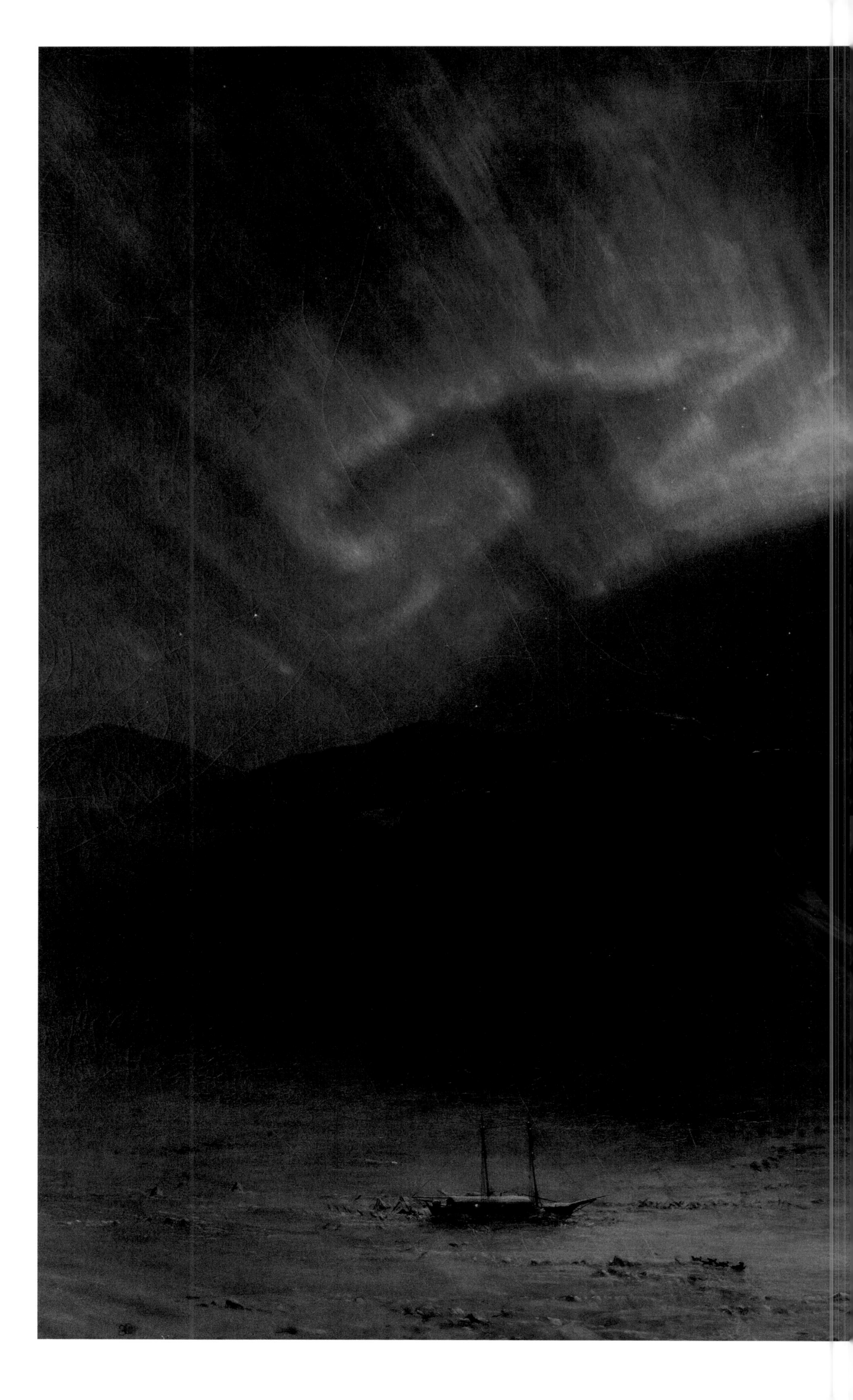

• 1865:

Titled *Aurora Borealis*, this luminous painting by Hudson River School painter Frederic Edwin Church was based in part on arctic explorer Isaac Hayes's own sketches, and also his written descriptions of the aurora. The northern lights ripple in a majestic arc over the ship carrying Hayes's 1860–61 arctic expedition. The alien landscape, the eerie lights in the sky, and the distant ship of exploration with its isolated beacon all seem to foreshadow a later sci-fi iconography.

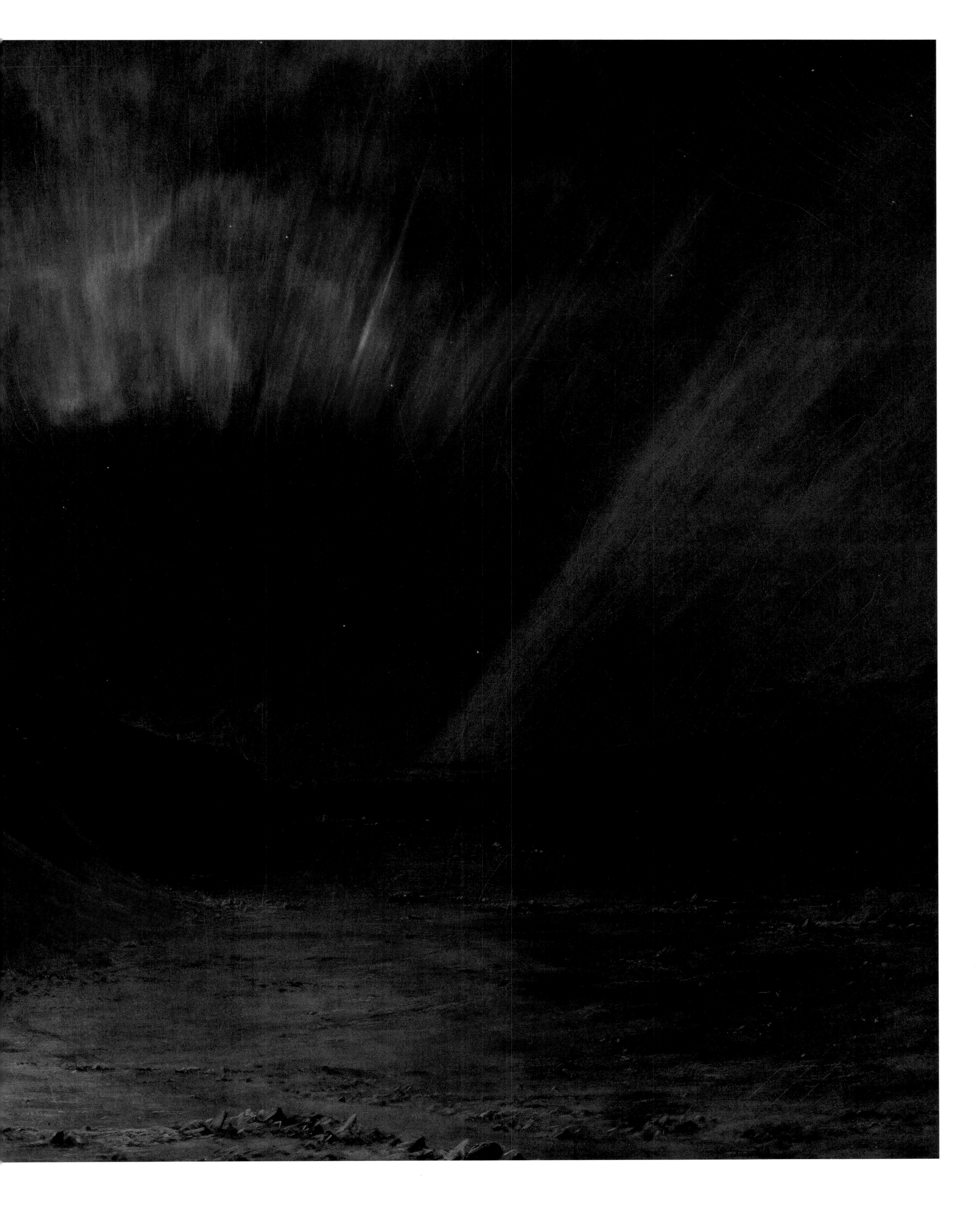

MOONLIGHT PHENOMENA AT THE BEGINNING OF THE POLAR NIGHT, November 1893.
A vertical axis passes through the moon, with a strongly-marked luminous patch where it intersects the horizon. A suggestion of a horizontal axis on each side of the moon: portions of the moon-ring with mock moons visible on either hand.

• 1896:

Norwegian arctic explorer and Nobel Peace Prize recipient Fridtjof Nansen drew these pictures himself while leading the 1893–96 *Fram* expedition, an ambitious but failed attempt to reach the North Pole via ship and dogsled. They were published in his best-selling 1897 account of the expedition, *Farthest North*. **Above top:** Six months into the expedition, the crew of the *Fram* saw a lunar version of a sun dog—a "parluna." **Above:** Auroras are a plasma that forms when charged particles excite electrons within tenuous gases in the Earth's upper atmosphere; they shed the excess energy as light. As seen here and on page 309, when seen from directly below the rippling curtains formed by Earth's magnetic field lines, sometimes exceptionally strong blasts of solar wind can cause particularly radial, kinetic auroral displays called coronae.

• 1899–1901:

Between 1898 and 1904, Danish painter and author Harald Moltke participated in four arctic expeditions. Two of them were funded and led by the Danish Meteorological Institute and specifically dedicated to studying the northern lights. Moltke's paintings of auroras were an integral part of this scientific research. This depiction of an auroral sheet over a church in northern Finland came from the twenty-four paintings of the aurora borealis that Moltke made during those two expeditions.

Harald Moltke's electric skyscape depicts an undulating forked sheet of auroral flame joining an abstract planetary landscape to unspeakably powerful and wordless cosmic forces.

• 1315–21:

The term *chora* dates back at least as far as Plato, who used it to mean space, and specifically the space where forms materialize: the *visible universe*, in other words. This fresco by an unknown fourteenth-century artist is on the ceiling of one of the outstanding surviving churches of Byzantium: the Chora Church in Istanbul. After the Ottoman conquest in the fifteenth century, it was covered by plaster for five hundred years while Chora played the role of mosque. Starting in 1948, however, the church was carefully restored, and in 1958 it opened its doors as a museum. The fresco is titled *The Angel of the Lord Rolling Up the Scroll of Heaven at the End of Time*.

Acknowledgments

I have consistently been surprised and grateful by how many people immediately grasped the essence of this rather specific project and helped me secure access to rare lodes of archival material.

Of the many people who have contributed to the realization of this book, noone has had more impact than Dr. Owen Gingerich of Harvard (and author of the foreword). Apart from allowing me to photograph a number of his rare books (nine images from which appear in *Cosmigraphics*), Dr. Gingerich patiently submitted to my questions concerning various aspects of astronomy. These ranged from what we can say with any degree of certainty concerning such objects from deep time as the Nebra Sky Disc, to many important details concerning the Copernican Revolution and its aftermath, to questions regarding the way we have structured our knowledge of the universe. He also read everything I wrote, contributed good-humored comments, pushed back against some interpretations (to my great benefit), and in general was indispensable. Having said that, any errors of interpretation or fact are of course entirely my own.

Dr. Gingerich also was kind enough to forward various of my inquiries to the invaluable HASTRO history of astronomy discussion group list. I want to thank Adam Jared Apt, John W. Briggs, Steve Ruskin, Charles Wood, Michael Hoskin, and William Tobin for their detailed responses to my questions. Particular thanks go to list contributors Stephen McCluskey and Randall Rosenfeld, who went well beyond the call of duty in their nuanced analyses of medieval representations of celestial motion and the Nebra Sky Disc.

In Somerville, Massachusetts I'm near one of the nation's oldest astronomy magazines, *Sky and Telescope*. When I contacted Senior Editor Alan MacRobert in pursuit of Czech astronomer Antonin Bečvář's *Skalnate Pleso Atlas*, he immediately invited me in to *Sky and Telescope*'s offices, and made the atlas and a scanner available. I soon met a number of people who make the magazine a valuable international resource, including Sean Walker, Dennis Di Cicco, and Gregg Dinderman. I'm grateful for all of their hospitality.

MacRobert advised me to contact another Cambridge organization that was too good to be true: the American Association of Variable Star Observers, or AAVSO. There archivist Dr. Michael Saladyga and senior technical assistant Elizabeth Waagen invited me to document the holdings of their extremely rich library. Over ten images in *Cosmigraphics* come from AAVSO, to the great benefit of this project. I owe them big thanks, and look forward to the day when I can contribute this book to their library.

I am also very grateful to the staff of the John G. Wolbach library at the Harvard-Smithsonian Center for Astrophysics, including head librarian Christopher Erdmann and librarians Maria McEachern and Mary Haegert, who excavated rare holdings from their archives and allowed me to set up a camera in their stacks.

I'm particularly grateful to astrophysicist Dr. R. Brent Tully of the University of Hawaii, and Daniel Pomarede, visualizations specialist at the Center for Computational Astrophysics in Saclay, France, who kindly gave me access to their supercomputer visualizations of galaxy formation and the dynamics of galaxy groups. Barry Ruderman of Ruderman Antique Maps, Inc in La Jolla kindly allowed me to use several scans in this book. Dr. Olga Shonova of South Bohemia went far out of her way in helping me gain access to the superb planet studies of pioneering Czech space artist Ludek Pesek; I hope his presentation here revives interest in his work. Astronomy historian Ton Lindemann contributed images and ideas to this project at a late stage. I owe them my thanks.

I'd also like to salute those enlightened online repositories of public domain materials that make these available for free to the world community: the astonishing David Rumsey Historical Map Collection; the Library of Congress; the Walters Art Museum; the Getty; the British Library; the National Library of France; the Rijksmuseum; the University of Michigan Library; the National Library of Poland; Wikimedia, Wikipedia, and WikiPaintings; the Yale Beinecke Library; the Boston Public Library; the University of Oklahoma History of Science Collections; NASA, USGS, NOAH, and the U. S Army Corps of Engineers; the Canadian Geological Survey; and the University of Cambridge, Institute of Astronomy Library.

Other friends of *Cosmigraphics* include, in alphabetical order, planetary cartographer Ralph Aeschlimann of ralphaeschliman.com; Cameron Beccario of earth.nullschool.net; cosmologist Dr. Francesco Bertola; David Bossard, who supplied his scans of arctic explorer Fridtjof Nansen's drawings; galaxy mapper Winchell D. Chung Jr., Tom Crouch of the Smithsonian National Air and Space Museum; Rik Declercq of the University Library, Ghent; astronomer Dr. John Dubinski; James Faber of Day & Faber; Daniel Fabrycky of the Kepler science team; Elena García-Puente of the Biblioteca Nationale Spain; the University Library, Ghent; Princeton astrophysicist Dr. J. Richard Gott; Planetary scientists and cartographer Dr. Henrik Hargitai, whose site planetologia.elte.hu/ipcd/ contains an invaluable repository of scanned planetary maps, including rare material from the USSR and the former Warsaw Pact countries; the Hayden Planetarium's Cynthia Angel, Dr. Carter Emmart, Sharon Stulberg, and Vivian Trakinski; Carlton Hobbs LLC; Kevin Jardene of galaxymap.org; Eric Jones of the Apollo Lunar Surface Journal; Karen Howes of interiorarchive.com; Dr. Homa Karimabadi; Dr. Daniel Lewis, the Dibner Senior Curator, History of Science, at the Huntington Library; Burlen Loring; Kerry V. Magruder, Curator, History of Science Collections, University of Oklahoma Libraries; John Overholt, Curator, the Donald and Mary Hyde Collection of Dr. Samuel Johnson Early Modern Books and Manuscripts, the Houghton Library, Harvard; space artist Ron Miller; the Old Print Gallery in Washington, DC; Alex Parker; Dr. Brigitte Pfeil of the Universitätsbibliothek Kassel; Bill Rankin, from radicalcartography.net; Matthias Rempel; the Saxon State and University Library, Dresden; author Peter Stauning for the Harald Moltke aurora paintings; Apollo landing site mapmaker Thomas Schwagmeier; Stephen A. Skuce of the MIT Library; the Sternberg Astronomical Institute; Felice Stoppa of atlascoelestis.com; Abel Mendez Torres; Arthur Woods of thespaceoption.com; and Michael Zeiler of eclipse-maps.com.

Many books and online sources contributed to this book; because this is not an academic study, they are not footnoted. These include but are not limited to the Quarterly Journal of the Royal Astronomical Society, *Star Maps* by Nick Kanas; *Star Struck* by Ronal Bashear and Daniel Lewis; *The Cosmographical Glass* by S. K. Heninger (from which I extracted the quotation in the first paragraph of chapter 2); *The Copernican Revolution* by Thomas Kuhn; *Seeing and Believing* by Richard Panek; *Longitude* and *A More Perfect Heaven: How Copernicus Revolutionized the Cosmos* by the inimitable Dava Sobel; *The Sun in the Church* by J. L. Heilbron; *The Transits of Venus* by William Sheehan and John Westfall; and *Rainbows, Halos, and Glories*, by Robert Greenler. I was particularly intrigued by the many ideas in architectural theorist Dalibor Vesely's *Architecture in the Age of Divided Representation*. I also derived information about the hand-made Mariner 4 Mars image on page 197 from Dan Goods' website.

Over the last few years my world-view has been productively bent into new shapes by discussions with Carter Emmart of the Hayden Planetarium, David Gersten of Cooper Union (who first clued me in to Vesely's work), and Chris Rose of RISD. Their ideas have impacted my work in innumerable ways, and I thank them for being ongoing wellsprings of inspiration. I also want to thank Wonder Cabinet maestro Ren Weschler for well over a decade of dialogue on many subjects.

This book wouldn't have been possible without the encouragement and support of my editor at Abrams and good friend, Eric Himmel, who revived the idea in the spring of 2013. *Cosmigraphics* also benefited from the professionalism of my managing editor at Abrams, David Blatty, who patiently fielded last minute incoming chapter essays and revisions with aplomb. I also am in great debt, as always, to my agent Sarah Lazin, who initiated my connection with Abrams many years (and five books) ago.

I also want to thank my family, starting with my parents, Shirley and Raymond Benson, and my sister and brother, Carolyn and Nick Benson, for their support. Finally, there's no way I could have done this project without my daily therapeutic discussions with my wife Melita, to whom this book is dedicated, and my son Daniel.

Image Credits

Title Page: USGS/NASA; map by Paul Spudis and James Prosser. **Chapter 1:** Page 19-21: Courtesy of U. of Oklahoma History of Science collections; Pg. 22-23: Courtesy the Huntington Library; Pg. 24-25: Courtesy Walters Art Museum; Pg. 26-29: Biblioteca Nacional de España; Pg. 30: Courtesy the David Rumsey Historical Maps Collection; Pg. 31: Courtesy Joanna Ebenstein; Pg. 32: Courtesy the U. of Oklahoma History of Science collections; Pg. 33: Image copyright © The Metropolitan Museum of Art. Image source: Art Resource, NY; Pg. 34-35: Courtesy ESA/Planck. **Chapter 2:** Page 39: State Library of Lucca; Pg. 40: Courtesy the U. Library, Ghent; Pg. 41, top: Isabella Steward Gardiner Museum, Boston; Pg. 41, bottom: Courtesy Dr. Owen Gingerich; Pg. 42: Bibliothèque nationale de France; Pg. 43: Courtesy the Huntington Library; Pg. 44: Wikimedia; Pg. 45: Courtesy the Rijksmuseum; Pg. 46: U. of Michigan Library/color data courtesy Ton Lindenmann; Pg. 48-51: Courtesy the National Library of Poland; Pg. 52: Courtesy the Yale Beinecke Library; left: Courtesy Barry Lawrence Ruderman Antique Maps, Inc.; Pg. 53, left: Courtesy Barry Lawrence Ruderman Antique Maps, Inc.; right: Courtesy the U. of Oklahoma History of Science collections; Pg. 54-55: Courtesy the Boston Public Library; Pg. 56-57: Courtesy the David Rumsey Historical Maps Collection; Pg. 58-59: Courtesy the U. of Oklahoma History of Science collections; Pg. 60-61: Courtesy the Library of Congress Geography and Map Division Washington, D.C.; Pg.s 62-63: Courtesy the U.S. Army Corps of Engineers Engineering Geology and Geophysics Branch; Pg. 64-65: Courtesy the David Rumsey Historical Maps Collection; Pg. 66: Courtesy Bill Rankin, from radicalcartography.net. Map based on data by Peter Bird, USCLA, and earthquake data from NEIC and USGS; Pg. 67: Courtesy the Canadian Geological Survey; Pg. 68: Courtesy NASA Goddard Space Flight Center Scientific Visualization Studio. Image by Greg Shirah (NASA/GSFC) (Lead) and Horace Mitchell (NASA/GSFC) based on data interpreted by Hong Zhang (UCLA) and Dimitris Menemenlis (NASA/JPL CalTech); Pg. 69: Courtesy Cameron Beccario, earth.nullschool.net. **Chapter 3:** Page 73: Wikimedia; Pg. 74, top: Courtesy the Getty Museum; Pg. 74: bottom: Courtesy the British Library; Pg. 75: Courtesy the British Library; Pg. 77: Courtesy the Huntington Library; Pg. 77, top: Courtesy Ton Lindenmann; Pg. 77, bottom: Courtesy the Royal Astronomical Society; Pg. 79: Courtesy the Library of Congress; Pg.s 80-81: MFA Images, Museum of Fine Arts, Boston; Pg. 82-83: Courtesy the Wolbach Library, Harvard; Pg. 84: U. of Michigan Library/color data courtesy Ton Lindenmann; Pg. 85: Courtesy the National Library of Poland; Pg. 86: Bibliothèque de l'Observatoire de Paris; Pg. 87: Dipartimento di Fisica e Astronomia, Università di Bologna; Pg. 88: Courtesy the Saxon State and U. Library, Dresden; Pg. 89: Courtesy the Getty Museum; Pg. 90-91: Courtesy the American Association of Variable Star Observers (AAVSO); Pg. 92-95: Courtesy the Wolbach Library, Harvard; Pg. 96-97: Courtesy the Library of Congress; Pg.s 98-99: Courtesy the U. of Michigan Library; Pg. 100: Courtesy Carlton Hobbs LLC; Pg. 101: Courtesy Anne Verdillon, Nain.de.Jardin blog; Pg. 102-103: Courtesy Olga Shonova; Pg. 104, top: Courtesy AAVSO; Pg. 104-105: Courtesy the Sternberg Astronomical Institute, via Henrik Hargitai, planetologia.elte.hu/ipcd/; Pg. 106-107: Apollo maps, Apollo Lunar Surface Journal, Thomas Schwagmeier; Pg. 108: Courtesy USGS/NASA/USAF; map by S. R. Titley; Pg. 109: Courtesy USGS/NASA/USAF; map by Don Wilhelms and John McCauley; Pg. 110: Courtesy USGS/NASA; map by David Scott, John McCauley, and Mareta West. Pg. 111: Courtesy USGS/NASA. **Chapter 4:** Page 115: Courtesy the U. Library, Ghent; Pg. 116: Courtesy the British Library; Pg. 117: Rylands Medieval Collection, U. of Manchester; Pg. 118: Courtesy the Library of Congress; Pg. 119: Courtesy the Huntington Library; Pg. 120: Courtesy the British Library; Pg. 121: Courtesy Owen Gingerich; Pg. 122-123: Courtesy the U. of Michigan Library; Pg. 124-125: Courtesy the National Library of Poland; Pg. 126: Image copyright © The Metropolitan Museum of Art. Image source: Art Resource, NY; Pg. 127: Courtesy the Wolbach Library, Harvard; Pg. 128-129: Courtesy the U. of Michigan Library; Pg. 130-131: Courtesy AAVSO; Pg. 132: Courtesy NCAR-Wyoming supercomputing facility ©UCAR, image courtesy Matthias Rempel, NCAR; Pg. 135: Courtesy Homa Karimabadi and Burlen Loring. **Chapter 5:** Page 141: Courtesy the U. Library, Ghent; Pg. 142-143: Courtesy the Bibliothèque nationale de France; Pg. 144, top and bottom: Courtesy the British Library; Pg. 145: The Ashmolean Museum Yousef Jameel Centre for Islamic and Asian Art; Pg. 146: Courtesy Owen Gingerich; Pg. 147-148: U. of Michigan Library/color data courtesy Ton Lindenmann; Pg. 149: Courtesy Owen Gingerich; Pg. 150-151: Courtesy U. of Oklahoma History of Science collections; Pg. 152, top: Courtesy U. of Oklahoma History of Science collections; Pg. 152, bottom: Courtesy the Rijksmuseum; Pg. 153-154: Courtesy the Wolbach Library, Harvard; Pg. 155: Courtesy the Old Print Gallery; Pg. 156: Courtesy the U. of Cambridge Institute of Astronomy Library; Pg. 157: Wikipedia; Pg. 158: Courtesy the Wolbach Library, Harvard; Pg. 159: Courtesy Dr. Francesco Bertola; Pg. 160-161: Courtesy R. Brent Tully; Pg. 162: Courtesy John Dubinski; Pg. 163: Courtesy Daniel Pomarede, COAST Project (CEA-Saclay/Irfu); Pg. 164-165: Courtesy J. Richard Gott and Mario Juric; Pg. 166: Simulation performed by Frédéric Bournaud using the RAMSES code developed by Romain Teyssier, visualization using the SDvision code developed by Daniel Pomarède, as part of the COAST Project (CEA-Saclay/Irfu); Pg. 167: Simulation performed by Lauriane Delaye and Frédéric Bournaud using the RAMSES code developed by Romain Teyssier, visualization using the SDvision code developed by Daniel Pomarède, as part of the COAST Project.; Pg. 168-171: Courtesy AMNH-Hayden Planetarium, from Dark Universe, directed by Carter Emmart, produced by Vivian Trakinski; Pg. 172-173: Courtesy R. Brent Tully, Daniel Pomarède, Helene Courtois (U. Lyon 1) and Yehuda Hoffman (Hebrew U.) **Chapter 6:** Page 177: Courtesy the U. Library, Ghent; Pg. 178-179: Courtesy the British Library; Pg. 180: Courtesy the Huntington Library; Pg. 181: Library of Congress; Pg. 182-183: U. of Michigan Library/color data courtesy Ton Lindenmann; Pg. 184: Dipartimento di Fisica e Astronomia, Università di Bologna; Pg. 185: Courtesy Ton Lindenmann; Pg. 186-187: Courtesy the Library of Congress; Pg. 188-191: Courtesy the Public Library of Cincinnati and Hamilton County; Pg. 192: Courtesy AAVSO; Pg. 193: Courtesy the U. of Cambridge Institute of Astronomy Library; Pg. 194: Top row and bottom right: Courtesy the Wolbach Library, Harvard; Pg. 194: Bottom left: Courtesy AAVSO; Pg. 195: Bonestell LLC; Pg. 196: Courtesy Olga Shonova; Pg. 197: Courtesy NASA/JPL/Dan Goods. Image by Richard Grumm; Pg. 198-199: Courtesy Olga Shonova; Pg. 200-201: USGS/NASA; map by Paul Spudis and James Prosser; Pg. 202: USGS/NASA; based on maps provided by the USSR Academy of Sciences; Pg. 203: USGS/NASA; map by Alexander Basilevsky; Pg. 204: USGS/NASA; map by Kenneth Tanaka and David Scott; Pg. 205: USGS/NASA; map by Kenneth Tanaka and Corey Fortezzo; Pg. 206: USGS/NASA; map by Baerbel Lucchitta; Pg. 207: Courtesy Ralph Aeschliman; Pg. 208-209: USGS/NASA; maps by Scott Murchie and James Head; Pg. 210-211: Courtesy Daniel Fabrycky, Kepler science team; Pg. 212: Courtesy Abel Mendez Torres, U. of Puerto Rico at Arecibo; Pg. 213: Courtesy Alex Parker **Chapter 7:** Page 217: Courtesy Owen Gingerich; Pg. 218: top: Library of Congress; Pg. 218: bottom: Courtesy the British Library; Pg. 219: Courtesy the British Library; Pg. 220: Forschungbibliothek Gotha; Pg. 222-223: Courtesy the Bibliothèque nationale de France; Pg. 224: Wikipedia France; Pg. 225: Courtesy the Bibliothèque nationale de France; Pg. 226: Courtesy Sotheby's; Pg. 227: Courtesy the Huntington Library; Pg. 228: Courtesy Owen Gingerich; Pg. 229: Courtesy Barry Lawrence Ruderman Antique Maps, Inc.; Pg. 230-231: Library of Congress; Pg. 232: Courtesy Ton Lindenmann; Pg. 233-235: U. of Michigan Library/color data courtesy Ton Lindenmann; Pg. 236-237: Courtesy the Wolbach Library, Harvard; Pg. 238: Courtesy Owen Gingerich; Pg. 239: Image copyright © The Metropolitan Museum of Art. Image source: Art Resource, NY; Pg. 240: Courtesy AAVSO; Pg. 241: Courtesy the Public Library of Cincinnati and Hamilton County; Pg. 242: Courtesy AAVSO; Pg. 243: Courtesy the U. of Cambridge, Institute of Astronomy Library; Pg. 244-245: Courtesy Sky and Telescope Magazine/Sky Publishing, F+W Media, Inc.; Pg. 246-247: Courtesy Winchell D. Chung Jr.; Pg. 248: NASA, JPL, Spitzer Space Telescope; map by Robert Hunt; Pg. 249: Courtesy Winchell D. Chung Jr.; Pg. 250-251: Courtesy Kevin Jardene, data Douglas Finkbeiner. **Chapter 8:** Page 255: Courtesy the Bibliothèque nationale de France; Pg. 256, top: Courtesy the British Library; Pg. 256: bottom: Courtesy Owen Gingerich; Pg. 257: Rylands Medieval Collection, U. of Manchester; Pg. 258-259: Courtesy Day & Faber; Pg. 260-261: Bayerische Stattsbibliothek; Pg. 262: Courtesy the Huntington Library; Pg. 263: Courtesy the Getty Museum; Pg. 264: Courtesy the Library of Congress; Pg. 265: Courtesy the U. of Cambridge, Institute of Astronomy Library; Pg. 266: Library of Congress; Pg. 267-269: Courtesy AAVSO; Pg. 270-271: Courtesy the Public Library of Cincinnati and Hamilton County; Pg. 272-275: Courtesy Michael Zeiler, www.eclipse-map.com. **Chapter 9:** Page 279: Wikimedia; Pg. 280-281: Courtesy Day & Faber; Pg. 282-284: Courtesy Universitätsbibliothek Kassel - Landesbibliothek und Murhardsche Bibliothek der Stadt Kassel; Pg. 285-286: Courtesy the Yale Beinecke Library; Pg. 287: Courtesy the National Library of Poland; Pg. 288-289: Courtesy the Yale Beinecke Library; Pg. 290-291: Royal Collections Trust; Pg. 292: Courtesy Dennis Di Cicco, Sky and Telescope magazine; Pg. 293: Courtesy the U. of Cambridge, Institute of Astronomy Library; Pg. 294, top: Courtesy NOAA; Pg. 294, bottom: Courtesy AAVSO; Pg. 295: Courtesy the U. of Michigan Library; Pg. 296: AAVSO; Pg. 297-299: Courtesy the U. of Michigan Library. **Chapter 10:** Page 303: Wikimedia Commons; Pg. 304-307: Courtesy Day & Faber; Pg. 308: Courtesy Owen Gingerich; Pg. 309: Courtesy AAVSO; Pg. 310-311: Smithsonian American Art Museum, via Wikipedia; Pg. 312: Courtesy © David C. Bossard, 19thcenturyscience.org.; Pg. 313-314: Courtesy Peter Stauning; Pg. 316: Courtesy Karen Howes, the Interior Archive. **Case:** USGS/NASA.